Plant Structure – A Colour Guide

Second Edition

BRYAN G. BOWES
Formerly Botany Department
University of Glasgow, UK

JAMES D. MAUSETH
Integrative Biology
University of Texas, USA

CRC Press
Taylor & Francis Group
Boca Raton London New York

CRC Press is an imprint of the
Taylor & Francis Group, an **informa** business

First published 2008 by Manson Publishing Ltd

CRC Press
Taylor & Francis Group
6000 Broken Sound Parkway NW, Suite 300
Boca Raton, FL 33487-2742

ISBN 13: 978-1-138-40761-9 (hbk)
ISBN 13: 978-1-84076-092-7 (pbk)

Visit the Taylor & Francis Web site at
http://www.taylorandfrancis.com

and the CRC Press Web site at
http://www.crcpress.com

Cover: Ayala Kingsley
A CIP catalogue record for this book is available from the British Library.

Contents

Preface

'Why should they care about the histogenesis of the leaf, or adventitious roots? ... The public wants heart transplants, a cure for AIDS, reversals of senility. It doesn't care a hoot for plant structures, and why should it? Sure it can tolerate the people who study them ... They're relatively inexpensive too. It costs more to keep two convicts in Statesville than one botanist in his chair'.

More Die by Heartbreak, Saul Bellow

While Bellow's character Kenneth Trachtenberg may convincingly relegate the study of plant structure to a backwater, it is a commonplace that we are all ultimately dependent on green plants for our survival on earth. Horticultural successes in increasing crop yields and developing new plant varieties emphasize the importance of plant physiology, biochemistry, and molecular biology, for all of which the study of the green plant's internal form and internal structure is a prerequisite. With the steadily increasing content of undergraduate and graduate biology courses, the proportion of a student's time devoted to plant morphology is inevitably reduced. There is no longer time for detailed study of the excellent and exhaustive texts in plant anatomy, and many find that plant structures are most easily understood when mainly described by annotated photographs and drawings. Such is the concept behind the present guide: knowledge of plant structure is fundamental to the study of plant science, and that knowledge has to be imparted clearly, briefly, and precisely. Following an introductory chapter on the morphology of the vascular plant, there are seven chapters each dealing with a major aspect of plant structure. A comprehensive glossary of botanical terms used in the guide is also included. The text for each chapter sets out the essential characteristics of the plant features described and makes extensive reference to appropriate illustrations in the particular chapter and elsewhere in the book. Each illustration is accompanied by a legend and salient features are numbered (not labelled) for maximum clarity, referencing the structures to a boxed key. It is hoped that these aspects of the guide, together with the photographs and drawings, will prove attractive and useful to many readers.

The guide is intended for use in different ways by different readers. For the university or college student, the guide is intended to be read either as a concise introductory text or as a revision guide in preparation for exams. For the professional instructor or for the researcher in academic life or in industry, it is hoped the guide will provide a source of rapid reference. For the artist or the amateur student of natural history, the intrinsic beauty of many plant specimens, in external form and under the microscope, is clearly shown in the photographs and drawings, arranged in sequence after the text in each chapter.

The overall intention has been to provide a concise and highly illustrated summary of present knowledge of the structure of vascular plants, with particular emphasis on flowering plants.

Bryan G. Bowes
James D. Mauseth

Abbreviations

ER endoplasmic reticulum

F-F freeze-fracturing

G-Os fixation in glutaraldehyde followed by osmic acid

KMn fixation in potassium permanganate

LM light microscope

LS longitudinal section

RER rough endoplasmic reticulum

RLS radial longitudinal section

SEM scanning electron microscope

SER smooth endoplasmic reticulum

TEM transmission electron microscope

TLS tangential longitudinal section

TS transverse section

Acknowledgements

JDM thanks the Teresa Lozano Long Institute of Latin American Studies at the University of Texas and the Cactus and Succulent Society of America for generous financial support.

Although BGB and JDM together wrote and also photographed the vast majority of the images used in this book, we have to thank all the talented team at Manson Publishing for integrating both text and figures so successfully in this guide. We especially thank Paul Bennett and Ruth Maxwell for their skillful, patient contributions to this project.

Dedications

From *Some Fruits of Solitude* by William Penn, 1693

The humble, meek, merciful, just, pious, and devout souls are everywhere of one religion; and when death has taken off the mask they will know one another, though the divers liveries they wear here make them strangers.

From *For A' That And A' That* by Robert Burns, 1795

Then let us pray that come it may,
As come it will for a' that,
That sense and worth, o'er a' the earth,
May bear the gree*, and a' that.
For a' that, and a'that,
Its comin yet for a' that,
That man to man, the world o'er,
Shall brothers be for a' that.

*have first place

From *My First Summer in the Sierra* by John Muir – first published in 1911, but based on his journals written in the early 1870s

(September 2nd: The Tuolumne Camp) One is constantly reminded of the infinite lavishness and fertility of Nature – inexhaustible abundance amid what seems enormous wastage. And yet when we look into any of her operations that lie within reach of our minds, we learn that no particle of her material is wasted or worn out. It is eternally flowing from use to use...

BGB

To my partner, Tommy R. Navarre, for 24 years of support and encouragement.

JDM

Biographies

Bryan Bowes was Senior Lecturer for over 30 years in the Department of Botany, Glasgow University in Scotland but has also been a Research Fellow in the Federal Technical University, Zurich (ETH, Switzerland), Harvard University (USA) and Armidale University (NSW, Australia). His research interests encompassed a broad field in angiosperm structure and development including plant anatomy and ultrastructure, plant regeneration and in vitro morphogenesis. He was the author of the first edition of *Colour Atlas of Plant Structure* (1996) and later edited and contributed Chapters to *Colour Atlas of Plant Propagation and Conservation* (1999) and *Colour Atlas of Trees* (2008), which were all published by Manson Publishing.

Although retired, he continues to take a keen interest in plant morphology (particularly veteran trees) and now records his botanical observations as digital images – comprising many thousands to date – taken with his camera on extensive travels of Scotland and abroad.

James Mauseth has been a Professor of Botany at the University of Texas in Austin, USA since 1975, after graduate studies at the University of Washington in Seattle. His research has centred on evolutionary modifications in plant structure that occurred as cacti became increasingly adapted to desert habitats. Field research has taken him throughout South America, Mexico, the Caribbean and the south-western USA, as well as botanical gardens at Kew, Zurich, Monaco, The Huntington in Pasadena and The Desert Botanical Garden in Phoenix. Parasitic plants have also been one of his research topics, particularly those in which most aspects of ordinary anatomy and morphogenesis have been highly modified. He has also published two textbooks (*Botany, An Introduction to Plant Biology*, third edition [2003], Jones and Bartlett Publishers; *Plant Anatomy* [1988], Addison/Wesley Publishers) and a popular book on plant biology (*A Cactus Odyssey: Travels in the Wilds of Bolivia, Peru and Argentina*, Mauseth JD, Kiesling R, and Ostolaza C [2002], Timber Press, Portland, Oregon).

Other interests are long-distance bicycle touring, kayaking, and snow-shoeing.

CHAPTER 1

Introduction

THE ASSORTMENT OF LAND PLANTS

Flowering plants, or angiosperms (**1, 2**), dominate large areas of the land surface and represent the climax of vascular plant evolution. They occupy a wide range of habitats and about a quarter of a million species have been recognized so far. However, many more, particularly from tropical regions, await scientific description. Angiosperms are very diversified in their form, and range in size from a few millimetres in diameter in the aquatic *Lemna* (**3**) to 90 m or more in height in *Eucalyptus* (**4**). Some complete their life cycle in less than 2 months while some specimens of *Quercus* (oak) may live nearly a thousand years (**5**).

Flowering plants provide the vast majority of those consumed by humans (**6, 7**) or utilized for domestic animal fodder. Likewise, angiosperms provide various very important commercial hardwood timbers such as *Acacia, Carya, Eucalyptus, Fagus, Juglans,* and *Quercus* (**4, 8, 9**), fibres (e.g. *Corchorus, Linum*; **10**) and drugs (e.g. *Papaver, Coffea*; **7**). Most decorative garden plants are grown to provide floral displays (**1, 2**).

Although the flowering plants are now dominant in many habitats, remnants of earlier evolved vascular plant groups are still present in the flora. There are about 700 species of gymnosperms whose seeds are naked (**11**); these are mostly conifers (**12**) but there are also some 289 tropical/sub-tropical cycad species (**11**) and a few others. Their naked seeds distinguish them from the flowering plants where the seeds are enclosed within a fruit (**13**). The conifers dominate the vast tracts of boreal forest which occur in North America and northern Europe and Asia, and many conifers provide very valuable softwood timbers for a multitude of purposes such as construction, paper pulp, and fencing. The spore-bearing ferns and their allies (**14–16**) number about 12,000 species.

As well as over 260,000 species of vascular plants, the land flora includes the nonvascular, spore-bearing bryophytes (**17, 18**). These small plants comprise about 1500 species of moss (**18**) but far fewer liverworts (**17**) and hornworts. They lack cuticular covering to the epidermis and are usually confined to moist locations. Hornworts and liverworts are often simply-organized thalloid structures without leaves, but mosses (and some liverworts) are more complex and have leafy green shoots.

The stems of many mosses and a few liverworts show a central strand of tissue, apparently concerned with the movement of water and soluble foodstuffs. However, except in a few taxa such as *Polytrichum* (**19**), this does not have the structural complexity of the xylem and phloem tissues (**20**) of vascular plants. The latter tissues are concerned with the rapid, long-distance transport of water and soluble foodstuffs (**21**).

Lichens are not true plants but rather symbiotic associations of fungi and algae; these however, often show a complex plant-like form (**22**).

TRANSPIRATION AND TRANSLOCATION IN VASCULAR PLANTS

The root system absorbs water, together with dissolved mineral salts, from the soil. This passes across the cortex and endodermis of the young root to the central xylem (**23**). The dead tracheary elements of this tissue have strong thickened walls (**20**) and their lumina are filled with columns of water moving upwards into the shoot (**21**).

This transpiration stream is powered by the evaporation of water vapour from the shoot surface, and mainly occurs through the stomata in the leaf epidermis. These small pores (**24**) normally remain open in the day and allow the entry of carbon dioxide, which is essential for photosynthesis in the green foliage. The sugars thus elaborated are translocated (**21**) in solution in the living sieve elements of the phloem (**20**) to the stem and root where they are either stored (**23, 25**) or metabolized.

GENERAL MORPHOLOGY OF ANGIOSPERMS

The young shoot of the generalized flowering plant (**26**) bears a number of leaves and normally a lateral bud occurs in the axil of each leaf. The leaf is attached to the stem at the node, while the internode lies between successive leaves. The leaf is usually flat and often is borne on a leaf stalk (petiole, **27**). In a horizontal leaf the adaxial surface (which was nearest to the shoot apex while within the bud, **28**) lies uppermost and the abaxial side forms the lower surface.

A simple leaf may be dissected or lobed, and a compound leaf shows several leaflets (**29**); leaflets do not subtend axillary buds. In the lamina (leaf blade) a network of veins is present (**30**) which links to the vascular system of the stem. The axillary (lateral) buds may remain dormant but normally develop into side shoots, or form flowers. At the base of the main stem the cotyledons (first leaves formed in the embryo; **31**) demarcate it from the hypocotyl; the latter represents a transition zone between stem and root.

Until recently, angiosperms were believed to consist of just two groups: the dicotyledons (dicots) and the monocotyledons (monocots). However, cladistic studies of DNA and other features have proposed that this fundamental diversification of angiosperms did not occur until after about eight small basal orders (containing less than 3% of all angiosperm species) had already become distinct. Four of these basal orders are now grouped together as the 'Magnoliids,' another is distinct as the water lilies (Nymphaeales). The rest of the angiosperms are divided into the monocots (containing about 22% of all angiosperm species), a separate order Ceratophyllales, and finally the 'eudicots' (with about 75% of all angiosperm species). In most structural features these all still form two fundamental groups: those with monocot-like structure (the monocots, of course) and those with dicot-like structure (almost everything except the monocots). There is no term to describe 'all the angiosperms other than monocots' other than the old term 'dicot;' we will therefore use the term 'dicot' here except in those instances in which one of the basal angiosperms differs from both monocots and eudicots.

Monocots and dicots show distinct morphological, anatomical, and floral characteristics (**32**). The dicots (crucifers, begonias, willows, oaks) contain most of the flowering plant species and the great majority shows some degree of secondary (woody) thickening (**4, 5, 8**). Monocots (e.g. bananas, grasses, lilies) do not undergo secondary thickening in the same way as dicots, but in some genera large trees may develop (**33, 39**).

Dicot leaves are commonly petiolate (**27**) and normally show a narrow attachment to the stem. In monocots the leaf is frequently sessile (without a petiole; **34**) and the leaf base often encloses a large sector of the stem (**35, 36**). Leaves of dicots are varied in shape and arrangement of their major veins but normally show a reticulate pattern of the small interconnecting veins (**30**). In monocots the leaf is typically elongate (**33, 34**), with the main veins paralleling its length. Their relatively rare lateral connections are normally unbranched (**37**).

In dicots, the radicle (seedling root) is normally persistent and the older region often increases in diameter by secondary thickening (**38**). By contrast, in monocots the radicle dies off and an adventitious root system develops from the base of the enlarging shoot. A number of larger monocots produce adventitious prop roots which stabilize their heavy upright shoots (**39**).

VASCULAR ANATOMY OF ANGIOSPERMS

The primary vascular systems of monocots and dicots generally differ considerably (**32**). In a transverse section of the monocot stem (**40**) there are many scattered vascular bundles, while in dicots a smaller number of bundles is usually arranged in a cylinder outside a pith (**41**). The roots of monocots frequently show a central pith-like ground tissue with a large number of strands of alternating xylem and phloem on its periphery (**42**). In the dicots, however, a star-shaped core of xylem commonly occurs with strands of phloem lying between its several arms (**23**).

In the majority of dicots and gymnosperms, a fascicular cambium separates the primary xylem and phloem of both the stem and root (**38, 41, 43**). If secondary thickening occurs the normally discrete strands of cambium become linked, and the continuous ring of vascular cambium produces secondary xylem internally and secondary phloem externally (**10, 43**). The vast majority of monocots is herbaceous; however, a number of palms grow into tall trees supported by numerous fibres. Others such as *Dracaena, Yucca*, and *Cordyline* produce new (secondary) vascular bundles from a secondary thickening meristem and may form large trees (**33**; see also Chapter 6 Anomalous Secondary Growth).

FLORAL AND REPRODUCTIVE FEATURES OF ANGIOSPERMS

In angiosperms the seeds are enclosed (**45B**) in contrast to gymnosperms whose seeds are naked (**11, 45A**). In monocots the floral parts (sepals, petals, stamens, and carpels) commonly develop in threes (**1, 32**), whereas in dicots these frequently occur in fives or fours (**2, 32**). However, a large and indefinite number of floral parts is present in many other dicots (**44**). The mature carpel (female part of the flower; **44, 45**) consists of several parts: the terminal stigma which receives the pollen (**1, 2, 44**), an intermediate style (**1, 2**), and the basal ovary (**45B**). In most taxa the carpels are fused (syncarpy; **13**) rather than free from each other (apocarpy; **44**). Within the ovary, one to numerous ovules are present and each contains an egg cell at the micropylar end of the ovule (**45B**). The pollen grain germinates on the stigma and the pollen tube grows down the style to enter the ovule (**45B**) where it liberates two haploid sperm nuclei.

One of these fertilizes the egg which forms the diploid zygote, while the other nucleus fuses with the two centrally located polar nuclei (**45C**) to give rise to the endosperm which nourishes the embryo. As the embryo develops from the zygote it enlarges and the surrounding tissues of the ovule expand to form the mature seed. The ovary concomitantly increases in size to form the mature fruit (**13**). In dicots two cotyledons are present on the embryo (**31, 32, 46**), but in monocots only a single one occurs (**32**).

THEME OF THE COLOUR GUIDE

This book is concerned with the development and mature form of the vascular plant and attention is focused on its structure at an anatomical, histological, and fine-structural level. As previously emphasized, the angiosperms possess the greatest number of evolutionarily modified features and dominate a varied range of habitats. They are the most numerous members of the land flora and provide nearly all of the plants, except for conifers which yield softwoods, exploited economically by humans. It is therefore appropriate that the examples discussed in ensuing chapters concentrate on the varying manifestations of the anatomy of the flowering plants on which we are all so dependent.

This colour guide is intended to serve both as an integrated series of clearly described illustrations and a concise text; where reference is made to a feature displayed by a particular genus this is normally intended as generally representative of a number of plants rather than exclusive to the plant quoted.

In this colour guide plants are described by their scientific names (i.e. genus and sometimes species), but some common names are additionally given for more generally familiar plants.

It is believed that this presentation will facilitate the appreciation and understanding of basic plant anatomy and it is not intended to be encylopaedic in its coverage. It is intended for use by biology undergraduates, as well as by graduates who have no previous knowledge of plant structure but are undertaking research in plant physiology, biochemistry, horticulture, or related fields. The colour guide will also be relevant to biological studies at advanced school and college levels, while the abundant and fully annotated illustrations should be of general interest to a wider audience.

A short bibliography details various texts which supplement the essentials of plant anatomy presented in this work, while the glossary will be of especial use to the reader where illustrations and legends are consulted without direct reference to the text.

1 Hermaphrodite flowers of *Lilium* 'Destiny' (lily). The floral parts are grouped in threes: six orange-spotted yellow perianth members, six stamens with orange anthers (1) and a three-lobed orange stigma (2). Floral parts in threes and elongate, narrow leaves typify monocots (cf. 32A).

1 Anthers 2 Stigma

2 Flowers of the dicot *Rhododendron falconeri*. Each shows a prominent corolla consisting of 5 fused petals, 10 stamens with 'strings' of pollen (1) extruded from terminal pores on their anthers and a central style (2, formed by five fused carpels) terminated by a prominent receptive stigma (3).

1 Pollen strings 3 Stigma
2 Style

3 Mature plants of the smallest known angiosperm, *Lemna minor*. These are hundreds of fully mature plants, each only a few millimetres across. Flowers are microscopic.

4 Gigantic trunk of the dicot *Eucalyptus diversicolor* (Karri tree). This is a native of southwest Australia and specimens up to 87 m tall and with a diameter of 4 m have been recorded. These trees have been extensively logged for their very strong timber which is used in the construction industry.

5 Ancient (now neglected) specimen of *Quercus petraea* (sessile oak) which was originally pollarded (in the 13th C in Scotland at Cadzow) at about 2 m above ground level (to prevent pastured cattle grazing on its foliage). Its spreading branches were used for boat building and general construction.

6 Most of the plants we eat are angiosperms, flowering plants, such as this wheat (*Triticum*). This structure is a set of numerous fruits which are surrounded by dry remnants of flower parts. Each fruit contains a single seed developed from a single inconspicuous flower.

7 Ripening fruits of *Coffea arabica*. Within each berry lie two coffee seeds ('beans') which are roasted and ground to make a delicious infusion. So-called instant coffee (which has, however, lost most of its distinctive coffee identity) is commercially prepared by spray- or freeze-drying various coffee brews.

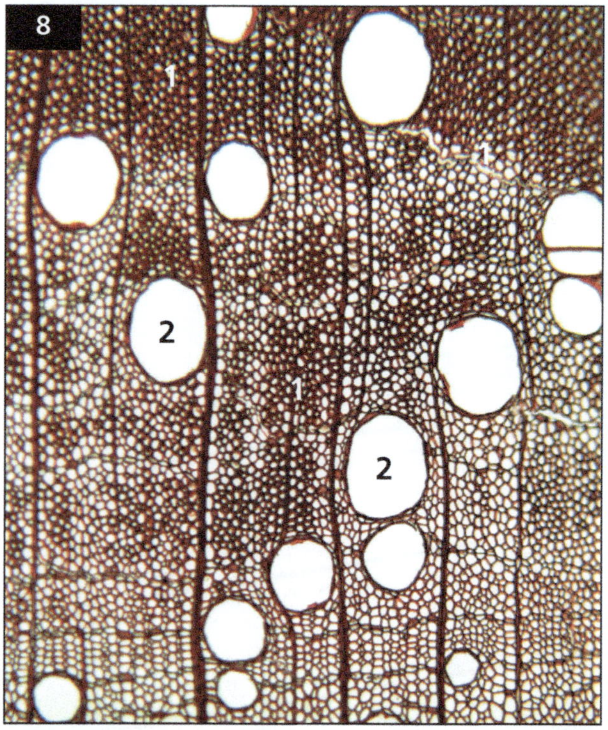

8 TS of *Juglans cinerea* (butternut) wood. Angiosperm trees provide hardwoods, many of which have beautiful colours and textures that make them suitable for fine woodworking. They typically have abundant fibres (1) that give strength and durability, as well as vessels (2) that conduct water. (LM.)

1 Wood fibres
2 Water conducting vessels

9 Original 14th C *Quercus robur* (English oak) beams supporting the roof of the main hall at Stokesey Castle, England.

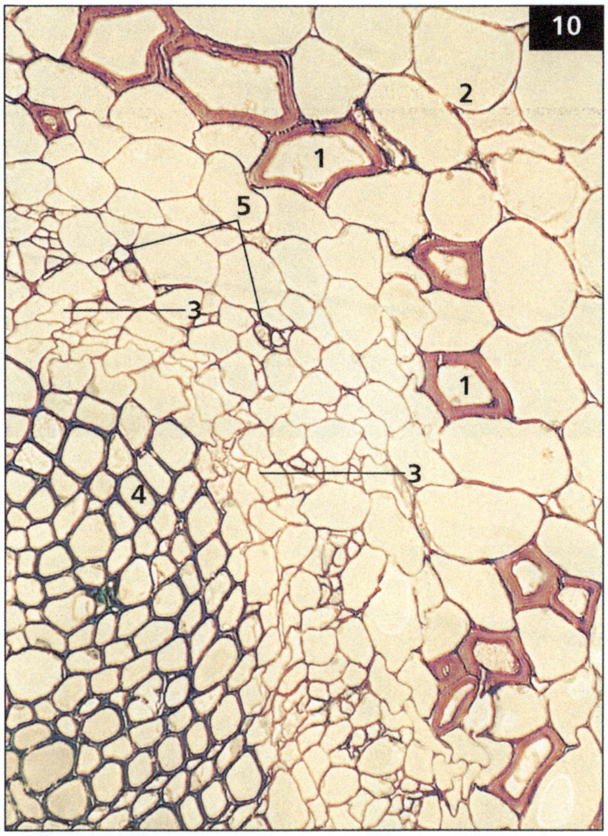

10 TS of the stem of the dicot *Linum usitatissimum* (flax) showing numerous large and thick-walled fibres (1) which are maturing adjacent to the thin-walled cortical parenchyma (2). Within the vascular cylinder a narrow cambial layer (3) is giving rise to secondary xylem (4), while groups of narrow elements (5) represent the translocating elements of the phloem. (G-Os, LM.)

1 Thick-walled fibres
2 Cortical parenchyma
3 Cambial layer
4 Secondary xylem
5 Translocating phloem

11 Specimen of the cycad gymnosperm *Cycas circinalis*, bearing numerous freely-exposed seeds. Unlike angiosperms, these are not enclosed within a fruit (cf. 46).

12 *Pinus ponderosa* (Western yellow pine) bearing female cones and needle-shaped leaves grouped in threes.

13 *Lycopersicon esculentum* (tomato) fruit. In TS the numerous seeds (1) are seen attached to axial placentas (2) and embedded in juicy tissue derived from placental tissue. The four locules indicate that the fruit developed from an ovary of four fused carpels. In the external view the remains of the flower stalk and the green sepals (3) are visible at its base; this fleshy dicot fruit developed from a single superior ovary which was situated above the perianth.

1 Seeds
2 Axial placentas
3 Sepals

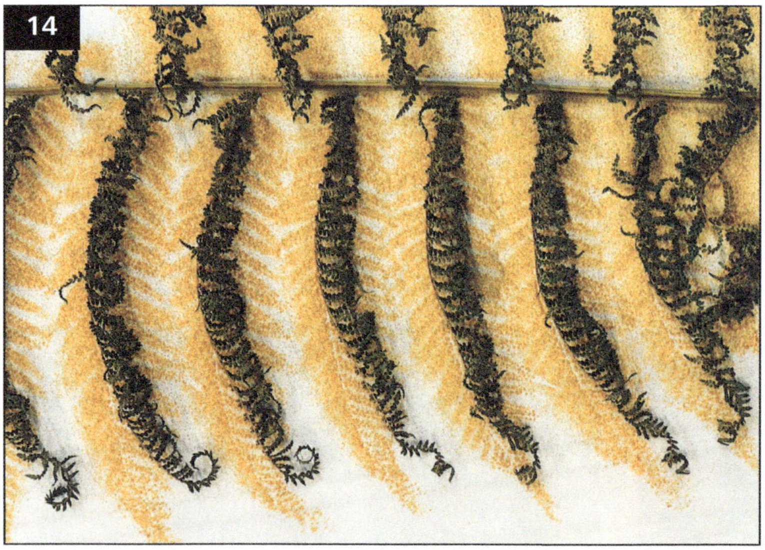

14 Portion of a withered fertile frond (leaf) of the subtropical tree fern *Dicksonia antartica.* The multitudinous sporangia on its abaxial surface have dehisced and shed millions of yellow, uniform-sized, haploid spores onto the bench surface.

15 These plants of *Equisetum arvense* (horsetails or scouring rushes) have vascular tissues but reproduce without seeds, thus they are vascular cryptogams similar to ferns. This is a single plant, spreading by underground rhizomes, with new buds emerging on the right.

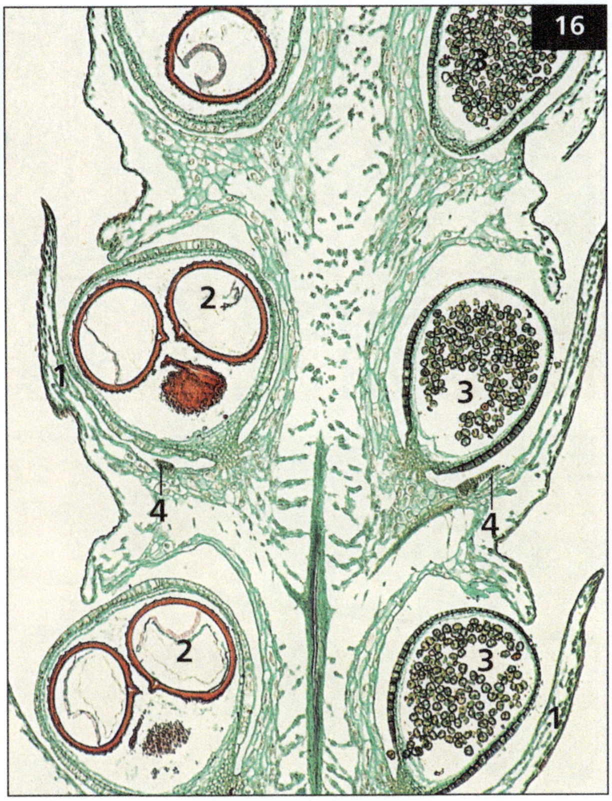

16 LS of the heterosporous cone of *Huperzia selago* (a fern ally). The sporangia are borne in the angle between the cone axis and the sporophylls (1). Within the sporangia the spore mother cells undergo meiosis: the megasporangia contain a few large and thick-walled haploid megaspores (2), while the microsporangia contain abundant small and thin-walled microspores (3). Note the small ligules (4) on the adaxial surface of the sporophylls; this feature shows the affinity of present day *Selaginella* with the fossil, *Lepidodendron*, the dominant tree of the Carboniferous coal measures. (LM.)

1 Sporophyll
2 Megaspores
3 Microspores
4 Ligule

17 A: Closely crowded gametophytic plants of the liverwort, *Pellia epiphylla*. Each green, flattened and lobed thallus bears male and female sex organs and after fertilization the embryo develops into the diploid sporophyte dependent on the gametophyte. A single mature sporophyte is visible, consisting of a seta (1) and terminal capsule (2). B: Dehisced capsule with some yellow spores still evident.

1 Seta
2 Capsule

18 The green plants at soil level are the closely crowded gametophytes of the moss *Polytrichum*. They grow as small, upright leafy haploid stems several centimetres tall in bright woodlands, and they bear male and female sex organs. Once an egg is fertilized, it develops into a diploid sporophyte with a stalk (1) and a sporangium (2; most are covered by a protective calyptra here). Although longer-lived and more complex than those of liverworts, the moss sporophytes are also dependent on the leafy gametophyte.

1 Sporophyte stalk
2 Sporophyte sporangium with calyptra

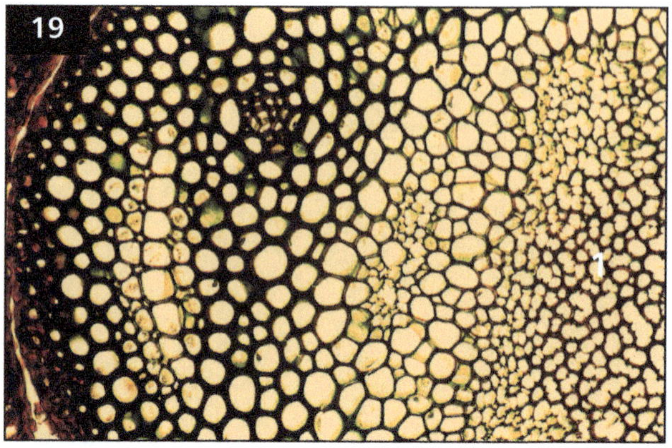

19 TS of the stem of the moss *Polytrichum* (hair moss) showing its considerable histological complexity with a central strand (1) of thicker-walled water conducting cells. (LM.)

1 Water conducting strand

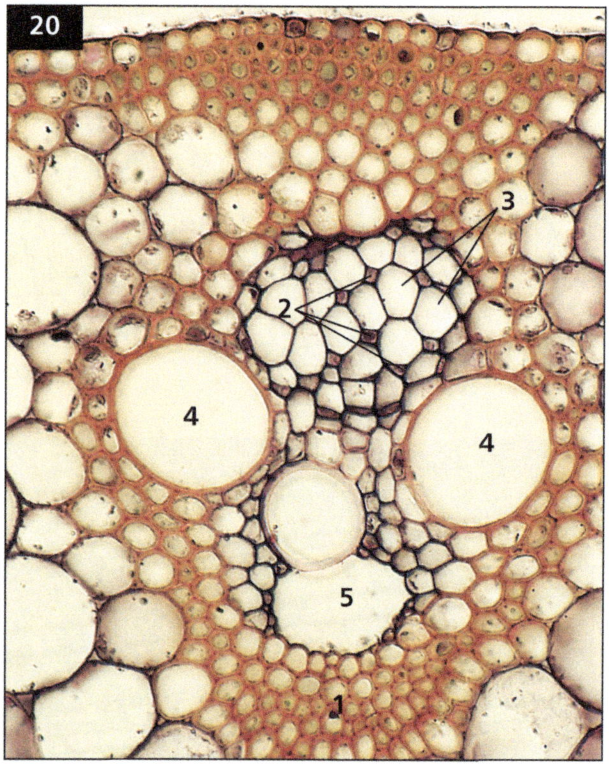

20 TS of the stem of the monocot *Zea mays* (maize) showing a peripheral vascular bundle. The bundle is invested by many small, thick-walled sclerenchyma fibres (1). The phloem lies on the exterior side of the xylem and its smaller, densely staining, companion cells (2) contrast with the larger sieve elements (3). The endarch xylem consists of several wide vessels (4) with thickened secondary walls. During internode elongation the innermost element has been overstretched, and its primary wall torn, and is now represented by a cavity (5). (LM.)

1 Sclerenchyma fibres
2 Phloem companion cells
3 Sieve elements
4 Xylem vessels
5 Cavity in xylem

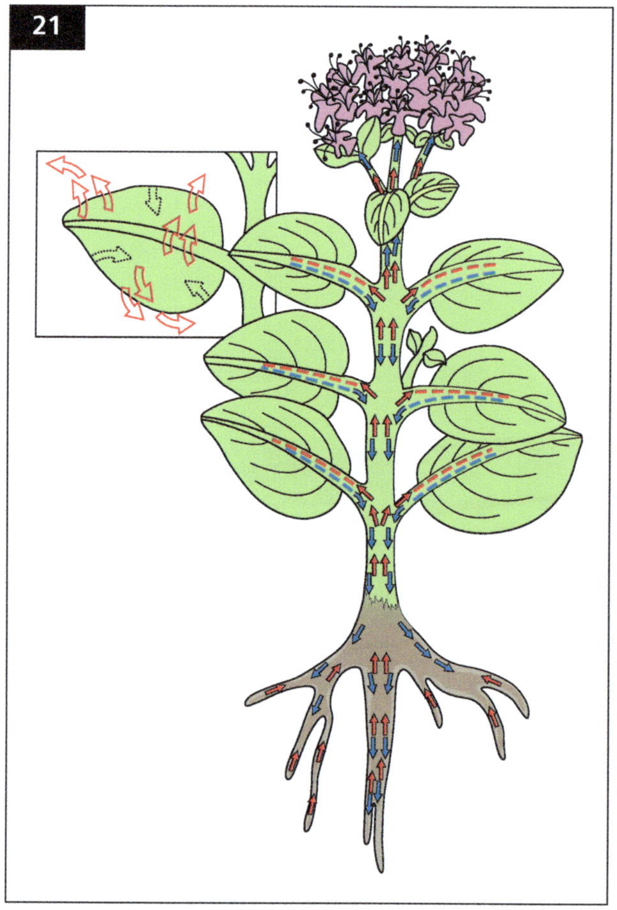

21 Diagram of a herbaceous flowering plant illustrating the following functions: 1. The flow of water in the xylem (red arrows) from the root to stem and transpiring leaves. 2. The translocation of sugars in the phloem (blue arrows) from the photosynthetic leaves to actively-growing organs or storage regions in the shoot and root. 3. The evaporation of water vapour (open arrows) into the atmosphere and the diffusion of carbon dioxide (stippled, open arrows) into the leaf via the open stomatal pores.

22 The broad lobed 'leafy' lichen *Lobaria pulmonaria* (tree lungwort). This is not a true plant but is composed of several symbiotic partners; embedded within a compact covering of fungal (ascomycete) hyphae lie photosynthetic green algal cells. In this lichen scattered colonies of prokaryotic blue-green algae also occur.

23 TS of the mature primary root of the dicot *Ranunculus* (buttercup). Note the parenchymatous cortex (1) packed with red-stained grains of starch and the central vascular cylinder which is surrounded by the endodermis (2). The thick-walled, dead xylem elements (3) conduct water through their lumina to the aerial shoot system. Strands of phloem (4) lie between the xylem arms. (LM.)

1 Parenchymatous cortex
2 Endodermis
3 Xylem elements
4 Phloem

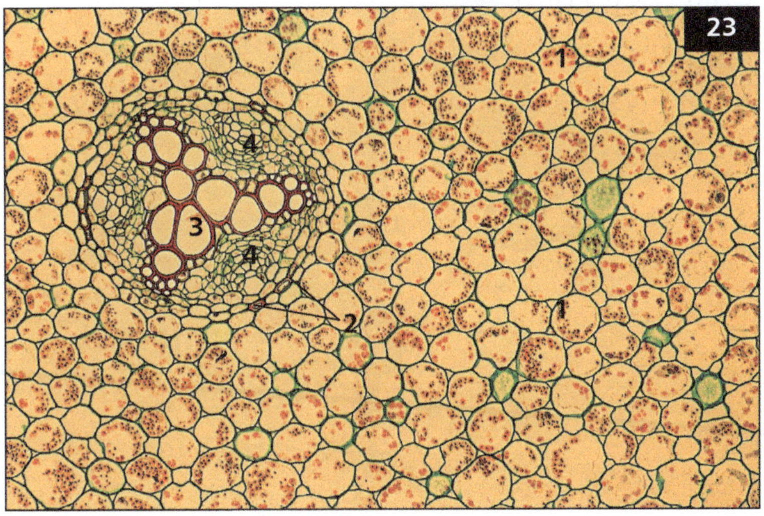

24 Cleared leaf of the gymnosperm *Taxus baccata* (yew). This is viewed from the abaxial surface and shows the epidermal stomata orientated parallel to the long axis of the leaf. (LM.)

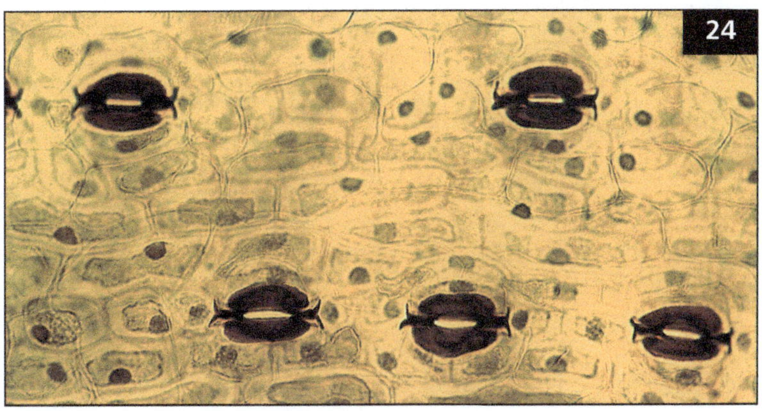

25

25 *Solanum tuberosum* (potato tuber). A group of amyloplasts showing the concentric lamellae present in the large starch grains; note also the small, yellow-green, amylochloroplasts (arrows). (Fresh section, LM.)

26

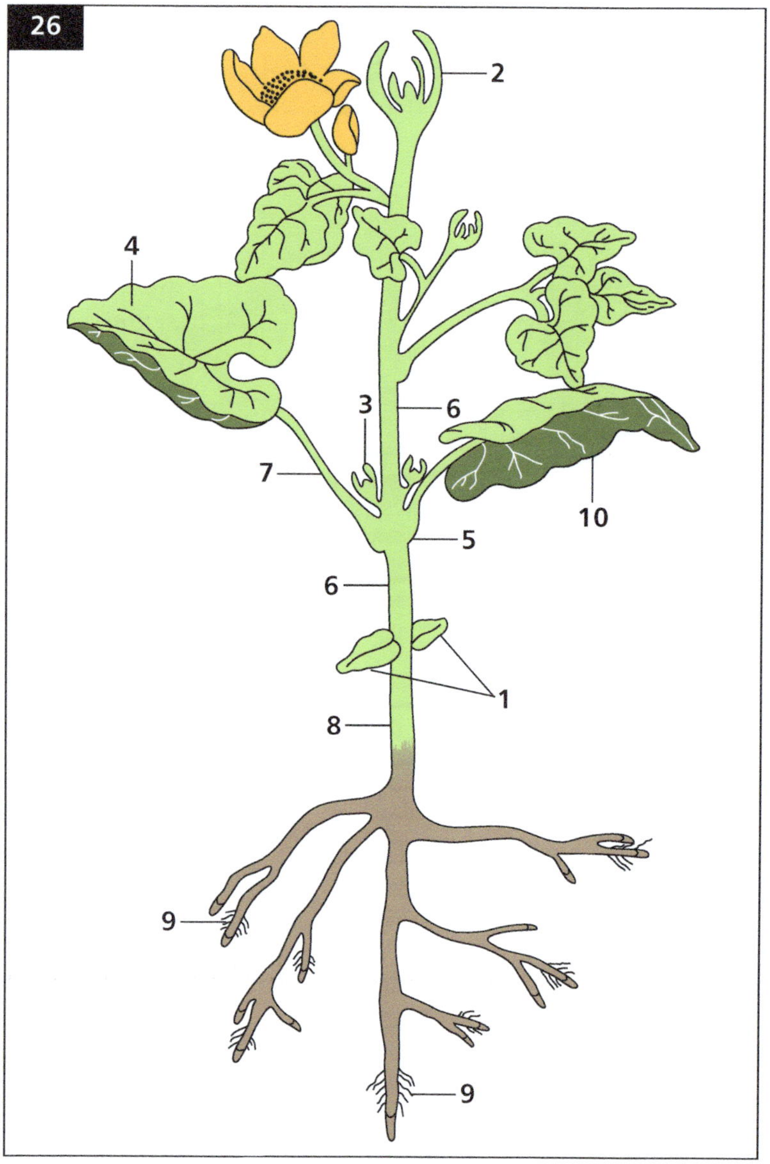

26 Diagram showing the morphology of a generalized herbaceous flowering plant. The two small cotyledons (1) at the base of the stem were the first leaves developed on the embryo and indicate that this is a dicot (cf. 32B). The shoot bears a terminal bud (2) while a number of lateral buds (3) lie in the axils between the stem and the adaxial face (4) of the leaves. The uppermost axillary bud has developed into a flower, but the remainder are vegetative. The leaves arise at the nodes (5) while the intervening regions of stem are the internodes (6). The leaf consists of the lamina borne on a petiole (7). In the hypocotyl (8) the vascular tissue changes from the arrangement in the stem to that of the root (cf. 32). The tap root and laterals bear numerous root hairs (9). Abaxial surface of leaf (10).

1 Cotyledons
2 Terminal bud
3 Lateral bud
4 Adaxial leaf face
5 Node
6 Internode
7 Petiole
8 Hypocotyl
9 Root hairs
10 Abaxial leaf surface

27 Simple leaves of the dicots *Kalanchoe diagromontiana* and *Begonia metallica*. Both are differentiated into lamina (1) and petiole (2). *Kalanchoe* (A) shows a single midrib in the lamina but in *Begonia* (B) several main veins are arranged palmately. Note the adventitious buds (3) which develop from the notched margins of the *Kalanchoe* blade; when these fall onto the soil they root and develop into new plants.

1 Lamina
2 Petiole
3 Adventitious buds

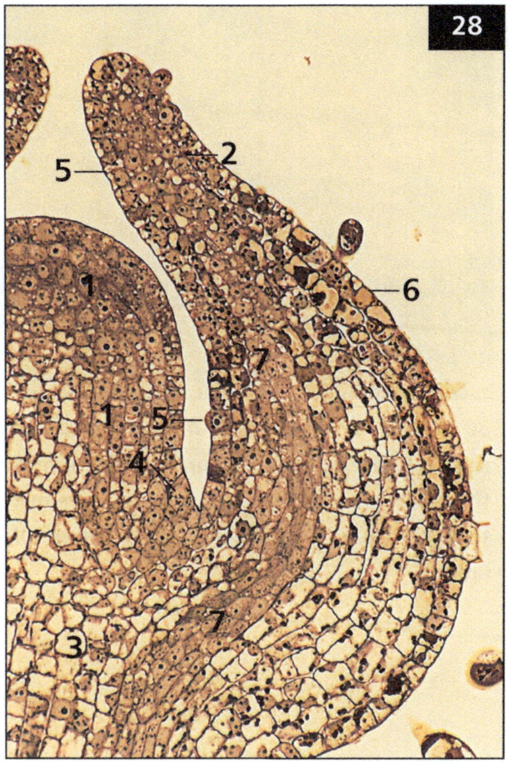

28 Median LS through the shoot tip of the dicot *Glechoma hederacea* (ground ivy). Note the hemispherical shoot apex (1) and leaf primordium (2) arising at its base. The core of the apex is somewhat vacuolated and will give rise to the pith (3) of the young stem. An axillary bud primordium (4) will develop at the adaxial base (5) of the leaf. The abaxial face (6) of the leaf shows considerable vacuolation and this is confluent with the cortex of the young stem. Procambium (7). (LM.)

1 Shoot apex
2 Leaf primordium
3 Pith
4 Axillary bud primordium site
5 Adaxial leaf face
6 Abaxial leaf face
7 Procambium

29 A compound leaf of the dicot tree *Sorbus aucuparia* (rowan, mountain ash) showing the numerous pinnately arranged leaflets which are, however, devoid of axillary buds; another leaf clearly reveals a bud (1) in its axil.

1 Axillary bud

30 Abaxial surface of the leaf blade of the dicot *Begonia rex*. Note the numerous lateral connections between the palmately arranged main veins. The smallest veins form a reticulum of polygons enclosing small islands of lamina.

31 Newly germinated seedlings of *Fagus sylvatica* (beech), each showing the two cotyledons (1) characteristic of a dicot, the small plumule (2), and the elongate hypocotyl (3) which links at soil level with the underground radicle. Note also the remnants of the fruit (nut) wall (4).

1 Cotyledons
2 Plumule
3 Hypocotyl
4 Fruit wall

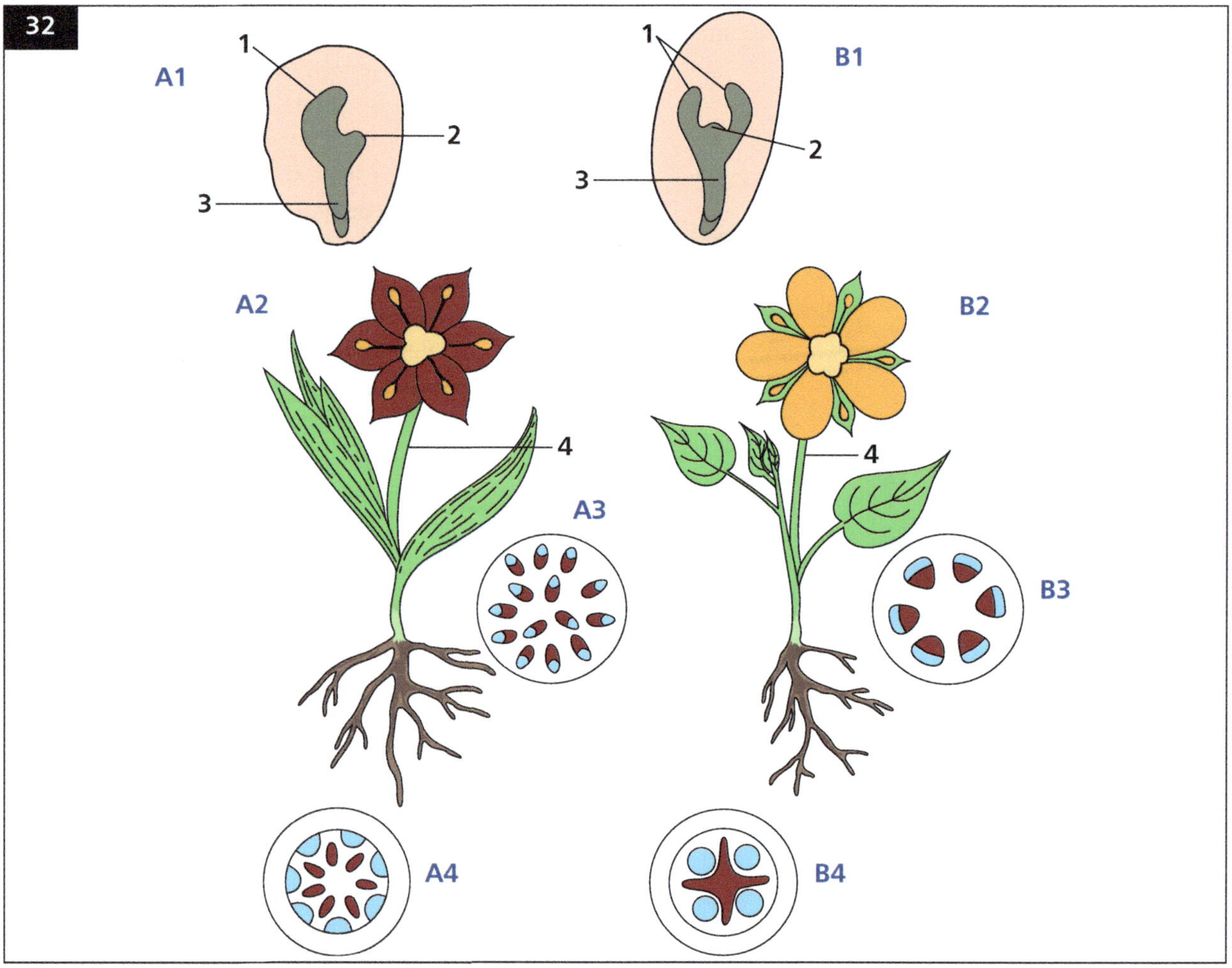

33 Tall trees of the tropical monocot *Yucca elephantipes*. In this and some other genera of arborescent monocots the secondary thickening meristem gives rise to discrete bundles of xylem and phloem which are embedded in a ground mass of fibres and parenchyma.

34 Leaves of *Iris* are sessile, lacking a petiole. They are also unusual for being extremely narrow from side to side but thick from top to bottom (described as sword-shaped). Like other monocots, they have a sheathing leaf base.

32 (Opposite) Diagrams showing the typical differences between monocots (series A) and herbaceous dicots (series B). A1, B1 are longitudinal sections of seeds; note the single cotyledon (1) in A1 but paired cotyledons (1) in B1. A2, B2 illustrate the following morphological differences: elongate leaves in monocots which are usually without petioles and with parallel veins; in dicots relatively shorter, usually petiolate, leaves occur which show branched and nonparallel veins. The floral parts are in threes in monocotyledons contrasting with fives (or fours or indefinite numbers) in dicots. A3, B3 and A4, B4 show the vascular systems in TS of the stem and root respectively, with xylem coloured red and the phloem blue. The monocot stem (A3) has many scattered vascular bundles consisting of inner xylem and outer phloem, whereas in dicot (B3) the stem bundles are usually arranged in a ring with vascular cambium present between the phloem and xylem. The roots of monocots (A4) are polyarch with numerous separate strands of phloem separated by xylem and a central pith is usually present. In dicots (B4) a pith is often lacking and from the xylem core several arms radiate outwards between the phloem strands. Plumule (2), radicle (3), pedicel (flower stalk, 4).

1 Cotyledon   2 Plumule   3 Radicle   4 Pedicel (flower stalk)

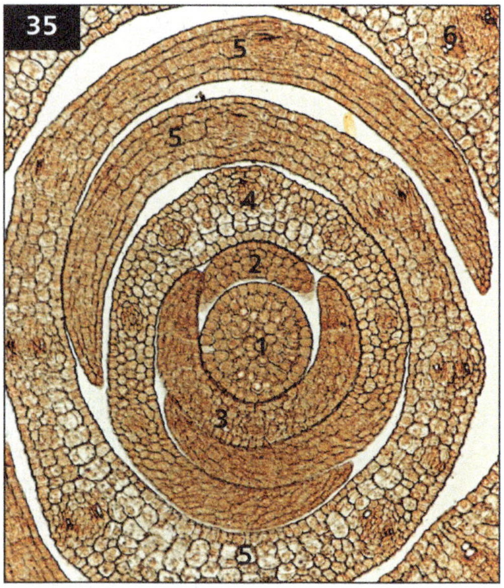

35 TS through the seedling shoot apex of the monocot *Zea mays* (maize). The shoot apex (1) is invested by progressively older leaf primordia (2–6) which are arranged in two alternate rows. These leaves are enclosed by the single coleoptile, a leaf-like cylindrical structure which protects the plumule before its emergence at germination. (LM.)

1	Shoot apex	2–6	Leaf primordia

36 Each leaf of this palm (*Pseudophoenix vinifera*) has a blade (not visible here), a petiole, and a sheathing base that encircles the bases of younger leaves as well as the stem itself; each leaf has a broad attachment to the stem and many vascular connections. If cut in TS, the concentric leaf bases would look like the rings of an onion or those of *Zea* (35). (Photographed at Montgomery Botanical Center, Florida.)

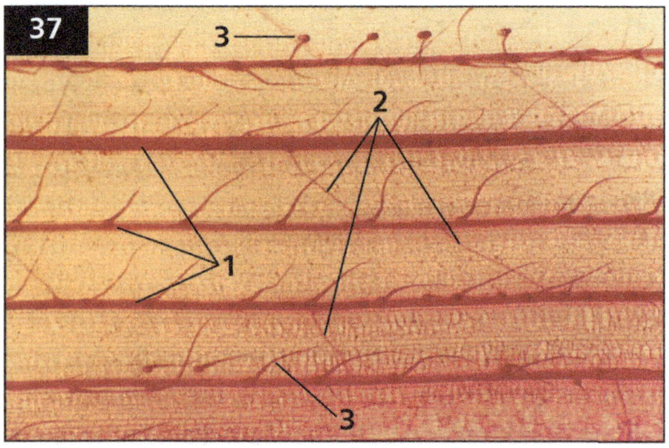

37 Cleared leaf blade of the grass *Bromus* (a monocot). The main longitudinal veins (1) are intermittently connected by fine oblique veins (2). Note the numerous trichomes (3) which arise from the epidermis above the main veins. (LM.)

1	Longitudinal veins	3	Trichomes
2	Oblique veins		

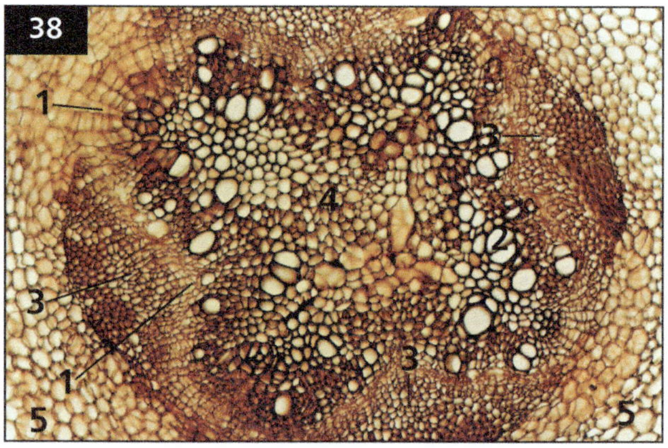

38 TS of the older root of *Vicia faba* (broad bean), a dicot. At this distance behind the apex secondary thickening has begun, and the convoluted vascular cambium (1) has formed considerable secondary xylem (2) and some secondary phloem (3). Primary xylem (4), cortex (5). (LM.)

1	Vascular cambium	4	Primary xylem
2	Secondary xylem	5	Cortex
3	Secondary phloem		

39 Trunks of several specimens of the arborescent monocot *Pandanus.* Note the numerous wide adventitious prop roots which run obliquely outwards from the trunks and grow down into the soil, thus stabilizing the tree.

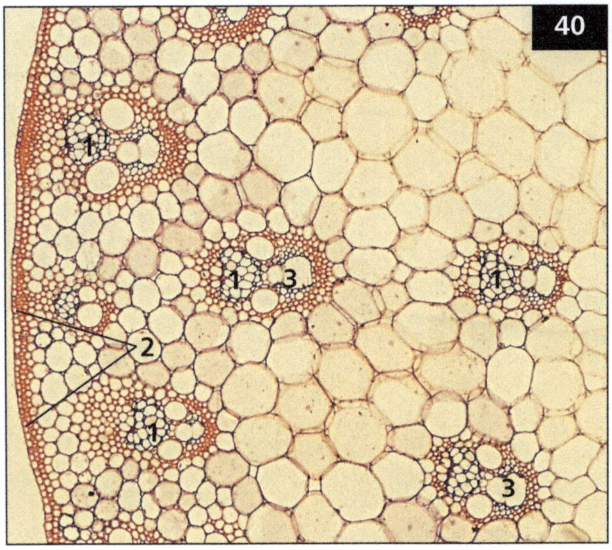

40 TS of the stem of the monocot *Zea mays* (maize). Numerous scattered collateral vascular bundles lie in parenchymatous ground tissue and in each the phloem (1) is situated nearest to the epidermis (2). Vascular cambium is absent between the xylem (3) and phloem (cf. 20). (LM.)

1 Phloem
2 Epidermis
3 Xylem

41 TS through the young stem of the dicot *Helianthus* (sunflower). The collateral vascular bundles lie in a peripheral cylinder with an extensive pith (1) situated internally and narrow cortex (2) lying externally. Each bundle is demarcated on the outside by a cap of phloem fibres (3) and a prominent fascicular vascular cambium (4) separates the xylem (5) from the phloem (6); but no interfascicular cambium has yet differentiated in the parenchyma between the bundles. (LM.)

1 Pith
2 Cortex
3 Phloem fibres
4 Fascicular vascular cambium
5 Xylem
6 Phloem

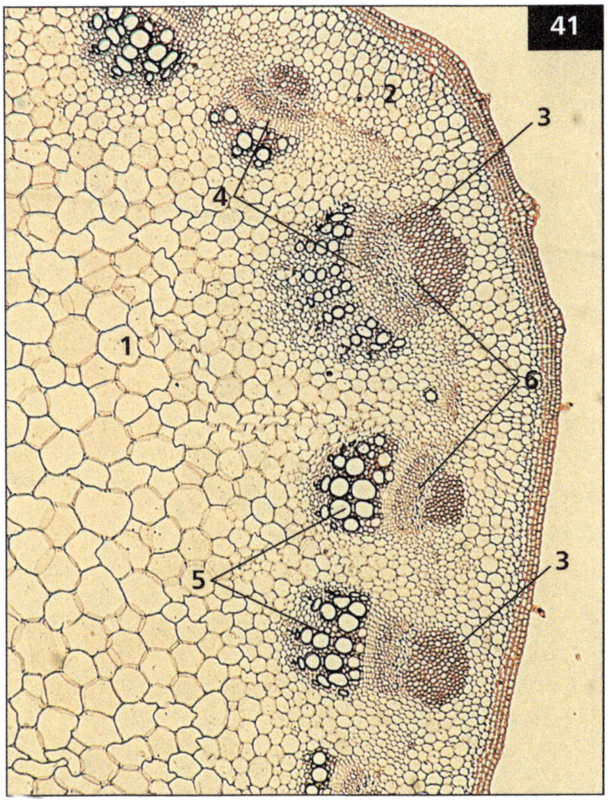

42

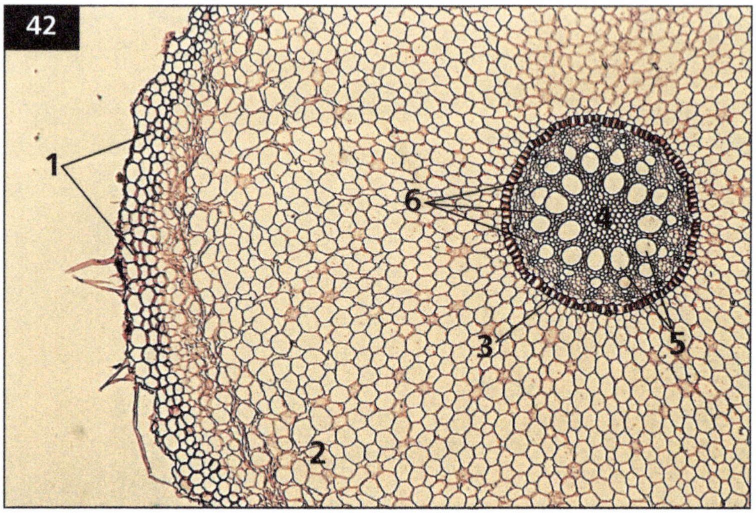

42 TS of the old root of the monocot *Iris*. Note the several-layered, thickened, exodermis (1), enclosing the wide parenchymatous cortex (2). A single layered, thickened endodermis (3) is also present and the vascular tissue enclosed within it shows a polyarch arrangement. The lignified ground tissue (4) is surrounded by radially aligned, wide xylem vessels (5) but with narrower tracheary elements adjacent to the endodermis, while phloem strands (6) occur between these narrower xylem elements. (LM.)

1 Exodermis
2 Parenchymatous cortex
3 Endodermis
4 Ground tissue
5 Xylem vessels
6 Phloem

43

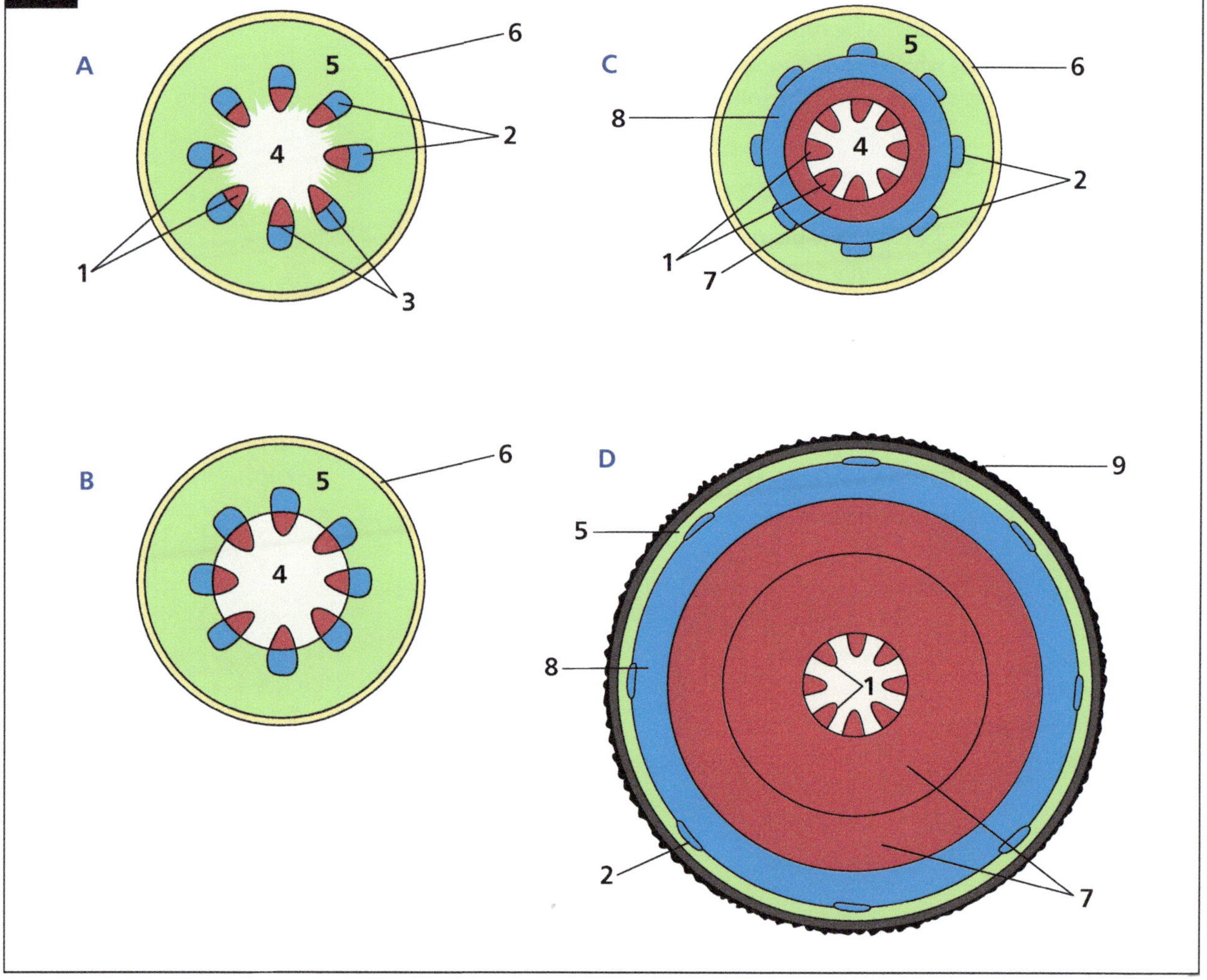

44 Inside view of a *Magnolia* flower (a magnoliid basal angiosperm). Some perianth members were removed to show the elongated receptacle which bears numerous spirally-arranged carpels (1) at its tip and numerous stamens below (2). The stigmatic surfaces of the carpels occur on the curved adaxial faces of the styles. The anther occupies the terminal two-thirds of each stamen.

1 Carpels 2 Stamens

43 (Opposite) Diagrammatic representation of secondary thickening in a woody stem in TS. In the young internode (A) a ring of vascular bundles occurs with primary xylem (1) lying internally, primary phloem externally (2), and a thin layer of fascicular vascular cambium (3) between them. The bundles are separated by ray parenchyma which extends from the pith (4) to the cortex (5). The epidermis (6) delimits the stem externally. In B, the vascular cambium has become a continuous cylinder and by the end of the first season's growth (C) it has formed a continuous ring of secondary xylem (7) internally and a thinner layer of secondary phloem (8) externally. At the end of the next season's growth (D), two growth rings are visible in the secondary xylem but the primary phloem and the first year's growth of secondary phloem is becoming disorganized by the expansion of the stem. The epidermis has been replaced by a layer of cork (9) which usually arises in the outer cortex.

1 Primary xylem
2 Primary phloem
3 Fascicular vascular cambium
4 Pith
5 Cortex
6 Epidermis
7 Secondary xylem
8 Secondary phloem
9 Cork

45

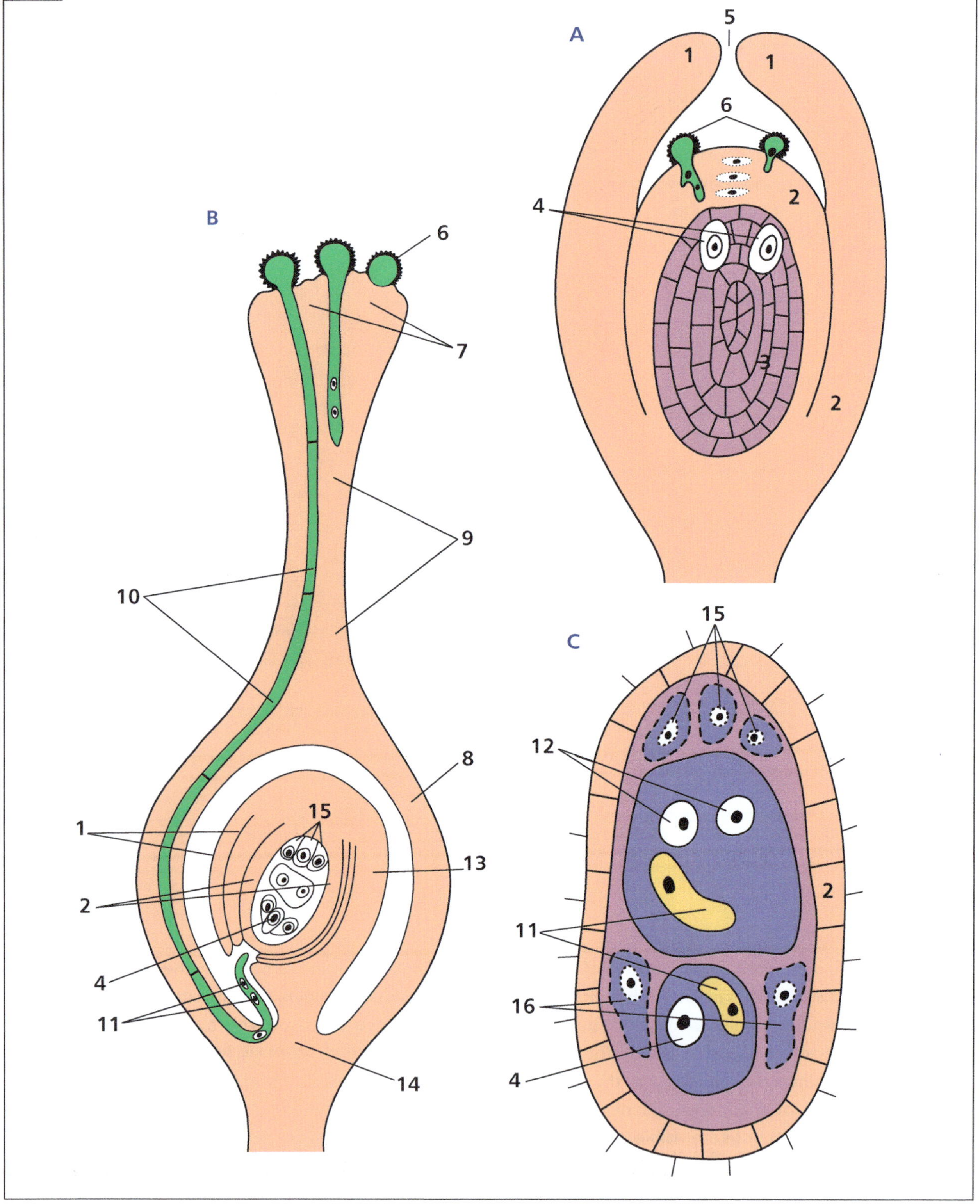

45 (Opposite) Diagrams of sexual reproduction in seed plants. In the naked ovule of a gymnosperm (A), an integument (1) invests the megasporangium (2); within, a single haploid megaspore has enlarged and divided to form the female gametophyte (3) which shows several large egg cells (4). The megaspore was one of four meiotic derivatives but the other three (dotted) have degenerated. The megasporangium connects with the outside of the ovule via a narrow micropyle (5) allowing entry of haploid pollen grains (6); these have now germinated on the megasporangium surface. A fertilized egg will give rise to a diploid embryo which will be nourished by the female gametophyte.

B shows the ovule of an angiosperm within a carpel. The latter comprises a receptive stigma (7) on which several pollen grains (6) have been deposited, a basal ovary (8), and the intermediately-located style (9). The single ovule within the ovary is composed of two integuments (1) which enclose the nucellus (megasporangium) containing a large embryo sac derived from the megaspore. Within the mature sac eight haploid nuclei typically occur, with the egg (lying between a pair of synergids) near to the micropyle. A long pollen tube (10, plugged at intervals by callose) has grown down the style and its tip (adjacent to the micropyle) contains two haploid male nuclei.

C shows detail of the embryo sac of an angiosperm at fertilization. Two male gametes have been liberated into the embryo sac and one male nucleus (11) is about to fertilize the egg nucleus (4) to form a diploid zygote, while the other will fuse with the two polar nuclei (12) to form triploid nutritive endosperm tissue. The other nuclei in the embryo sac normally degenerate. Funiculus (13), placenta (14), antipodals (15), synergids (16).

1 Integument
2 Megasporangium
3 Female gametophyte
4 Egg cell
5 Micropyle
6 Pollen grains
7 Stigma
8 Ovary
9 Style
10 Pollen tube
11 Male nuclei
12 Polar nuclei
13 Funiculus
14 Placenta
15 Antipodals
16 Synergids

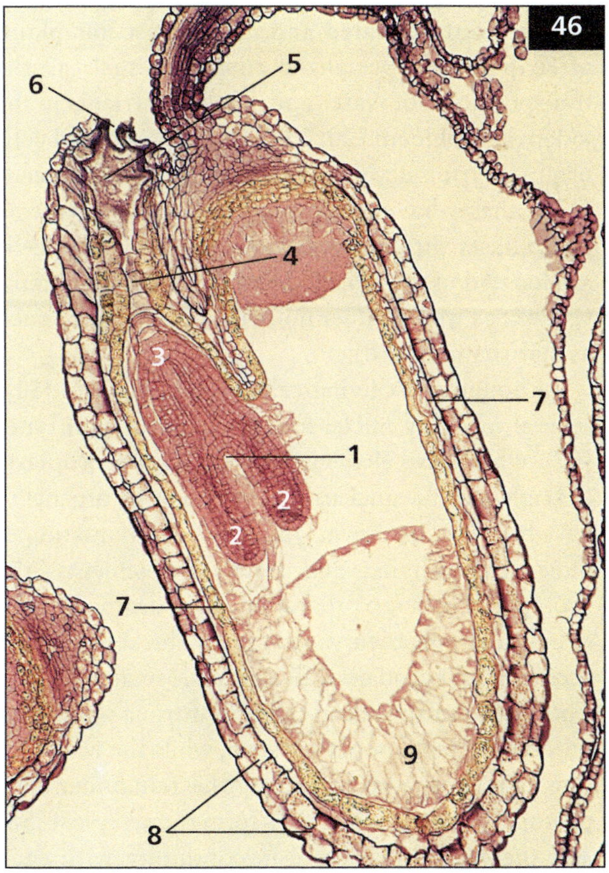

46 LS of an immature seed of the dicot *Capsella bursa-pastoris* (shepherd's purse). The apical pole of the enclosed embryo shows a rudimentary plumule (1) lying between a pair of prominent cotyledons (2), while the basal pole is terminated by the radicle (3). This is attached to a filamentous suspensor (4) which terminates in a large basal cell (5) at the micropyle (6). The embryo is enclosed by the nucellus (7) and integuments (8); within the embryo sac cellular endosperm (9) is forming. (LM.)

1 Rudimentary plumule
2 Cotyledons
3 Radicle
4 Filamentous suspensor
5 Basal cell
6 Micropyle
7 Nucellus
8 Integuments
9 Cellular endosperm

CHAPTER 2

The Plant Cell

INTRODUCTION

Even in a small green plant such as *Lemna* (**3**) there are millions of cells, each delimited from its neighbours by a cell wall. The vast majority of these cells are differentiated and, in the vascular plant, often perform specialized functions such as the transportation of water and soluble nutrients in the xylem and phloem (**20, 21**). All differentiated cells originate from actively dividing meristematic cells (**47**); these have thin primary walls investing protoplasts with dense cytoplasm (**48**). These cells are located in the apical (**28**) and lateral meristems (**10, 49**) as well as other more localized regions such as meristemoids (**50**).

Although most living cells are uninucleate (**51**), several or many nuclei may occur in certain types (**52**), while sieve elements (**20**) contain protoplasts (**53**) in which the nucleus and several other organelles have been broken down. During the differentiation of most sclerenchyma and tracheary elements, the protoplasts undergo programmed cell death (**54–56**); at maturity only their walls, greatly thickened by the accretion of secondary wall deposits, remain (**20, 57**). The lumina of such dead cells (**58**) form a significant part of the plant apoplast system, while the walls and intercellular spaces constitute the remainder. The protoplasts of the living cells form the symplast (**58**) and these protoplasts are in continuity with each other via the plasmodesmata (**58–60**). Between 1000 and 10,000 such protoplasmic connections occur per cell, but it seems that only relatively small molecules (with a molecular weight of less than 1000) are able to pass through them.

The fine structure of a partly differentiated cell is illustrated diagrammatically in **61**. The thin external primary wall is delimited from the protoplast by the membranous plasmalemma. Several large organelles are present including the nucleus, several vacuoles, and a number of chloroplasts. Normally under the light microscope (LM), only the wall and these larger organelles would be apparent (**47, 62**). However, with the greater resolution of the transmission electron microscope (TEM), mitochondria, endoplasmic reticulum, dictyosomes, ribosomes, microtubules, and plasmodesmata are also distinguishable (**61**). Most of these organelles are membrane-bounded (**51, 55, 59–61, 63**).

In this chapter, only the fine structural features of plant cells are considered, whereas the histological structure of differentiated cells and tissues is discussed in Chapter 3.

CELL MEMBRANES

Substances located exterior to the plasmalemma, or in the cytosol surrounding membrane-bounded organelles (**51, 61**), cannot mix freely with the materials localized internally because these membranes are semi-permeable. Membranes consist of a lipid bilayer (**63**), with the interspersed proteins and complexes forming the molecular pumps, enzymes, and other structural components. Most proteins are large and project onto the surface of the membrane (**63–65**). Differing types of organelles normally remain discrete within the cell since their membranes vary somewhat in individual structure.

The plasmalemma (**59, 60, 63**) and the membranes of mature dictyosome cisternae and vesicles (**55, 66**) are generally the thickest membranes of the cell and measure about 10 nm wide. When viewed in transverse section (in chemically fixed material), membranes usually show a tripartite appearance (**48, 66, 67**); but in freeze-fractured specimens the plasmalemma (**64, 65**) and other membranes (**68, 69**) show numerous particles which are protein complexes (**63**).

The plasmalemma adjacent to the plant cell wall sometimes reveals hexagonal arrays of particles which are possibly the sites of cellulose microfibrillar synthesis (**63**). Likewise in *Saccharomyces* (yeast), chitin microfibrils in the wall apparently link with particles in the plasmalemma (**65**).

NUCLEUS

The genetic material of the cell is primarily located in the nucleus (**47, 61, 70**). The nondividing nucleus is bounded by an envelope composed of the outer and inner membranes (**48, 61, 69**). These are separated by a perinuclear space approximately 20 nm wide, but are confluent at the margins of the abundant nuclear pores (**61, 68, 69**). These pores are approximately 70 nm wide but are apparently partly occluded by a complex fibrillar–particulate network. In the meristematic cell (**51, 70**) the nucleus may occupy one half or more of the volume of the protoplast, but this ratio rapidly decreases as the cell increases in size, with the individual small vacuoles (**51**) expanding and fusing to form a large central vacuole (**71**). The nucleus sometimes becomes highly lobed (**62, 70**) and in elongate, narrow cells may be spindle-shaped.

The chromatin (DNA complexed with histones) is not organized into chromosomes in the interphase cell of higher plants but regions of densely staining heterochromatin and lighter euchromatin are often visible (**61, 70, 71**). One to several nucleoli (**52, 72**) occur within the nucleoplasm; these contain stores of ribosome precursors (**73**) which apparently migrate into the cytoplasm via the nuclear pores. Prominent vacuoles are sometimes evident within the nucleoli (**62, 72, 73**). The nuclei of meristematic cells are usually diploid; however, DNA replication in interphase is not necessarily followed by nuclear division and in actively metabolizing tissues the cells are frequently polyploid.

From the onset of mitosis (prophase) the chromatin condenses into discrete chromosomes which become aligned at metaphase (**74**) on the equator of the mitotic spindle. Meanwhile the nucleolus has disappeared and the nuclear envelope fragmented. The fibres of the mitotic spindle, which are just visible at LM level (**75**), actually consist of aggregated bundles of microtubules. These are generated at the poles of the spindle and some fibres attach to chromosomes whereas others run between the two poles.

At metaphase each chromosome consists of two chromatids joined at their kinetochores, while microtubules are linked to each kinetochore (**74**). At anaphase the sister chromatids separate, pulled by the fibres to opposite poles of the mitotic spindle (**76**). Finally, at telophase, the chromatin becomes dispersed so that discrete chromosomes are no longer visible. Each nucleus becomes invested by an envelope, the nucleoli reappear, while a cell plate (new cell wall) separates the two progeny nuclei (**72**).

PLASTIDS

A variety of plastids with differing metabolic significance occurs in plants (**77**) but, within a particular cell, generally only one form is present (**71**). However, in dedifferentiating cells both the mature and juvenile types may occur (**25, 62**). All plastids are bounded by a membranous envelope (**77, 79**); the outer membrane is continuous but the inner membrane sometimes shows invaginations into the matrix (stroma). The latter contains ribosomes (70S) and circular DNA, which are both chemically distinct from those of the nucleus, while starch and lipid droplets (plastoglobuli) are frequently present (**59, 77, 79**). Internal membranes usually occur in the stroma and these sometimes form complex configurations (77, 79). Plastid interconversions are common (77); for instance, amyloplasts can turn green (80) and form chloroplasts, while the latter may divide to form young chloroplasts (77, 81) or senesce and give rise to chromoplasts (82).

Proplastids

These precursors of other plastids are usually 1–2 μm in maximal width and may be rounded or amoeboid and they contain few internal membranes (83). Proplastids occur in meristematic cells (48, 83) and in the root apex up to 40 are present per cell. Proplastids divide and their population remains more or less stable in the cells formed during the numerous divisions within the apical meristems. However, as the derivatives of the apical cells grow and differentiate (28) their proplastid populations generally increase, and the proplastids develop into the various types characteristic of different mature tissues and organs (77).

Chloroplasts

At maturity these are usually oval to lenticular and 5–10 μm in length (71, 79, 81). In the leaf mesophyll (71) up to 50 chloroplasts per cell are commonly present, but in some species they are even more abundant. The membranes of the envelope are separated by a space 10–20 nm wide (79) and the inner member sometimes shows connections with the thylakoid membranes in the stroma. These membranes are predominantly orientated parallel to the long axis of the plastid (79, 81) and are normally elaborated into a complex three-dimensional, photosynthetic system. This consists of grana interconnected by stromal lamellae or frets (79, 81, 84).

Each granum consists of 2–100 flattened and stacked membranous discs and each granal membrane encloses an intrathylakoidal space (79, 85). The sub-structure of the granal membranes is highly complex, with the chlorophyll molecules being integral components, and their functioning in the light reactions of photosynthesis is under intensive research. The major proteinaceous component in the stroma is the enzyme ribulose bisphosphate carboxylase.

Starch is commonly present in the stroma of the chloroplasts (71, 81) but represents a temporary store of excess carbohydrate. Plastoglobuli (79, 81) are frequent and contain pigments concerned in electron transport, while phytoferritin deposits (86), a storage form of iron, are sometimes observed. During leaf development the chloroplasts increase in number per cell by division (77, 81).

Etioplasts

In most flowering plants grown in the light the proplastids rapidly differentiate into chloroplasts in the young leaves (81), but in dark-grown plants etioplasts develop (77, 87). These contain an elaborate membranous prolamellar body with radiating lamellae but, on exposure to light, these rapidly form a granal–fretwork system and protochlorophyllide is converted to chlorophyll. In some plants, for example grasses, the leaf primordia are tightly ensheathed by the older leaves (35, 36); in such a darkened internal environment etioplasts initially differentiate in the mesophyll but the leaf blade turns green as it grows out free from the enclosing leaf bases.

Amyloplasts

In storage parenchyma cells long-term deposition of starch occurs in the amyloplasts (31, 78, 88) in which internal membranes are few, but one to numerous starch grains occur. The starch consists of varying proportions of amylose and amylopectin and is deposited in layers which may be visible at LM level (25). Amyloplasts in *Solanum tuberosum* (potato tuber) reach 20 μm in width and the development of several large starch grains within the plastid often causes the envelope to break and release the grains into the cytoplasm (78).

When starch is mobilized in the germinating cotyledon of *Phaseolus* (bean), the grain is initially digested from the centre and fragmented grains may be visible in the cytoplasm (89). In root cap cells, nodal regions and sometimes elsewhere (90), large sedimented amyloplasts occur and are apparently concerned with gravity perception. Amylochloroplasts (51) are common in the young shoot; these have thylakoids but also contain prominent deposits of storage starch. In plastids of *Helianthus*, carbohydrate is stored as insulin.

Chromoplasts

Yellow, red, and orange plastids are designated as chromoplasts and these accumulate a variety of carotenoid pigments (in globular or crystalline form) which colour many flowers and fruits (**13**). The changing foliage colours of deciduous trees prior to leaf fall (**91**) are caused by the degeneration of the thylakoid system of the chloroplasts, with the carotenoids accumulating in numerous plastoglobuli (**82**). However, many chromoplasts do not represent degenerate chloroplasts but develop directly from proplastids or via amyloplasts (**77, 92**).

Leucoplasts

In the epidermis of the green shoot (**93**) and the mesophyll of variegated leaves, nonpigmented leucoplasts often occur. These contain few internal membranes and little or no starch.

MITOCHONDRIA

Aerobic respiration occurs in these organelles and, as with plastids, mitochondria show nucleoid regions containing circular DNA (**94**) and ribosomes. Mitochondria are spherical, elliptical, or irregular in shape (**48, 94, 95**). They are delimited from the cytoplasm by an outer membrane that has a high lipid content. Their inner membrane shows frequent invaginations into the stroma to form irregular inflated tubules (**94**) or cristae (**95**) which contain enzymes of the electron transport chain. Krebs cycle enzymes mainly occur within the stroma. Mitochondrial cristae (**95**) are especially well developed in rapidly respiring tissues. Although a sectioned cell may show numerous small mitochondria which are one to several micrometres wide (**95**), such apparently separate organelles may represent segments of larger polymorphic individuals.

ERGASTIC SUBSTANCES

Plants cells store numerous food reserve compounds of which starch, synthesized in the amyloplasts, is the commonest (**62, 78, 88–90**) but protein bodies and fat (lipid) bodies also occur, especially in seeds and fruits (**96, 104**). Deposits of tannin (a phenol which protects plants against insect predators) commonly occur in bark, seed coats, and leathery leaves (**302**). Latex (rubber) is synthesized in the laticifers of *Hevea* and many other genera (see Chapter 3). Various forms of crystals are found in plant tissues and calcium oxalate deposits are the most common, as in elongate bundles of raphides (**97**) and star-shaped druses (**98**). Calcium carbonate cystoliths (**99**) are deposited on invaginations of the epidermal cell walls in the Moraceae and several other families. Silica bodies (silicon dioxide) frequently occur in the epidermal cells of various grasses while the cells of many desert plants contain copious mucilage which binds water and helps prevent desiccation.

ENDOPLASMIC RETICULUM (ER)

This system ramifies throughout the protoplast (**61**) and in plants it is commonly cisternal (lamellar) in form (**94, 100**). It is delimited by a single membrane (**94**) which is often confluent with the outer nuclear membrane and is also continuous with the central (desmotubular) component of plasmodesmata (**60, 61**). In rough endoplasmic reticulum (RER) the outer surface of the membrane is studded with ribosomes of 17–20 nm diameter (**94**) which are somewhat larger than the mitochondrial and chloroplast ribosomes.

The proteins synthesized by the ribosomes sometimes accumulate within the lumen of the associated ER and may be transported in the ER to other sites in the cell. Smooth endoplasmic reticulum (SER) lacks ribosomes and is particularly concerned with lipid synthesis. Extensive tracts occur in oily seeds. The ER system is believed to provide the essential proteinaceous and lipidic components for the other membranous systems of the protoplast.

GOLGI APPARATUS

This comprises several to numerous discrete membranous dictyosomes (Golgi bodies) per cell (**55, 66, 101**). Each consists of a plate-like stack (**61, 102**) of smooth cisternae 1–2 µm in diameter. Their margins are frequently branched into a tubular network (**101**) which proliferates a number of small vesicles. The cisternae are separated from each other by about 10 nm, but individual dictyosomes remain intact when isolated from the cell.

In the longitudinal view of a dictyosome a polarity is sometimes evident (**102**). At its forming face the cisternae are thought to be reconstituted from vesicles budded off from adjacent membranes of the

ER (**61, 102**). The cisternal membranes progressively increase in thickness across the dictyosome and at its maturing face the cisternae frequently become concave, with vesicles budding off from their margins (**61, 101, 102**).

The vesicles apparently migrate and fuse with the plasmalemma where the contents are voided into the apoplast. Within the dictyosome, proteins derived from the RER combine with various sugars; the secreted vesicles contain carbohydrates and glycoproteins concerned with cell wall synthesis (**55**), mucilage (**66**), or nectar secretion.

Dictyosomes are particularly abundant in cells actively undergoing extensive wall thickening or forming new dividing walls (**55, 103**); in root cap cells, where cisternae often become inflated with mucilage and sloughed off whole (**66**); and in glandular cells. In actively secreting cells a vast potential excess of membranous components (derived from the dictyosomes) arrives at the plasmalemma, but some of this material is apparently returned to the cytoplasm in the 'coated' vesicles (**61**).

VACUOLE

In the meristematic cell a number of small vacuoles occurs, each bounded by the membranous tonoplast (**48, 51, 60**). However, during cell growth these vacuoles massively enlarge and fuse so that up to 90% of the volume in a parenchyma cell is occupied by the vacuole, while the cytoplasm and other organelles are mainly peripheral (**71, 88**). The vacuole contains various solutes (normally at about 0.5 M concentration), and its consequent turgidity greatly contributes to the turgor of the whole protoplast.

The vacuole also contains a number of hydrolytic enzymes and the tonoplast breaks down during differentiation of sclerenchyma and tracheary cells (**54–56**). The enzymes which are liberated digest the protoplasts so that only the walls remain intact (**10, 20**). Vacuoles may contain anthocyanins, and other pigments and also become modified as protein bodies in a number of seeds (**89, 104**).

MICROBODIES

These small, membrane-bounded bodies are of two types. Peroxisomes occur in close proximity to chloroplasts (**84, 85**); they contain a variety of enzymes which oxidize the glycolic acid resulting from photorespiration and then return glyoxylic acid to the chloroplasts. Additionally, in another pathway glyoxylic acid is transformed into glycine and transferred to the mitochondria. Glyoxysomes occur in fatty seeds and contain enzymes catalysing fatty acid breakdown, which releases energy during germination, while the hydrogen peroxide produced is broken down by peroxidase.

RIBOSOMES

The cytoplasmic ribosomes are 17–20 nm wide and occur both floating free within the cytosol (cytoplasmic ground substance) as well as attached to the outer surfaces of the RER (**94, 100, 101**). Ribosomes contain RNA and protein and are composed of two sub-units which are synthesized in the nucleolus (**73**) but subsequently combine in the cytoplasm. Ribosomes are especially dense in cells which are rapidly synthesizing protein, where they frequently occur in clusters termed polysomes (**94**). The cytoplasmic ribosomes (80S) are slightly larger and biochemically distinct from those located in the plastids and mitochondria (70S).

MICROTUBULES AND MICROFILAMENTS

The microtubules are proteinaceous structures about 25 nm wide, with a hollow core 12 nm wide, and may be up to several micrometres long (**105**). In the nondividing cell they are normally located adjacent to the plasmalemma (**48, 105, 106**), but at nuclear division these become reassembled and aggregate into the fibres of the mitotic spindle (**75, 106**). When mitosis is completed, the microtubules apparently guide dictyosome vesicles to the equatorial region of the spindle where the vesicles fuse to form the cell plate (**103, 107, 108**). It has been suggested that the peripheral microtubules are concerned with the orientation of the cellulose microfibrils which are being formed in the young wall on the outside of the plasmalemma (**63, 64**); however, the evidence for this is equivocal.

Another smaller type of proteinaceous element has sometimes been observed in the plant cell. This is termed a microfilament and is about 7 nm in width.

In the green alga *Nitella*, the microfilaments are concerned with cytoplasmic streaming; in pollen tubes of flowering plants they apparently guide vesicles concerned with wall synthesis through the cytosol to the growing tip of the tube.

CELL WALL

Plant protoplasts are normally enclosed by a wall (**49, 51**) which gives rigidity and protection to the cell but, unless impregnated with hydrophobic materials such as cutin, suberin, and lignin, does not prevent water and solutes diffusing across it to the plasmalemma. The walls of adjacent cells are held together by a common wall layer (**58, 59**), termed the middle lamella, in which the pectin network is a major component, but there is relatively little cellulose. Because of being cemented together by this middle lamella, plant cells are immobile, although fibres and some other cells elongate by tip growth and intrude between the neighbouring cells.

Following cell division the progeny usually undergo vacuolation growth and their primary walls also expand. Commonly in parenchymatous tissue the middle lamellae partially break down at the sites where several cells connect to each other, and intercellular spaces develop (**109**). When expansion growth ceases some cell types undergo deposition of a secondary wall (**54–57**). The constitution of the secondary wall components and the orientation of its cellulose microfibrils (**110**) are markedly different from those of the primary wall.

Primary wall

Cytokinesis normally immediately follows mitosis and so the two progeny are divided by a common cell plate (**72**). The plate first appears at the equator of the mitotic spindle (**106**) and then advances centrifugally (**106–108, 111**) to fuse with the mother cell wall. The plate is formed from fusing dictyosome vesicles (**103**) and is delimited by a plasmalemma derived from the dictyosome membranes (**107, 108**). Strands of ER penetrate the plate and these form the central desmotubular component of the plasmodesmata (**60, 61**) which connect the protoplasts of adjacent cells. The unthickened cell plate constitutes the largely pectinaceous middle lamella common to both daughter cells (**107, 108**). Subsequently a thin primary wall is deposited on both surfaces of the plate (**59, 72**).

When a highly vacuolated cell divides, the growing margins of the cell plate are marked by dense cytoplasm (the phragmosome, **111**) in which clusters of short microtubules occur (**106–108**). In some tissues mitosis is not immediately followed by cell plate development: in the endosperm of many species the initially coenocytic cytoplasm (**52**) later becomes divided by freely-forming walls. These often develop in a tortuous pattern and a similar phenomenon occurs in callus tissue (**112**). In transfer cells labyrinthine ingrowths of the primary wall into the protoplast occur (**113**) and such modifications of parenchyma cells are common adjacent to vascular elements.

In thinner areas of the wall plasmodesmata are often clustered together (**48**), with up to 60 present per square micrometre of wall surface, to form pit fields (**114, 115**). These regions often remain unthickened (when secondary wall deposition occurs) and give rise to pits closed by a pit membrane composed of primary wall (**57, 104, 116**). Where lignification of the wall occurs, it largely restricts the passage of water and nutrients between protoplasts to the nonlignified pitted regions (**57**).

In the tracheary elements of the protoxylem (**56**) a lignified secondary wall is deposited internal to the primary walls in discrete rings or a spiral, but extensive tracts of nonlignified primary wall lie between the thickenings. These nonlignified regions are attacked by hydrolytic enzymes released from the vacuole of the degenerating protoplast, so that frequently only the cellulosic skeleton ('holey' wall) remains in the mature element to indicate the original position of the primary wall (**117**).

The somewhat thickened walls of sieve tubes (**53**), and the thick walls of many storage parenchyma cells (**104**), are primary and do not normally undergo lignification. During differentiation of the sieve elements their protoplasts largely degenerate (**118**), leaving intact the plasmalemma together with modified plastids (**119, 120**), mitochondria, ER, and deposits of P-protein (**118, 119**). The end walls of the sieve elements become modified as the amorphous polysaccharide, callose, is deposited within the wall around the plasmodesmata (**53**, **118**). The desmotubular component of the plasmodesma disappears and eventually a wide pore develops, ranging from 1–15 μm in diameter, but this usually appears plugged in sectioned material (**53**, **119**).

The primary wall (**49**, **54**) contains up to 80% of its fresh weight as water, while the other components are predominantly polysaccharide. Cellulose composes 25–30% by weight of the dried wall while hemicelluloses constitute a further 15–25%, pectic substances up to 35%, and glycoproteins 5–10%.The cellulose–hemicellulose framework (**121**) gives mechanical strength to the primary wall but the pore size is controlled by the pectin network. These two networks provide a mechanically strong cell wall which can nevertheless stretch and permit cell expansion.

Cellulose consists of long chains of ß-1,4-glucans, with 30 or more chains being aligned in parallel to form a microfibril (**64**, **121**) up to several micrometres long and 3.0–8.5 nm wide. Microfibrils possess high tensile strength and reinforce the wall in a form analogous to steel rods in reinforced concrete. These microfibrils are held together by numerous hydrogen bonds, resulting in a structure that is crystalline in its core, and partially crystalline at the exterior. In the primary wall newly formed microfibrils are often aligned parallel to each other (**64**), but randomly orientated microfibrils are also common (**110**).

However, microfibrils do not bind laterally to one another, so further matrix components are required to hold them in place. These are primarily hemicelluloses (in particular xyloglucan, but glucuronoarabinoxylans are also present in smaller amounts) having a backbone with a secondary structure similar to cellulose, which hydrogen-bond to adjacent microfibrils. The strength of the cross-bridges is probably regulated by proteins (termed expansins) which are able to break the hydrogen bonding between hemicelluloses and cellulose, thus allowing cell expansion.

Additionally, a pectin network surrounds the cellulose–hemicellulose network but has relatively few bonds to it (**121**, **122**). It is likely that the pectin network controls the porosity of primary cell walls and, due to its predominance in the middle lamella, also intercellular adhesion. Pectin consists of highly hydrated polymers rich in galacturonic acids but how the different pectins are linked together is unclear.

Wall matrix substances are synthesized in the ER and dictyosomes and transported via dictyosome vesicles to the newly-forming wall. It seems, however, that the cellulose microfibrils form *de novo* at the plasmalemma (**63**). In yeast the chitin microfibrils of the wall apparently originate from rosettes (aggregates of particles) in the plasmalemma (**65**). A similar mechanism has been suggested for the formation of cellulose microfibrils in higher plants in which enzyme rosettes, composed of cellulose synthase molecules (**63**), receive activated glucose from the protoplast.

Secondary wall

The secondary walls (**8**, **54**, **55**, **110**) are typically much less hydrated than the primary wall, and a higher proportion of the polysaccharide content is cellulose; however, hemicelluloses also occur (mainly glucuronoxylans in flowering plants and glucomannans in gymnosperms), but very little pectin is present. Additionally, the hydrophobic networks of lignin, cutin, and suberin present in many secondary walls provide support against mechanical stresses and desiccation while also providing highly effective barriers to invasion by micro-organisms.

Lignin is a hydrophobic highly cross-linked polymer of phenylpropanoid units. Between 15–35% of the dry weight of tracheary elements and sclerenchyma fibres (**56**, **57**, **117**) is composed of lignin which greatly increases the mechanical strength of their cell walls. These elements can resist the mechanical stresses imposed by gravity and wind, and also the hydraulic forces generated in the water-conducting tracheids and vessels. By displacing water in these cell walls, lignin prevents relative movement of the polysaccharides and stops cell extension. The loss of most of the water means that diffusion of enzymes into the wall is almost impossible and most micro-organisms are, therefore, unable to degrade lignified woody tissue.

The differential orientation of the microfibrils in a fibre or tracheary element is an important factor in the strength of their walls. Normally in a mature fibre the microfibrils in the oldest (outermost) layer of the secondary wall (S1) and the youngest (innermost) layer (S3) show microfibrils orientated more-or-less transversely to the fibre's long axis (**110**). In the thicker middle layer (S2) they are aligned nearly parallel to this axis. Since the microfibrils provide a high degree of tensile strength along their longitudinal axis, this differential layering gives all-round strength to a fibre similar to that found, but on a different scale, in plywood. Nearly all fibres are lignified, but in *Linum* (flax, **54**) this rarely occurs and their multilamellate secondary walls contain about 90% cellulose dry weight.

Cutin and suberin

The outer surfaces of the young shoot are covered by a hydrophobic cuticle (**123**). The outermost layer of the cuticle consists of hydrocarbon waxes and beneath lies a layer of cutin which forms the bulk of the cuticle. Cutin is made up of C16 and C18 hydroxylated fatty acids, linked together by ester bonds to form an insoluble polymeric network. The cork surface covering many older stems and roots (**5**) contains suberins which are composed of cross-linked, hydroxylated fatty acids of 20 or more carbon atoms in length, and some cross-linked phenolic groups. Suberin is also found in internal hydrophobic walls of the endodermis (**42**).

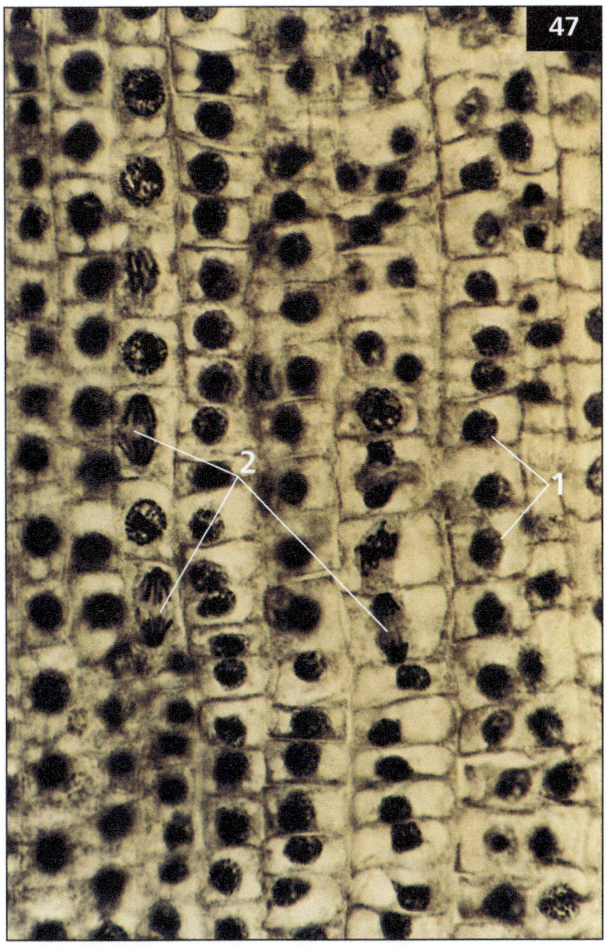

47 LS of *Phaseolus vulgaris* (French bean) root just behind the apex. This section shows longitudinal files of incipient cortical cells which are still actively dividing. Note the densely staining interphase nuclei (1) while a number of other nuclei are in various stages of mitosis (2). (LM.)

1 Interphase nuclei
2 Nuclei in mitosis

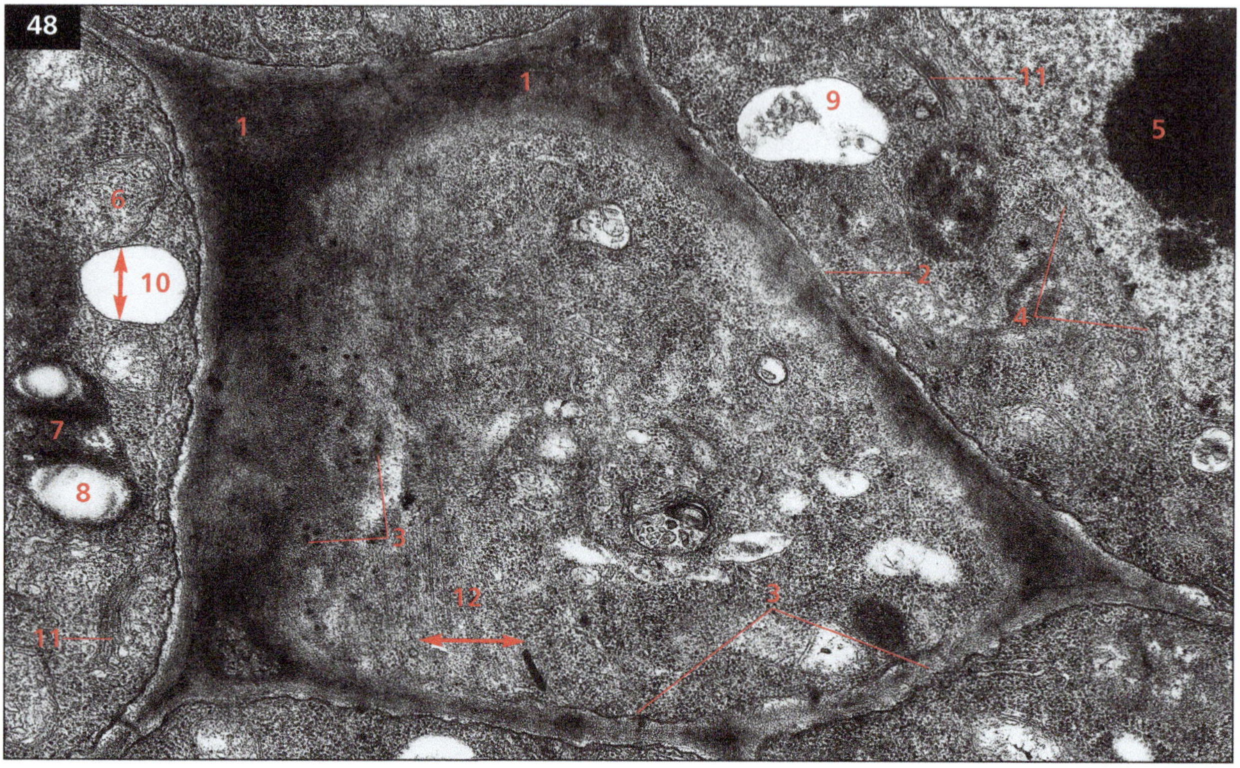

48 Meristematic cells from a bud of *Glechoma hederacea* (ground ivy). The protoplast is separated from the external cell wall (1) by the membranous plasmalemma (2), while plasmodesmata (3) penetrate the wall to connect adjacent protoplasts. The nucleus is invested by the two membranes of the nuclear envelope (4) and contains a densely staining nucleolus (5). The mitochondria (6) and proplastids (7) are also enclosed by membranous envelopes; these latter organelles are difficult to distinguish from each other in meristematic cells but starch (8) is diagnostic of a plastid. The small vacuoles (9) are delimited from the dense cytoplasm by a single tonoplast membrane (10). Several membranous dictyosomes (11) occur but little endoplasmic reticulum can be distinguished. A number of proteinaceous microtubules (12) are present. (G-Os, TEM.)

1 Cell wall
2 Plasmalemma
3 Plasmodesmata
4 Nuclear envelope
5 Nucleolus
6 Mitochondrion
7 Proplastid
8 Starch
9 Vacuole
10 Tonoplast
11 Dictyosome
12 Proteinaceous microtubules

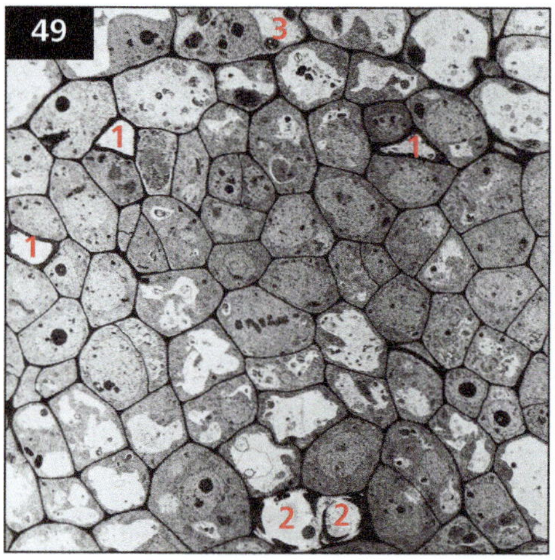

49 TS of a procambial strand in a bud of *Glechoma hederacea* (ground ivy). The narrow procambial cells are thin-walled, densely staining, and contain relatively large nuclei. These meristematic cells contrast with both the thick-walled, apparently empty, protophloem sieve tubes (1) and the mature protoxylem tracheary elements (2) which show irregularly-thickened secondary walls. Cortex (3). (G-Os, TEM.)

1 Protophloem sieve tube
2 Protoxylem tracheary element
3 Cortex

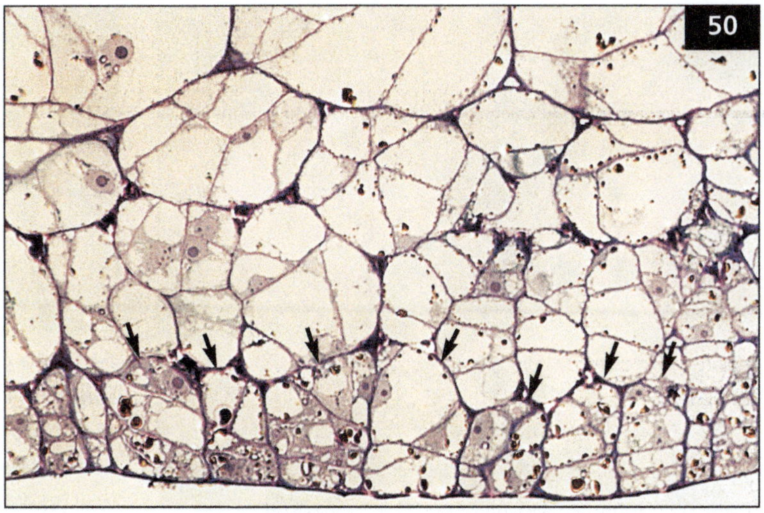

50 TS of the abaxial epidermis of a *Begonia rex* leaf segment cultured *in vitro* for several weeks. Note the dedifferentiation of the epidermal cells (arrows indicate the inner limits of the original epidermis) into a mass of primary meristematic cells which collectively constitute a meristemoid. (G-Os, LM.)

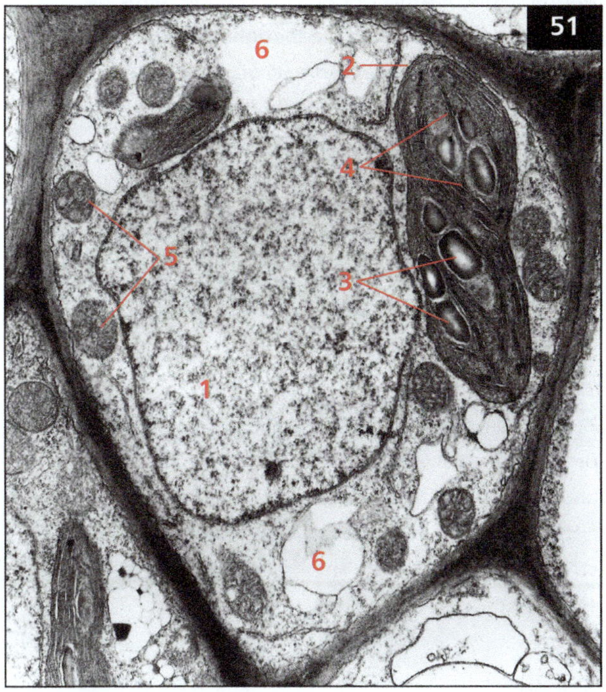

51 Bundle sheath cell from the leaf of *Phaseolus vulgaris* (bean). Note the large nucleus (1), prominent amylochloroplast (2) with numerous starch grains (3) and a well developed thylakoid system (4), mitochondria (5), and several small vacuoles (6). (G-Os, TEM.)

1 Nucleus
2 Amylochloroplast
3 Starch grains
4 Thylakoid system
5 Mitochondrion
6 Vacuole

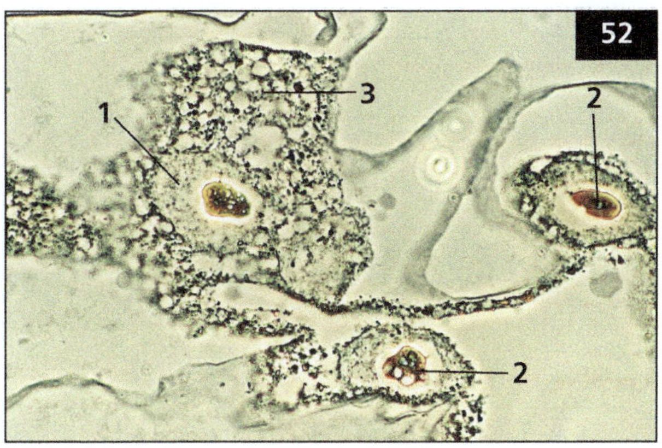

52 Coenocytic endosperm surrounding the young embryo of *Phaseolus vulgaris* (bean). Several nuclei (1) with prominent nucleoli (2) lie embedded in vesiculate cytoplasm (3) but no cell walls are present. (G-Os, Phase contrast LM.)

1 Nucleus
2 Nucleolus
3 Vesiculate cytoplasm

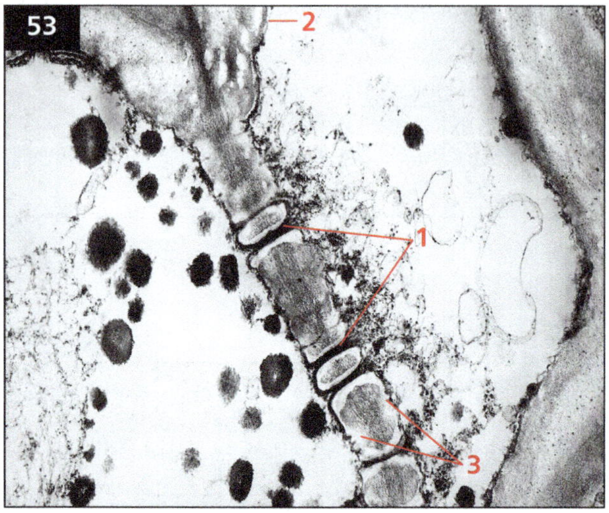

53 LS through a sieve plate of *Sorbus aucuparia* (rowan tree) showing narrow pores (1). Although these are apparently blocked, they are thought to be open in life. The protoplasts of the sieve elements are enucleate and the other organelles are degraded, but the plasmalemma (2) is still intact. Callose (3). (G-Os, TEM.)

1 Sieve pore
2 Plasmalemma
3 Callose

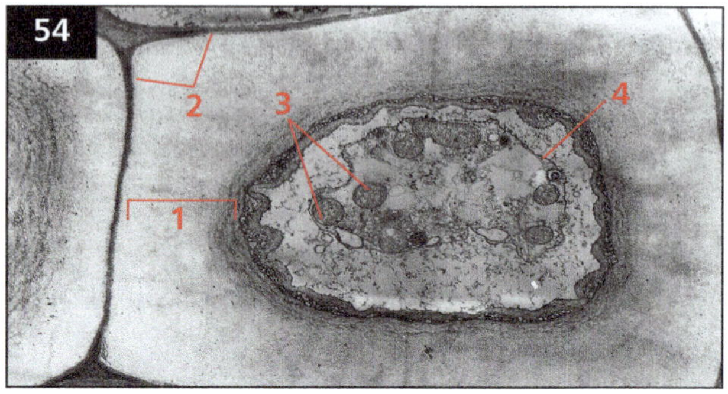

54 TS of a phloem fibre from the young hypocotyl of *Linum usitatissimum* (flax). This region of the fibre is nearing maturity and shows a massively thickened, but nonlignified, secondary wall (1) inside the thin primary wall (2). Within the degenerating protoplast a number of mitochondria (3) are distinguishable but the plasmalemma (4) has become detached from the innermost layer of secondary wall. (G-Os, TEM.)

1 Secondary wall
2 Primary wall
3 Mitochondria
4 Plasmalemma

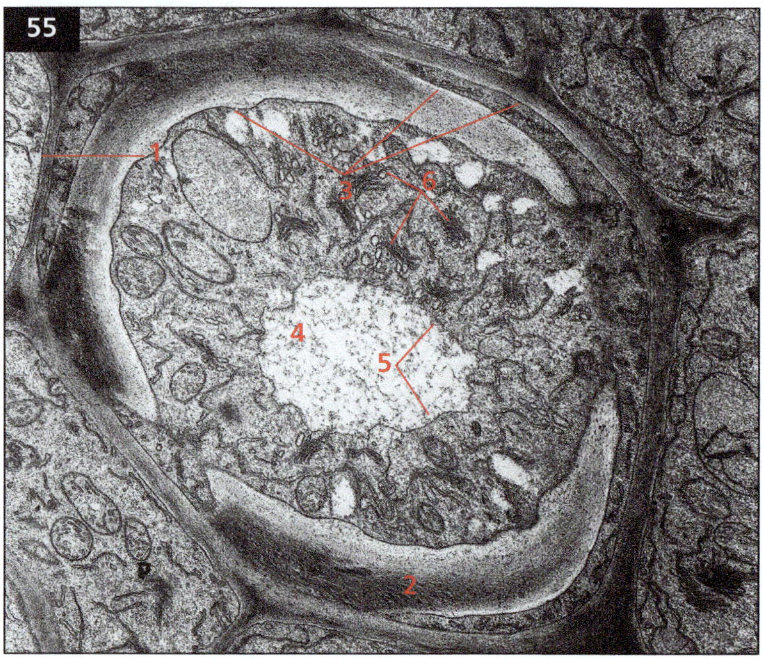

55 TS of a helically-thickened, differentiating protoxylem element of *Glechoma hederacea* (ground ivy). The thin primary wall (1) and the thickened, lignified secondary wall (2) are delimited from the protoplast by the plasmalemma (3). The protoplast contains a central vacuole (4) with its tonoplast (5) still intact. Numerous dictyosomes (6) are secreting vesicles into cytoplasm and these are presumed to transport noncellulosic polysaccharides to the growing wall. Numerous endoplasmic reticulum cisternae are evident but with this fixation the ribosomes are not preserved. (KMn, TEM.)

1 Primary wall
2 Secondary wall
3 Plasmalemma
4 Vacuole
5 Tonoplast
6 Dictyosomes

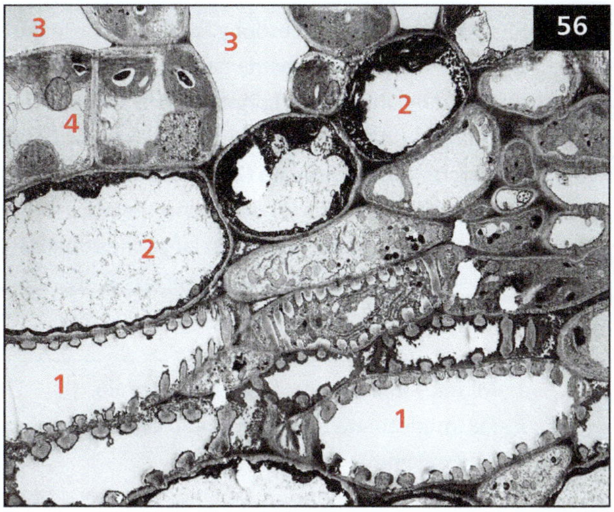

56 LS of a veinlet from a leaf of *Sorbus aucuparia* (rowan tree). Several annular protoxylem elements are already mature (1) and without protoplasts. Note the bundle sheath (2) parenchyma cells which separate the xylem elements from the air spaces (3) of the surrounding mesophyll tissue (4). (G-Os, TEM.)

1 Annular protoxylem
2 Bundle sheath
3 Air spaces
4 Mesophyll tissue

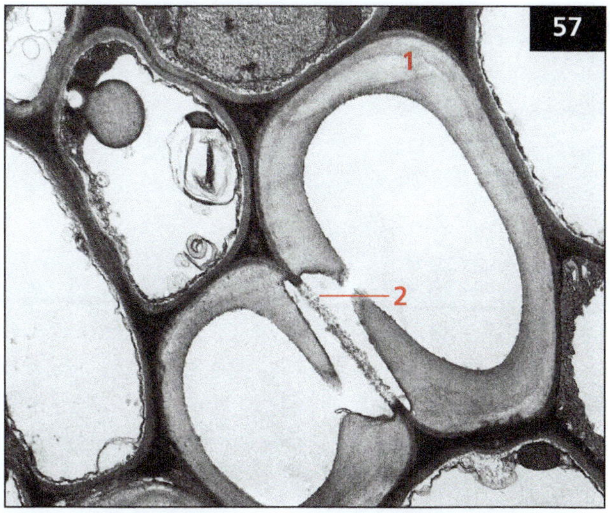

57 TS of xylem from the leaf midrib of *Sorbus aucuparia* (rowan tree). Note the several mature tracheary elements with thick secondary walls (1) and lumina devoid of protoplasts. A prominent bordered pit is visible and the primary walls and middle lamella (2), which form the pit membrane, are less dense where they are not overlain by secondary wall. (G-Os, TEM.)

1 Secondary wall
2 Pit membrane

58 Diagram to illustrate the symplast and apoplast. Each dead sclerenchyma fibre has an empty central lumen (1) which is enclosed by a thick secondary wall (2) lying within a thin primary wall (3). The living parenchyma cells possess primary walls only (3) and their protoplasts are interconnected by numerous plasmodesmatal channels (4). The protoplasts collectively constitute the symplast while the cell walls and dead cell lumens (plus intercellular space when present) form the apoplast. Nucleus (5), vacuole (6).

1 Central lumen
2 Secondary wall
3 Primary walls
4 Plasmodesmatal channels
5 Nucleus
6 Vacuole

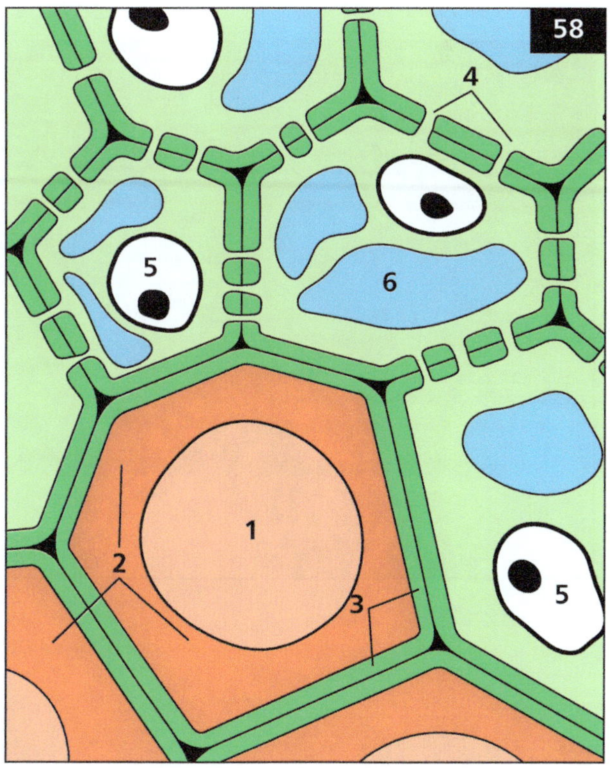

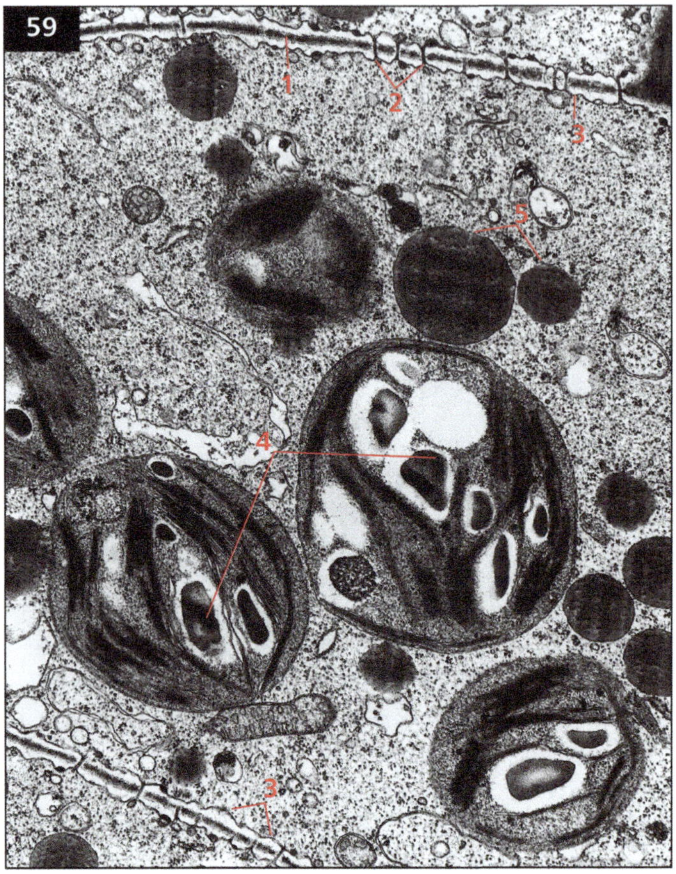

59 Fine structure of a mesophyll cell from the leaf of the moss *Polytrichum commune*. The cells are bounded by unthickened primary walls separated by a densely staining middle lamella (1); note the numerous plasmodesmata (2) traversing the walls and the plasmalemma (3) delimiting the wall from the protoplast. In the cytoplasm large chloroplasts (with starch grains, 4) and lipid vesicles (5) are especially prominent. (G-Os, TEM.)

1 Middle lamella
2 Plasmodesmata
3 Plasmalemma
4 Starch grain
5 Lipid vesicles

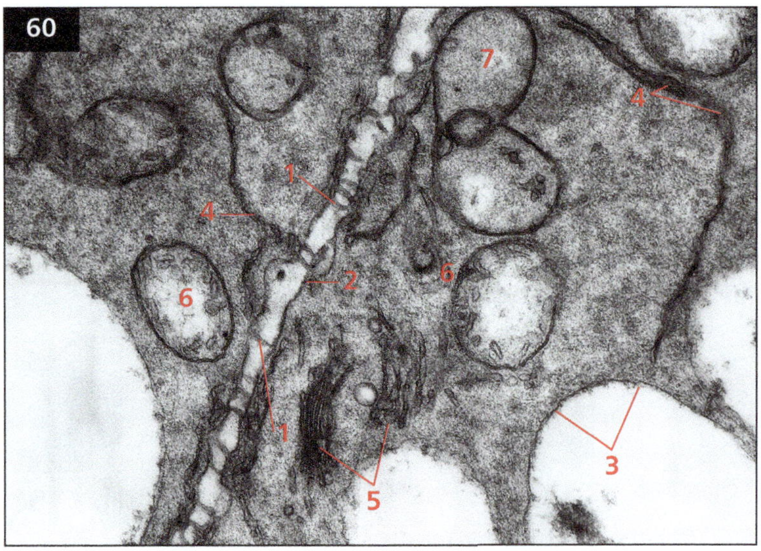

60 Root tip cells from *Allium cepa* (onion). Numerous plasmodesmata (1) cross the thin, unstained primary wall. The single membranes of the plasmalemma (2), tonoplast (3), endoplasmic reticulum (4), and dictyosomes (5) are well defined while the double membranes investing the mitochondria (6) and proplastids (7) can also be discerned. However, neither ribosomes nor microtubules are preserved with this fixative. (KMn, TEM.)

1 Plasmodesmata
2 Plasmalemma
3 Tonoplast
4 Endoplasmic reticulum
5 Dictyosomes
6 Mitochondria
7 Proplastid

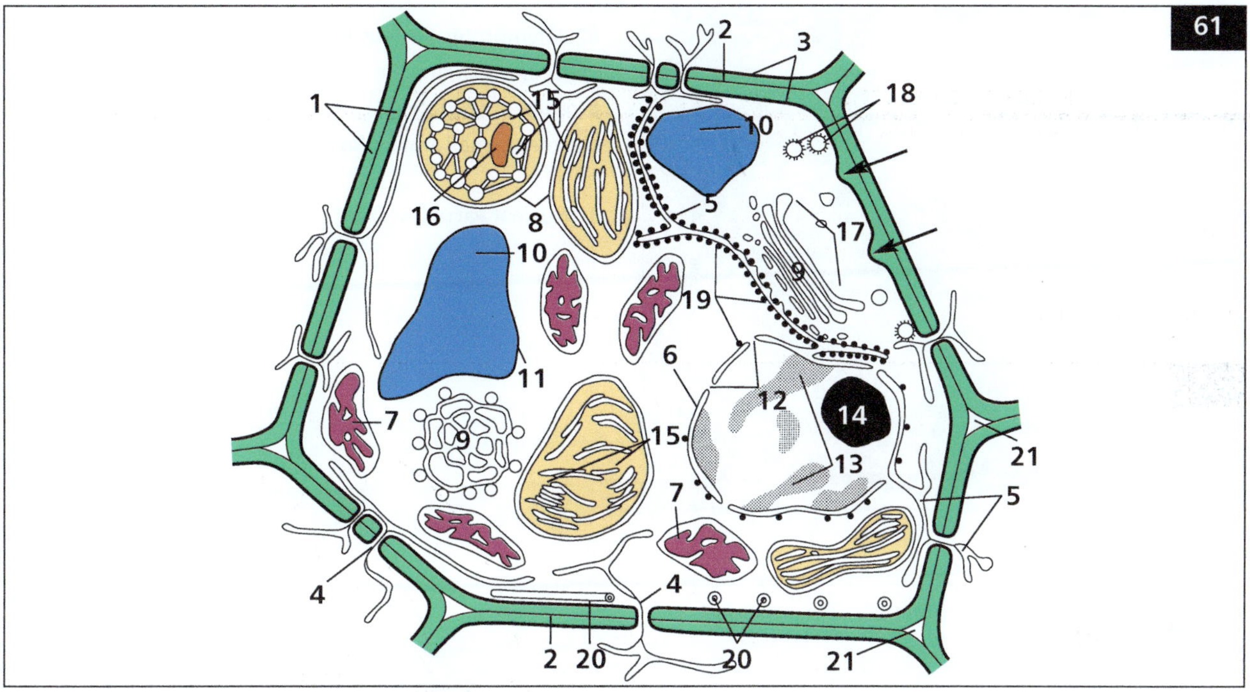

61 Diagram of the fine structure of a relatively undifferentiated plant cell. The protoplast is bounded by a thin primary wall (1) with a median middle lamella (2). The plasmalemma (3) encloses the protoplast and this membrane also lines the plasmodesmatal pores. Narrow desmotubules (4) traverse these pores and link the endoplasmic reticulum (ER, 5) of adjacent protoplasts. Although only partially indicated, the outer surfaces of the ER and outer nuclear membrane are normally covered by ribosomes; the numerous free cytoplasmic ribosomes are not shown in this diagram. The majority of cell organelles are membrane-bounded; two membranes enclose the nucleus (6), mitochondria (7), and chloroplasts (8), while the cisternae of the ER (5) and dictyosomes (9) are delimited by single membranes. The vacuoles (10) are also bounded by a single tonoplast (11). The nuclear envelope is linked to the ER, while at the numerous nuclear pores (12) the inner and outer membranes are confluent. Within the interphase nucleus denser DNA-rich areas of heterochromatin (13) occur and a large nucleolus (14) is evident. Large chloroplasts are present showing well-developed photosynthetic grana (15) and starch (16). The inner membrane of the mitochondrial envelopes shows convoluted tubular, or sometimes plate-like, invaginations. The dictyosomes (illustrated in both longitudinal and transverse views) show a polarity with the maturing face budding off numerous vesicles (17) which apparently migrate through cytoplasm to fuse with the plasmalemma (arrows). The coated vesicles (18) return surplus membranous material to be recycled by the protoplast. The principal nonmembranous organelles within the cell are the ribosomes (19) and the proteinaceous microtubules (20). At interphase the latter lie adjacent to the plasmalemma. Note that the middle lamella lying at the angles of the cell wall is beginning to break down to form intercellular spaces (21).

1 Primary wall
2 Middle lamella
3 Plasmalemma
4 Desmotubules
5 Endoplasmic reticulum
6 Nucleus
7 Mitochondria
8 Chloroplasts
9 Dictyosomes
10 Vacuoles
11 Tonoplast
12 Nuclear pores
13 Heterochromatin
14 Nucleolus
15 Grana
16 Starch
17 Dictyosome vesicles
18 Coated vesicles
19 Ribosomes
20 Proteinaceous microtubules
21 Intercellular spaces

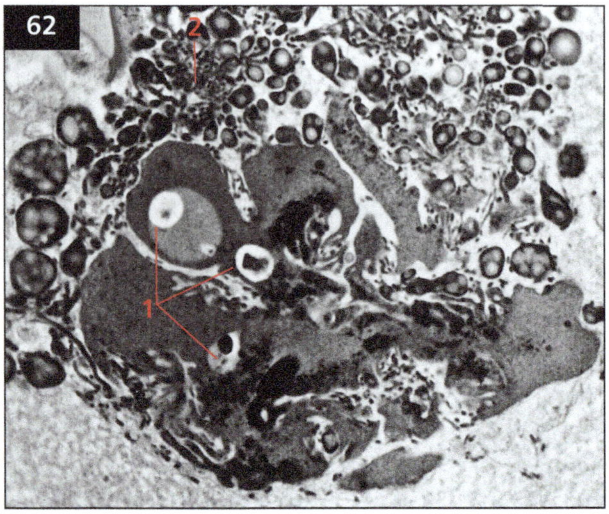

62 Polymorphic nucleus from a dedifferentiating cell of *Phaseolus vulgaris* (bean). The nucleus contains several prominent vacuoles (1) and is surrounded by a dense cluster of small amyloplasts (2). (G-Os, LM.)

1 Nuclear vacuoles
2 Amyloplasts

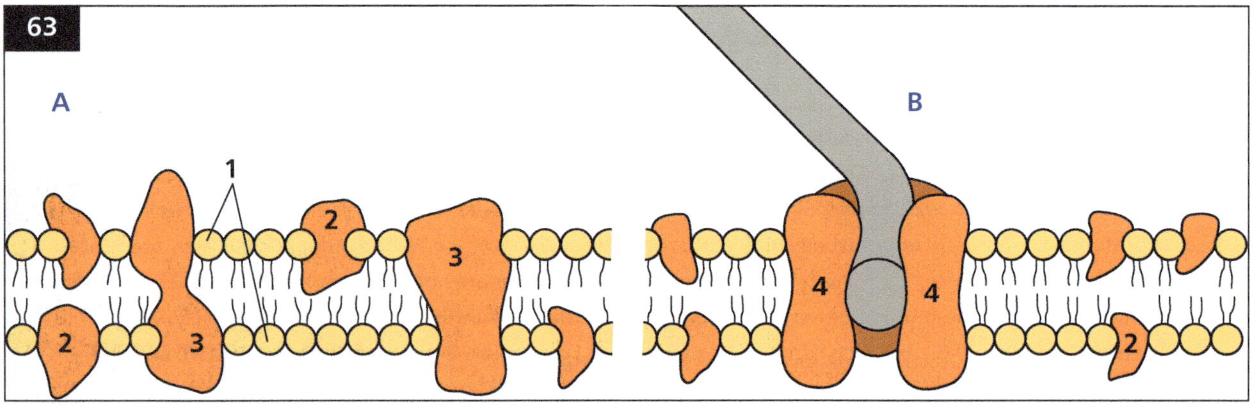

63 Model of a generalized plant membrane (A) and plasmalemma (B). A: Membrane is thought to consist of a bilayer of phospholipids with their hydrophilic heads (1) outermost, in which proteins are interspersed. Some proteins are confined to the membrane surfaces (2) while others (3) traverse the bilayer. At the plasmalemma (B) rosette protein complexes (4), composed of six cellulose synthase molecules, also span the membrane. Here cellulose precursor molecules are taken up from the cytoplasm, while cellulose microfibrils are extruded into the cell wall on the outer face of the plasmalemma.

1 Phospholipid heads
2 Surface proteins
3 Traversing proteins
4 Cellulose synthase molecules

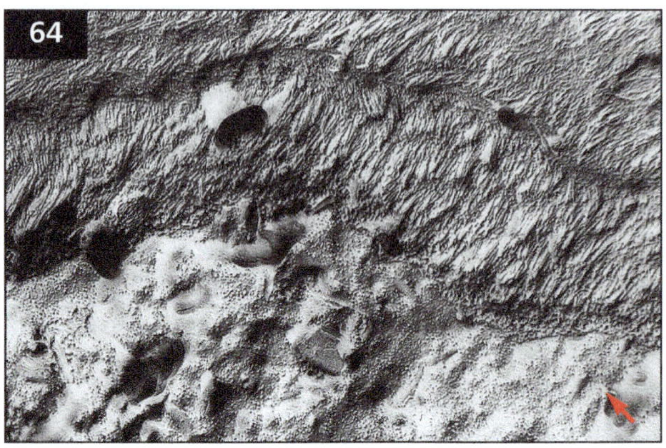

64 Interface between plasmalemma and cell wall in the root tip of *Lycopersicon esculentum* (tomato). The plasmalemma shows numerous small particles (presumably protein molecules) but no rosettes are apparent. Much of the water in the primary wall has sublimed away to reveal several layers of wall in which numerous, predominantly parallel, cellulose microfibrils occur. The specimen should be viewed from its direction of shadowing (arrow). (F-E, TEM.)

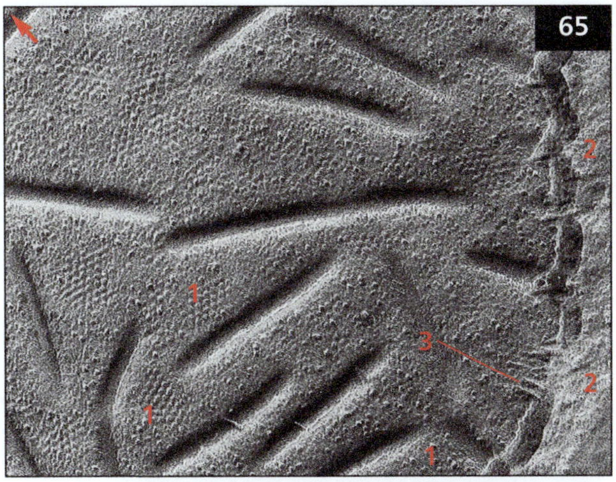

65 The plasmalemma of *Saccharomyces* (baker's yeast). Its fractured surface shows numerous intra-membranous particles, many of which are tightly grouped into rosettes (1). A portion of the chitinaceous cell wall (2) is visible and in some regions (3) wall microfibrils appear to originate from the rosette particles. Note also the variously orientated grooves in the plasmalemma: the specimen should be viewed from the direction of its shadowing (arrow). (F-F, TEM.)

1 Rosettes
2 Chitinaceous cell wall
3 Wall microfibrils

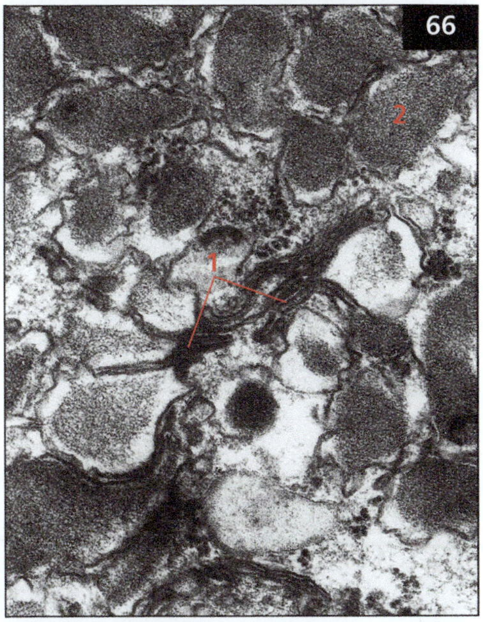

66 Detail from a root cap cell of *Zea mays* (maize) showing a dictyosome (1) with hypertrophied cisternae. These cisternae are bounded by a tripartite membrane and contain a finely granular dense material (2, probably mucilage) which is excreted into the apoplast after the cisternae fuse with the plasmalemma. (G-Os, TEM.)

1 Dictyosome
2 Granular material

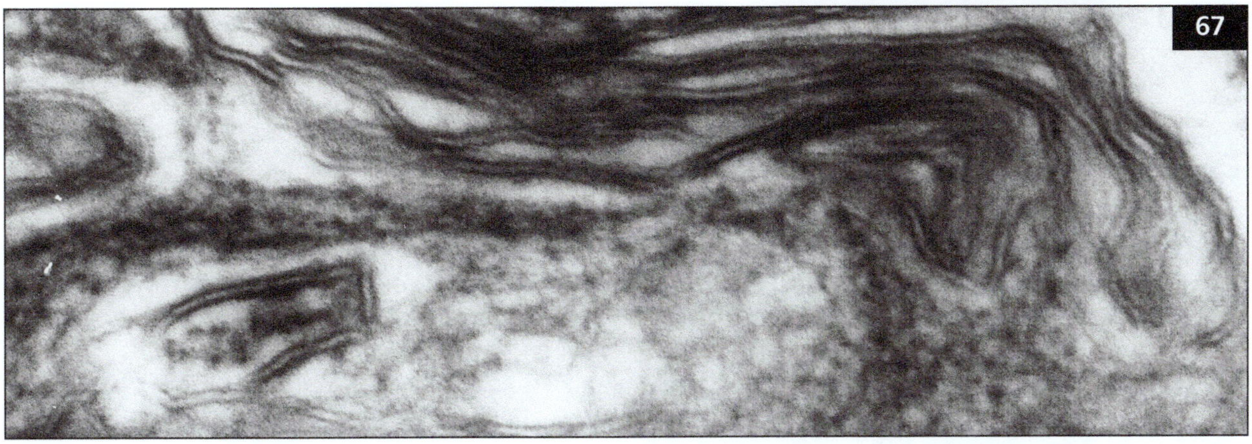

67 Membrane complex associated with the plasmalemma of *Andrographis paniculata*. Note the tripartite appearance of transversely-sectioned membranes which show a central translucent layer between two dense, but narrower, outer layers. The surrounding matrix is the cell wall. (G-Os, TEM.)

68

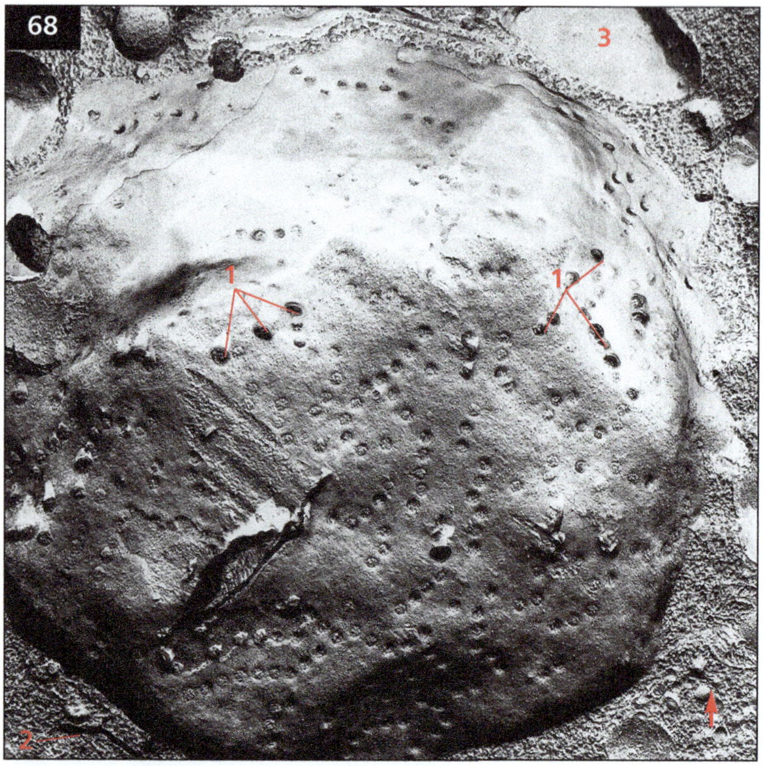

68 Surface view of a nucleus from the root tip of *Lycopersicon esculentum* (tomato). Note the numerous nuclear pores (1) connecting the nucleoplasm and cytoplasm. The specimen should be viewed from the direction of shadowing (arrow). Endoplasmic reticulum (2), vacuole (3). (F-F, TEM.)

1 Nuclear pores
2 Endoplasmic reticulum
3 Vacuole

69

69 Detail of a nucleus from the root tip of *Lycopersicon esculentum* (tomato). The inner (1) and outer nuclear membrane (2) both show numerous pores (3) but the small particles (probably proteins) are much scarcer in the inner membrane. The specimen should be viewed from the direction of shadowing (arrow). (F-F, TEM.)

1 Inner nuclear membrane
2 Outer nuclear membrane
3 Pores

70

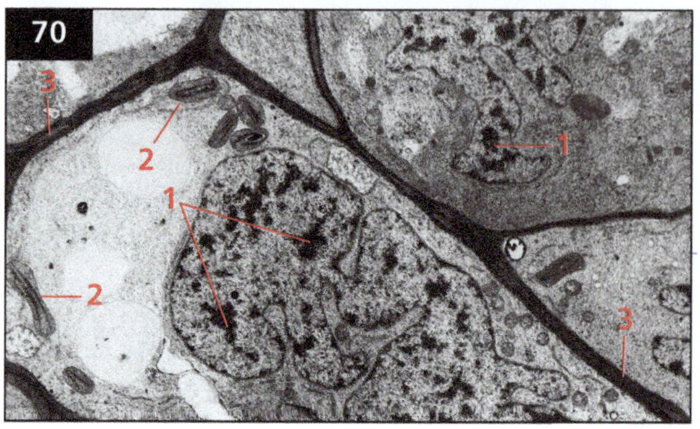

70 Large polymorphic nuclei from callused root tissue of *Pisum sativum* (pea). Note the dense areas of heterochromatin (1) visible internally. Plastids (2), cell wall (3). (G-Os, TEM.)

1 Heterochromatin
2 Plastid
3 Cell wall

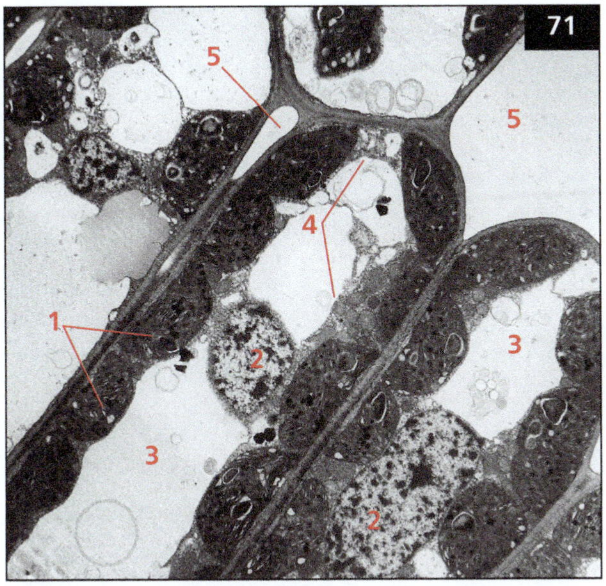

71 LS of the palisade mesophyll in the lamina of *Sorbus aucuparia* (rowan tree). Each thin-walled palisade cell shows a dense cytoplasmic layer packed with chloroplasts (1) and a single nucleus (2), while the extensive central vacuole (3) is delimited from the cytoplasm by the tonoplast (4). Note the dense heterochromatin in the nuclei and the prominent intercellular spaces (5). (G-Os, TEM.)

1 Chloroplasts
2 Nucleus
3 Central vacuole
4 Tonoplast
5 Intercellular spaces

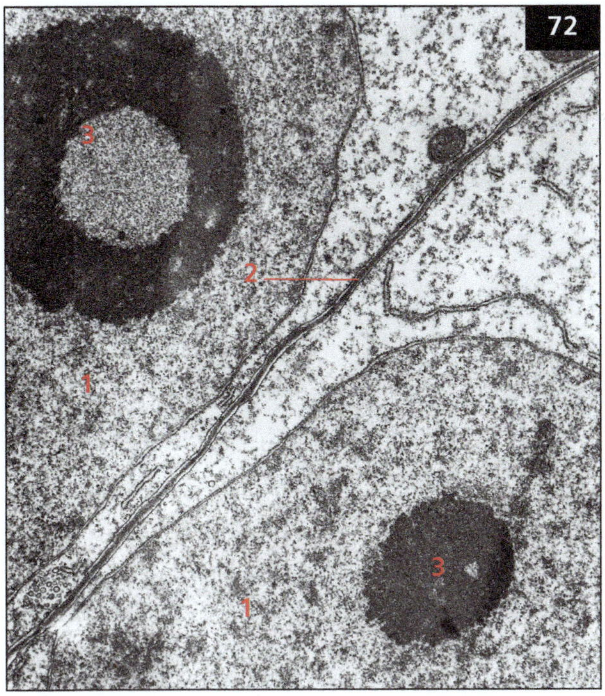

72 Dividing cells of the root of *Pisum sativum* (pea). Note the thin, newly-formed cell wall between the two progeny nuclei (1) resulting from mitosis. The densely stained middle lamella (2) separates the translucent primary walls of the daughter cells to either side. Nucleolus (3). (G-Os, TEM.)

1 Nuclei
2 Middle lamella
3 Nucleolus

73 Detail of a nucleolus from the root of *Pisum sativum* (pea). Note its large vacuole (1) whose empty appearance contrasts with the crowded particles (2) of the surrounding nucleolus. (G-Os, TEM.)

1 Vacuole
2 Crowded particles

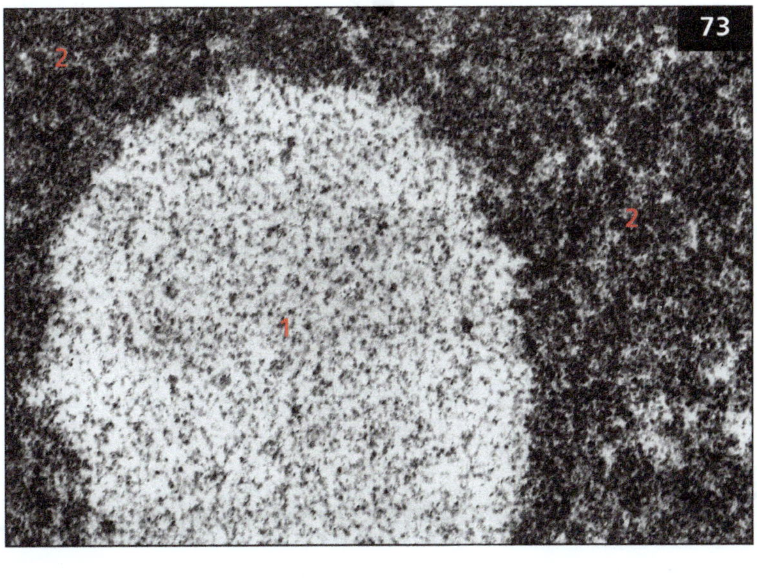

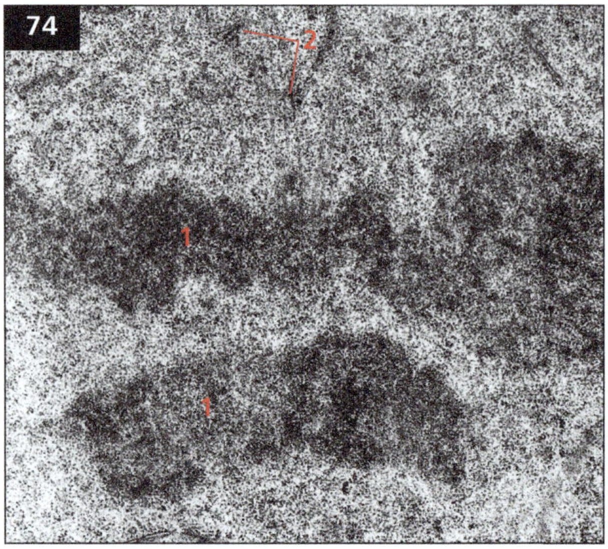

74 Detail of a dividing nucleus from *Pisum sativum* (pea) root. The mitotic spindle is sectioned longitudinally at metaphase and shows a pair of densely staining chromosomes (1) with attached kinetochore microtubules (2). (G-Os, TEM.)

1 Chromosomes
2 Kinetochore microtubules

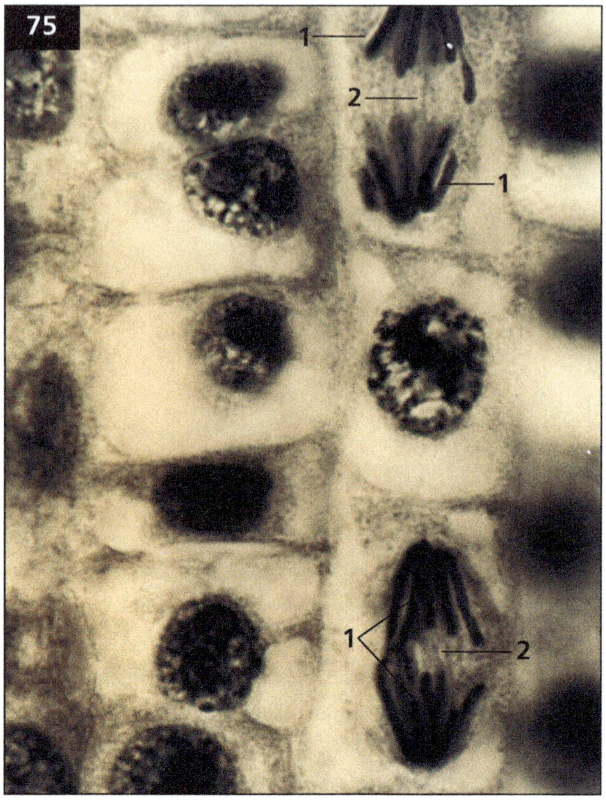

75 LS of *Phaseolus vulgaris* (bean) root just behind the apex. Note the two cells in which the nuclei are at the telophase stage of division with densely staining chromosomes (1) at either pole of the mitotic spindle. The spindle fibres (2) represent microtubules which are grouped in bundles and therefore are visible at light microscopic level. (LM.)

1 Chromosomes
2 Spindle fibres

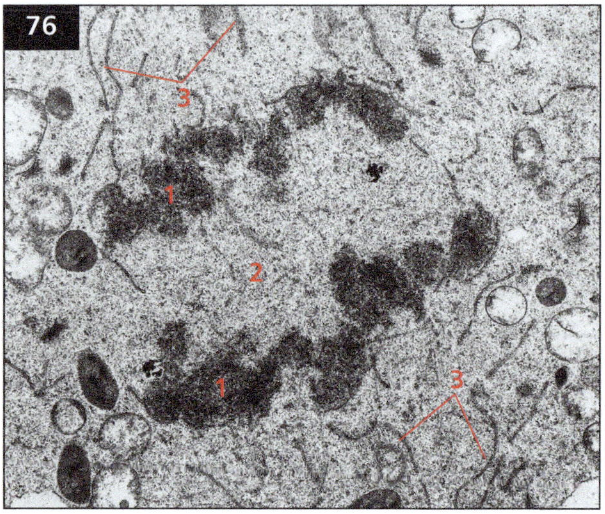

76 LS through a mitotic nucleus from *Pisum sativum* (pea) root. At anaphase the densely staining chromosomes (1) have separated leaving the equator region (2) devoid of large organelles. The nuclear envelope broke down at prophase, but numerous strands of endoplasmic reticulum (3) lie in the cytoplasm together with other organelles. (G-Os, TEM.)

1 Chromosomes
2 Equator region
3 Endoplasmic reticulum

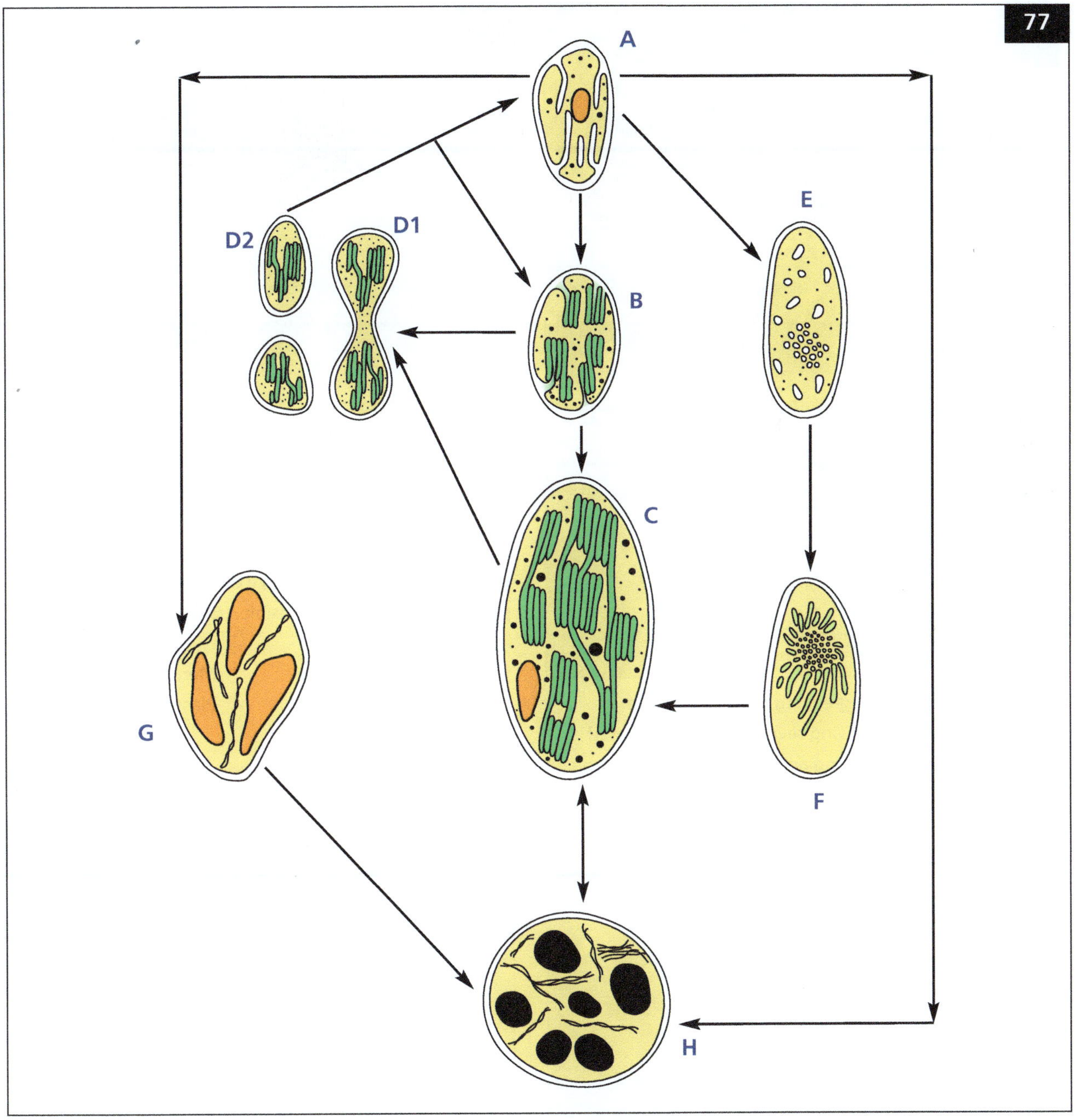

77 A–H: Diagrams showing the principal pathways of plastid ontogeny. In young green shoots the typical mode is from nonpigmented proplastids (A) to mature green chloroplasts (C) via young chloroplasts (B). Plastid replication may occur at any stage but is particularly common in proplastids and young chloroplasts (D1, D2). In dark-grown shoots (or in cells shielded from light by overlying tissue) etioplasts (E) form but on exposure to light these rapidly differentiate (F) into chloroplasts (C).

In roots (and in most epidermal cells of the shoot) nonpigmented leucoplasts develop from proplastids and amyloplasts (G). Chromoplasts (H) contain red, orange, or yellow pigments; these plastids either develop directly from proplastids or form from amyloplasts or degenerate chloroplasts. Both mature chloroplasts and amyloplasts may be induced to divide (and their derivatives sometimes revert to proplastids) by wounding the tissue in which they occur. The transformation of chromoplasts into chloroplasts has also been reported.

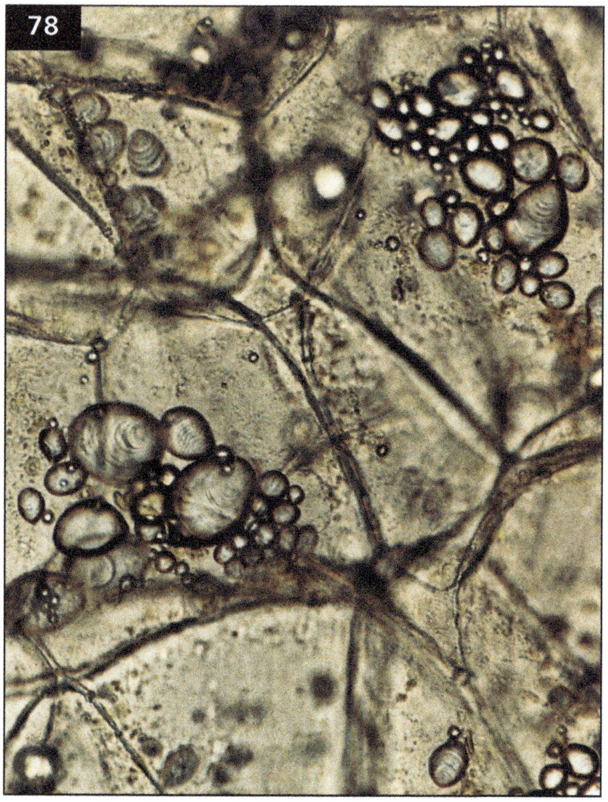

78 Storage tissue of the greening tuber of *Solanum tuberosum* (potato) showing several highly vacuolated parenchyma cells containing amyloplasts with very large starch grains. (Fresh section, LM.)

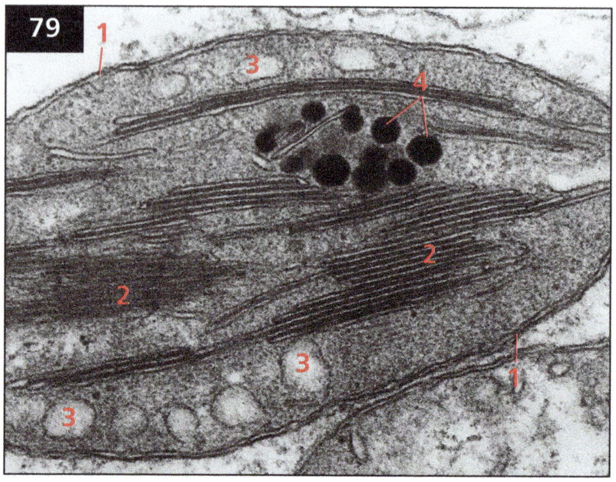

79 Young chloroplast from *Crambe maritima* (seakale) sectioned through its long axis. The two membranes of the chloroplast envelope (1) are distinct, while internally flattened membranous thylakoids are stacked into grana (2) interconnected by frets. A number of peripheral membranous vesicles (3) and a densely staining group of plastoglobuli (4) are also visible. (G-Os, TEM.)

1 Chloroplast envelope
2 Grana
3 Membranous vesicles
4 Plastoglobuli

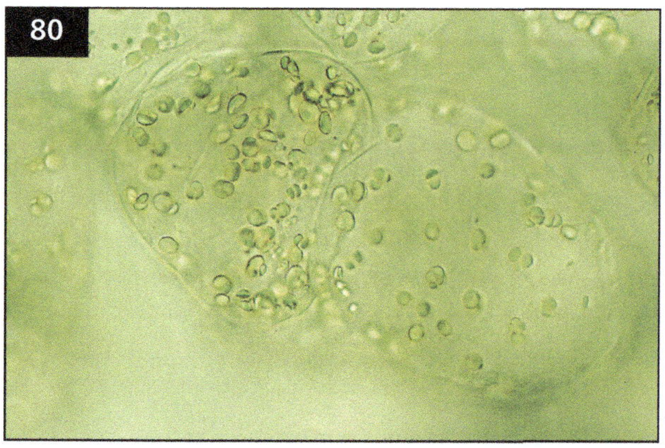

80 Storage tissue of the greening tuber of *Solanum tuberosum* (potato). Note the numerous green chloroplasts which have developed after prolonged exposure to light. (Fresh section, LM.)

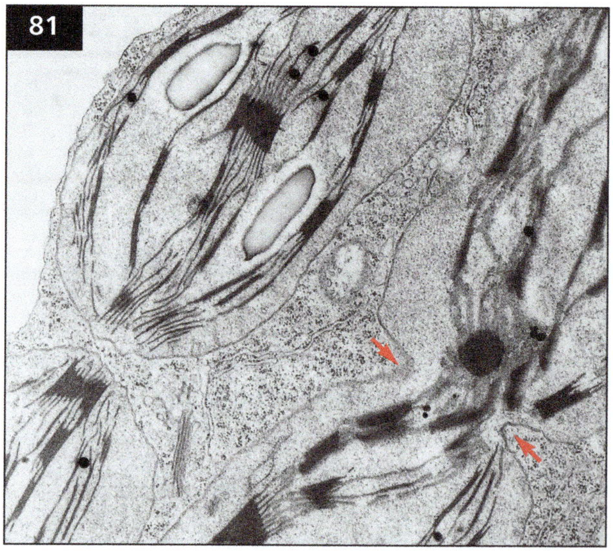

81 Dividing chloroplasts from mesophyll tissue in *Linum usitatissimum* (flax). One plastid has a median isthmus (arrows) which indicates the prospective plane of division, while the other two plastids have just separated. (G-Os, TEM.)

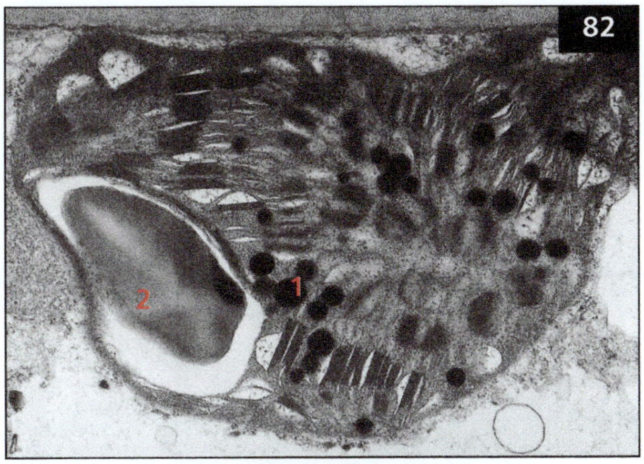

82 Chloroplast undergoing transformation into a chromoplast in the leaf of *Sorbus aucuparia* (rowan tree). The leaf was sampled in autumn; note the degenerating grana and the numerous plastoglobuli (1) in which carotenoid pigments accumulate to give the yellow-, orange- and red-coloured autumn foliage of deciduous species. Starch grain (2). (G-Os, TEM.)

1 Plastoglobuli
2 Starch grain

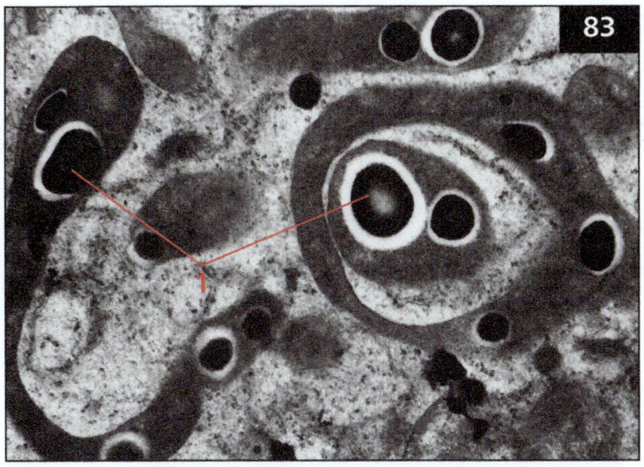

83 Proplastids from dedifferentiating cotyledonary tissue of *Phaseolus vulgaris* (bean). Few membranes are evident in the stroma but the numerous small starch grains (1) clearly define these as plastids. (G-Os, TEM.)

1 Starch grains

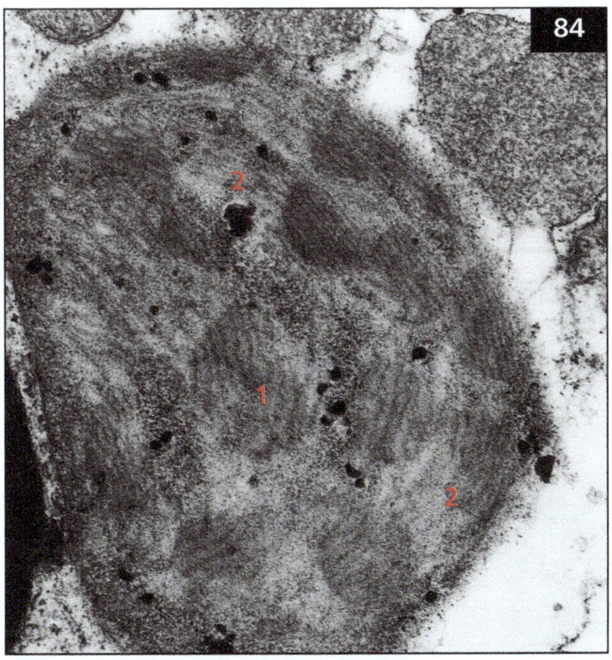

84 Small chloroplast from *Crambe maritima* (seakale) sectioned across its shorter axis. Note the discoidal shape of the granal membranes (1) and the tubular fret system (2) which interconnects the grana. (G-Os, TEM.)

1 Granal membranes
2 Fret system

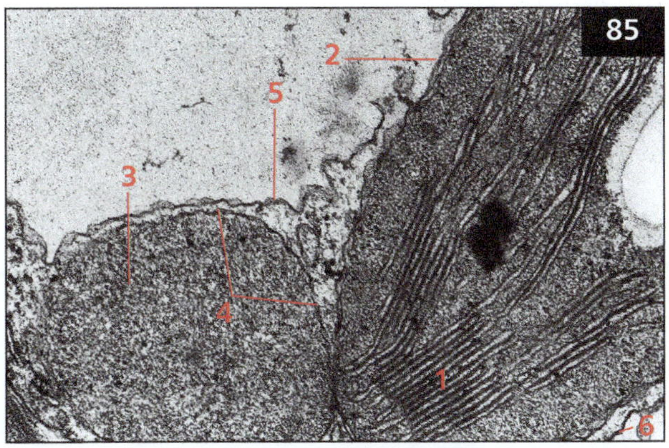

85 LS of chloroplast from *Linum usitatissimum* (flax). Note the well developed granal membranes (1) within the stroma and the two membranes of the investing envelope (2). Closely associated with the plastid is a large peroxisome, with a granular matrix (3), which is bounded by a single membrane (4). Tonoplast (5), plasmalemma (6). (G-Os, TEM.)

1 Granal membranes
2 Chloroplast envelope
3 Peroxisome
4 Single membrane of peroxisome
5 Tonoplast
6 Plasmalemma

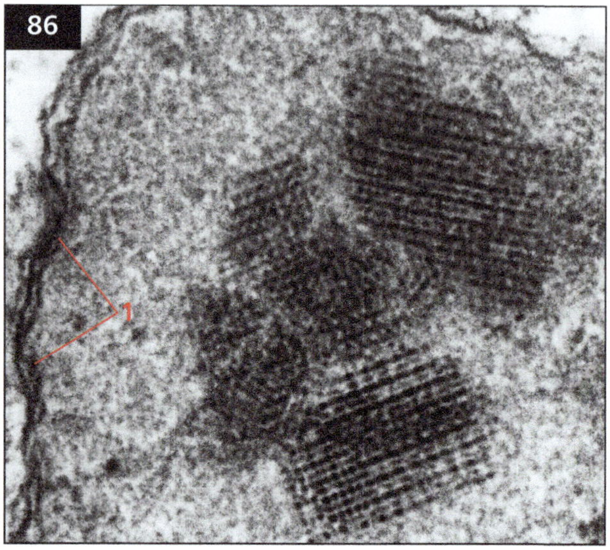

86 Plastid of *Crambe maritima* (seakale) showing crystalline phytoferritin within the stroma. The two membranes constituting the plastid envelope (1) are clearly evident. (G-Os, TEM.)

1 Plastid envelope

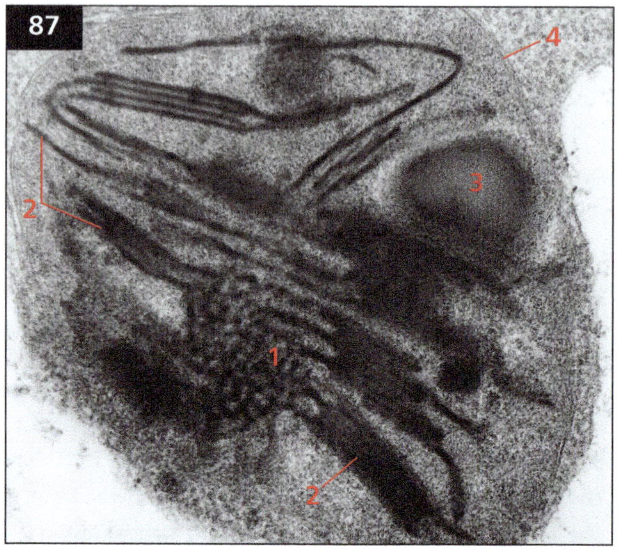

87 Greening etioplast in the stem of *Glechoma hederacea* (ground ivy). From the vestiges of the prolamellar body (1) a granal/fret membranous system (2) is developing. Starch (3), plastid envelope (4). (G-Os, TEM.)

1 Prolamellar body
2 Granal/fret membranous system
3 Starch
4 Plastid envelope

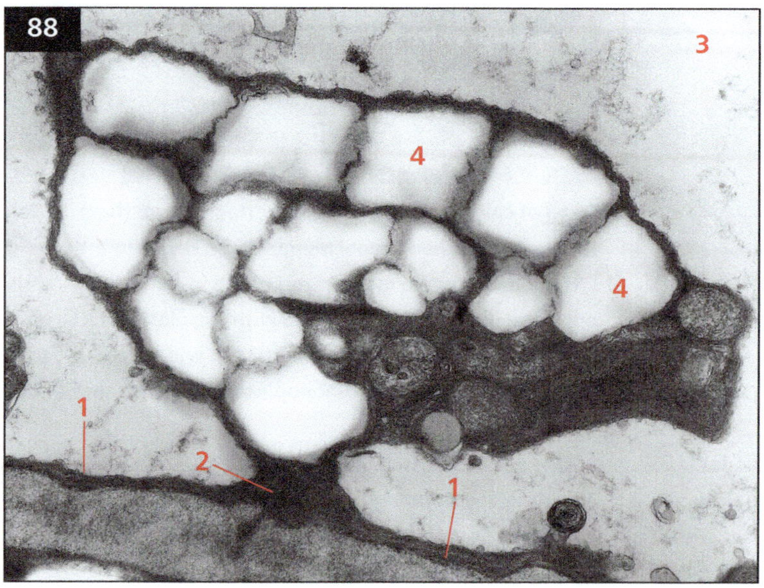

88 Large amyloplast from callus tissue of *Andrographis paniculata*. The plastid is connected to the cytoplasm lining the cell wall (1) by an isthmus of cytoplasm (2) which also invests the amyloplast and separates it from the large vacuole (3). Within the plastid few membranes are evident but numerous starch grains (4) occur. (G-Os, TEM.)

1 Cytoplasm lining cell wall
2 Connecting cytoplasm
3 Vacuole
4 Starch grains

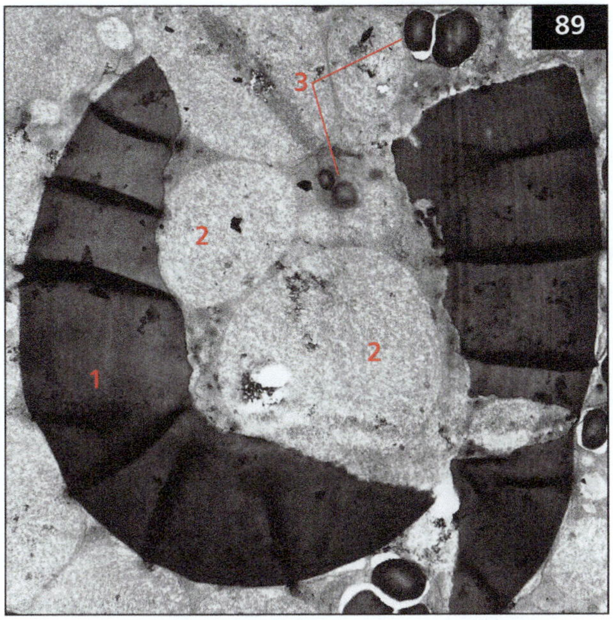

89 Starch grain fragments from the germinating cotyledon of *Phaseolus vulgaris* (bean). The densely staining starch grain has apparently been digested from the centre leading to its break-up into several pieces (1). Note also the membrane-bounded protein bodies (2), now largely empty, and the small amyloplasts (3). (G-Os, TEM.)

1 Starch grain fragments
2 Protein bodies
3 Amyloplasts

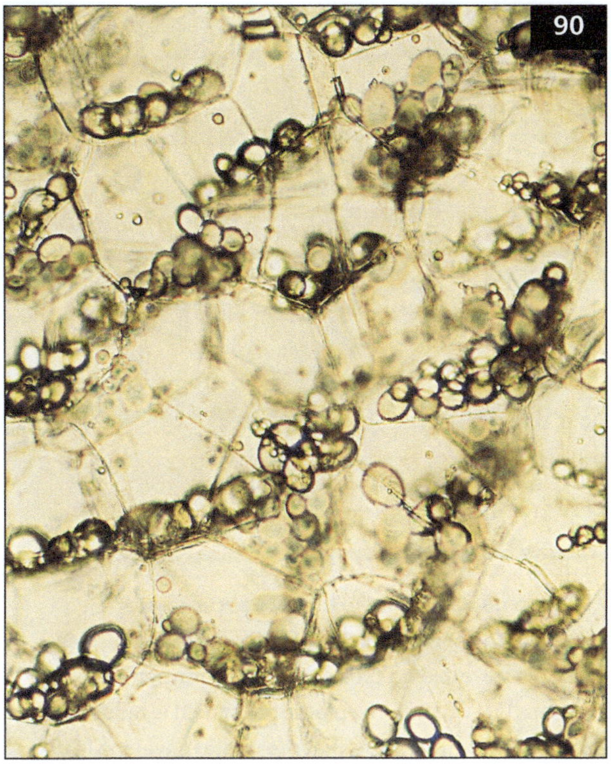

90 LS through an excised segment of *Solanum tuberosum* (potato) tuber. This was incubated in a vertical position for 4 weeks; note that the amyloplasts have sedimented and lie against the lower walls of the parenchyma cells. (Fresh section, LM.)

91 *Aesculus hippocastanum* leaves in autumn turning colour prior to leaf abscission. Note the range of leaflet colours, with a few which are still green through to orange and yellow; the latter colours are due to the accumulation of carotenoids within the degenerate chloroplasts.

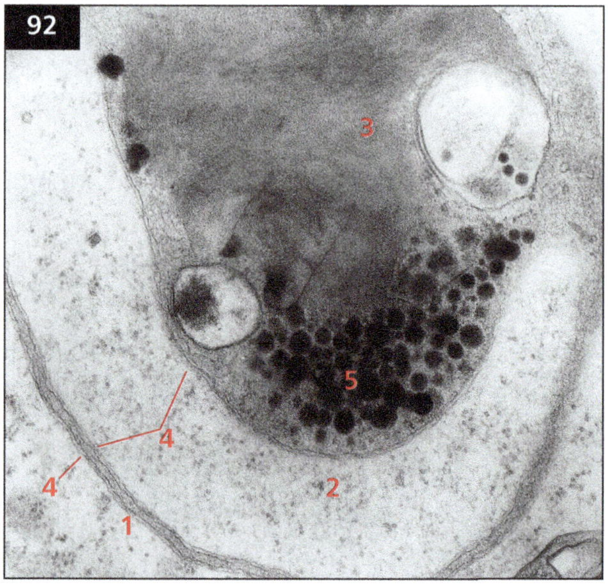

92 Chromoplast from the root of *Daucus carota* (carrot). The amoeboid plastid has an irregular surface and the surrounding cytoplasm (1) has formed an enclave (2) within it. The plastid stroma (3) is reduced here to a very narrow layer so that the plastid envelope (4) is apparently composed of four membranes. Plastoglobuli (5). (G-Os, TEM.)

1 Cytoplasm
2 Enclave of cytoplasm
3 Plastid stroma
4 Plastid envelope
5 Plastoglobuli

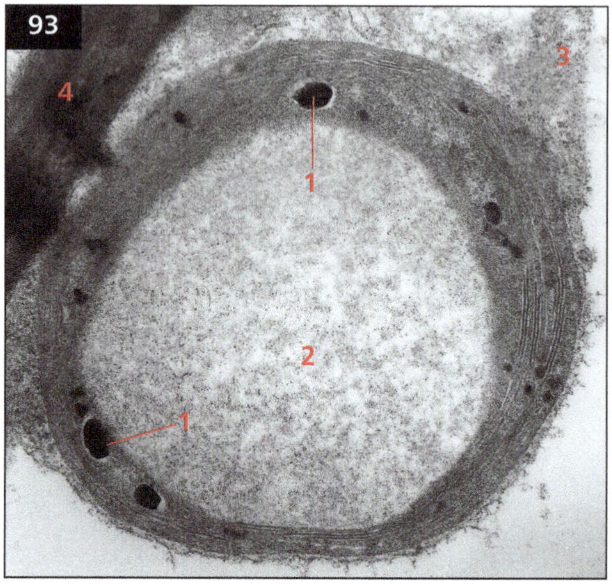

93 Epidermal leucoplast from the green leaf of *Glechoma hederacea* (ground ivy). In the peripherally located stroma the thylakoid system is only poorly developed but several small starch grains (1) are evident. The centre of the plastid is occupied by a large vacuole (2). Cytoplasm (3), cell wall (4). (G-Os, TEM.)

1 Starch grains
2 Vacuole
3 Cytoplasm
4 Cell wall

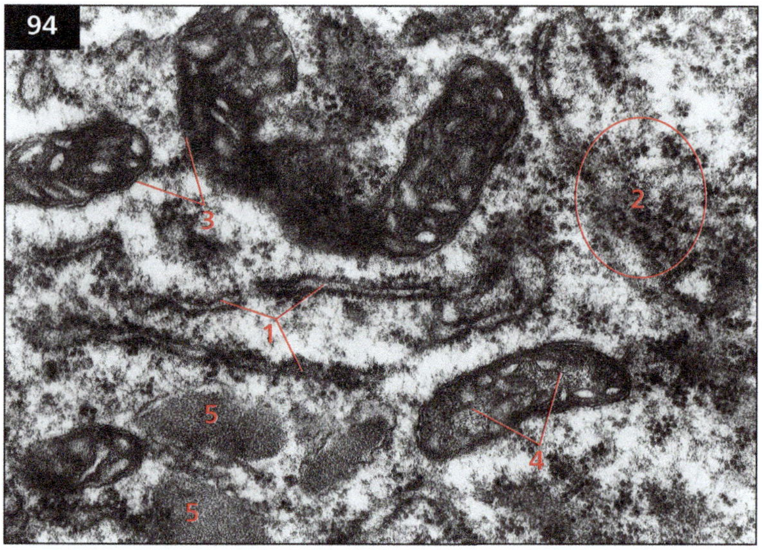

94 Densely cytoplasmic region from *Zea mays* (maize) root tip. Note the rough endoplasmic reticulum cisternae (1) covered with ribosomes which are often aggregated into polysomes (2). Each mitochondrion has an outer envelope (3) and numerous irregular cristae lie within the matrix. These are connected with the inner membrane of the envelope; note also the fibrillar nucleoid zones (4) in the mitochondrial matrix. Several mature hypertrophied cisternae (5) have been released from a dictyosome. (G-Os, TEM.)

1 Endoplasmic reticulum cisternae
2 Polysomes
3 Envelope of mitochondrion
4 Fibrillar nucleoid zones
5 Hypertrophied dictyosome cisternae

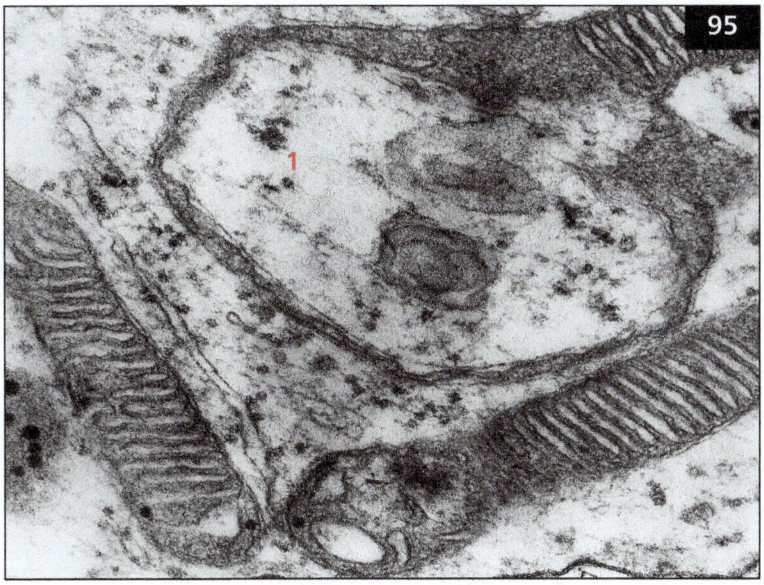

95 Mitochondria from callus tissue of *Taraxacum officinale* (dandelion). These probably represent segments of a single complex, polymorphic mitochondrion. The well-developed scalariform cristae are connected laterally with the inner membrane of the mitochondrial envelope; note also the large cytoplasmic enclave (1) in the upper mitochondrion. (G-Os, TEM.)

1 Cytoplasmic enclave

96 TS of the seed of the dicot *Ricinus communis* (castor oil) showing the numerous lipid (oil) bodies (stained red-yellow) confined within the protoplasts of the thin-walled cells of the cotyledon (1) and storage endosperm (2). Cell walls are stained green. (LM.)

1 Cotyledon 2 Endosperm

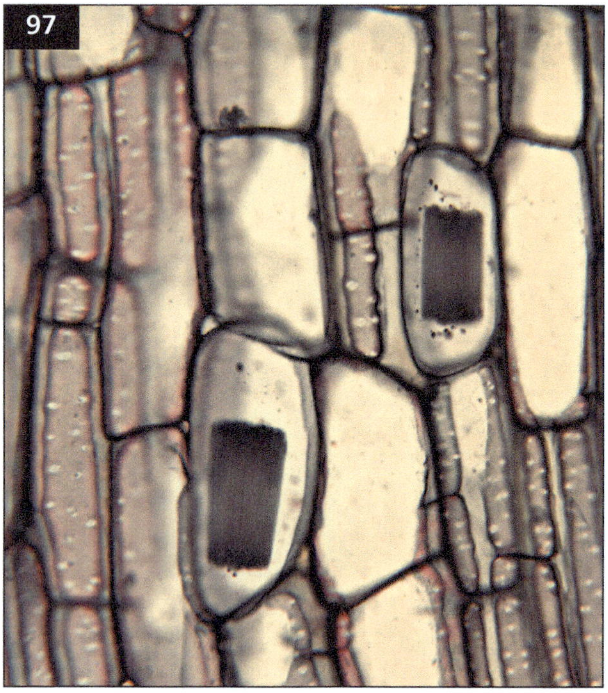

97 LS of root of *Dracaena*. Two cells contain bundles of raphide crystals. The raphides are tightly clustered in the vacuole in healthy cells and appear harmless, but if an animal damages the cell, the individual crystals are released and cause pain in either the mouth or the ovipositor. (LM.)

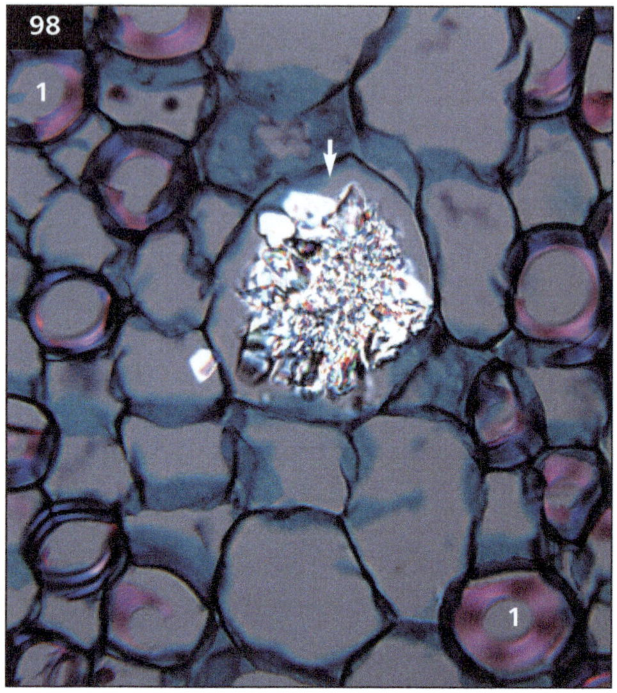

98 TS of wood of a cactus, *Discocactus alteolens*. One of the ray parenchyma cells contains a large druse (arrow) in its central vacuole. Its crystalline structure causes it to rotate polarized light, so it shines brightly whereas noncrystalline cell components are dark. Red circles (1) are the thick secondary walls of tracheids in the wood. (Polarized LM.)

1 Tracheids with thick cell walls

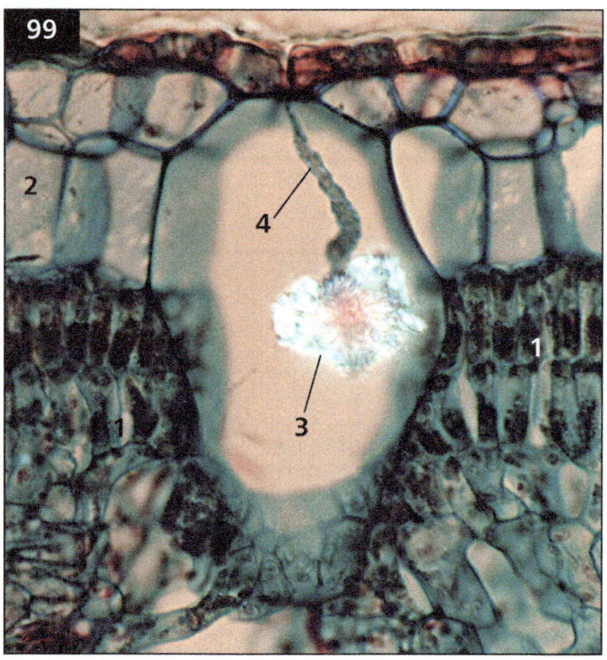

99 TS showing detail of the bifacial leaf of the dicot *Ficus elastica* (rubber-fig plant). The palisade mesophyll (1) is adaxially covered by a layer of large water storage cells (2) which represent the inner layers of a multiple epidermis derived by periclinal divisions of the protoderm. Note the cystolith which consists of a precipitate of calcium carbonate (3) encrusted onto a stalk of cellulose (4) connected to the wall of an enlarged epidermal cell. (Polarized LM.)

1 Palisade mesophyll
2 Water storage cells
3 Cystolith
4 Cellulose stalk

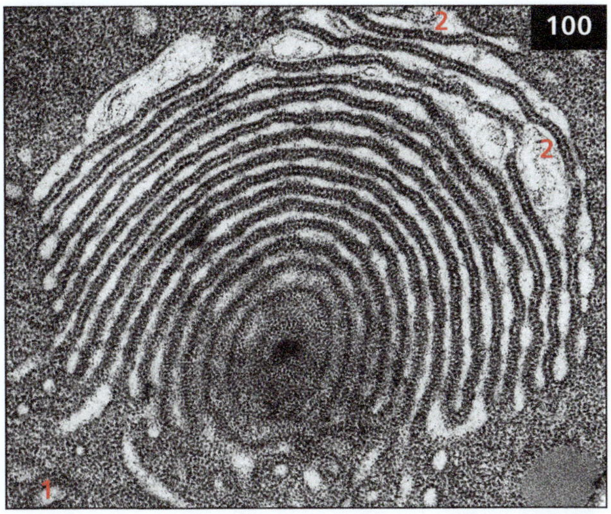

100 Concentric cisternae of rough endoplasmic reticulum from the root tip of *Allium cepa* (onion). The cisternal membranes are densely studded with attached ribosomes which are also packed free in the adjacent cytoplasm (1). The outer cisternal lamellae are inflated and contain fibrillar deposits (2) which probably represent protein synthesized on the associated ribosomes. (G-Os, TEM.)

1 Cytoplasm
2 Fibrillar deposits

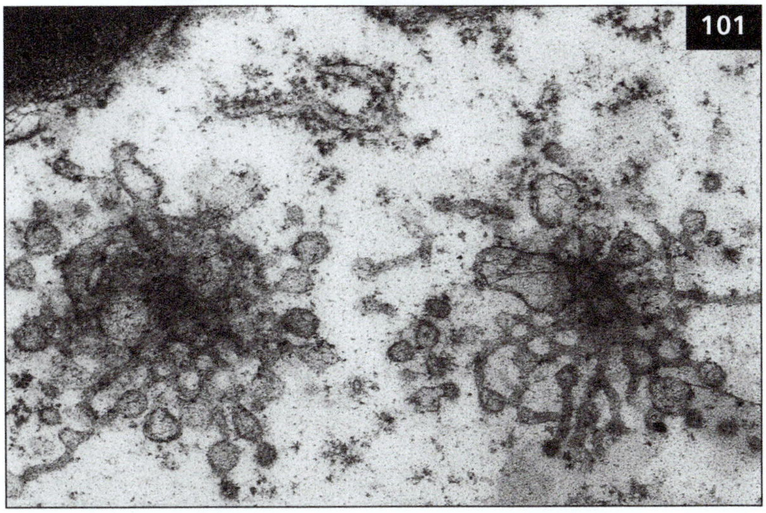

101 Dictyosomes from *Taraxacum officinale* (dandelion) callus sectioned in face view. Note that the cisternae are composed of anastomosing membranous tubules and that vesicles are forming at the margins. (G-Os, TEM.)

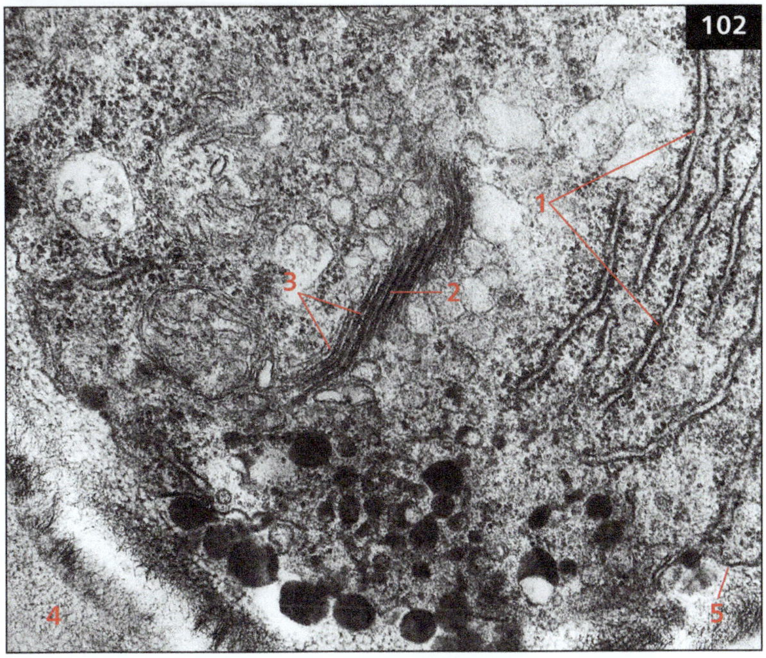

102 Peripheral cytoplasm of a cell from *Polytrichum commune* (hair moss). Note the parallel rough endoplasmic reticulum cisternae (1), numerous free ribosomes in the cytoplasm, and a prominent dictyosome. This shows some polarity, with the overall thickness of the cisternae progressively increasing from forming face (2) to mature face (3); in the latter cisternae a distinct lumen is apparent within the investing membranes. Cell wall (4), plasmalemma (5). (G-Os, TEM.)

1 Endoplasmic reticulum cisternae
2 Forming face of dictyosome
3 Mature face of dictyosome
4 Cell wall
5 Plasmalemma

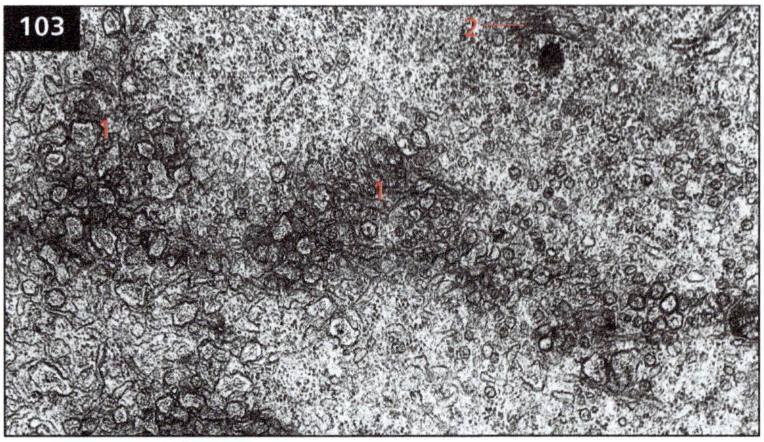

103 Obliquely sectioned cell plate forming between two daughter cells in *Linum usitatissimum* (flax). The cell plate forms from the abundant inflated vesicles (1) secreted by the dictyosomes (2). The vesicles migrate to the equator of the mitotic spindle and fuse with each other; thus the dictyosome membranes give rise to the plasmalemma. (G-Os, TEM.)

1 Dictyosome vesicles
2 Dictyosome

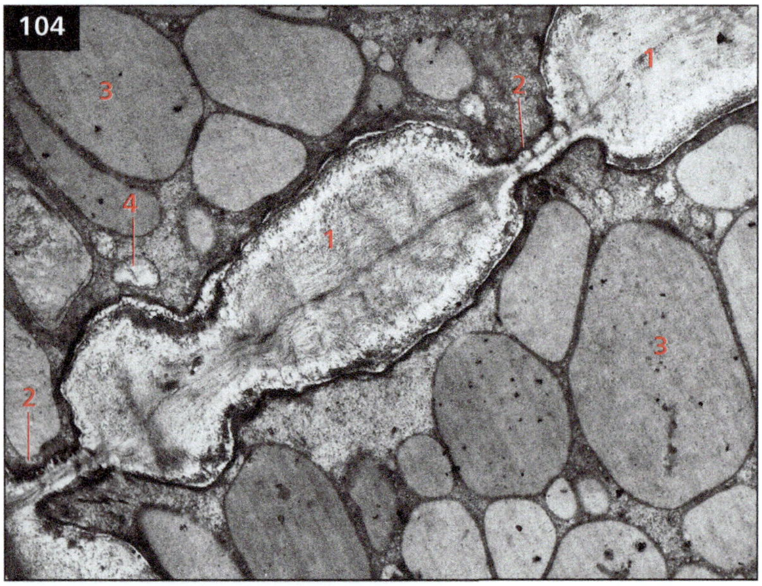

104 Partly hydrated cotyledonary storage cell from the germinating seed of *Phaseolus vulgaris* (bean). The greatly thickened wall (1) contains storage polysaccharides while the thinner regions represent simple pits (2) containing plasmodesmata. In the cytoplasm numerous large protein bodies (3) occur and several small mitochondria (4) can be distinguished. (G-Os, TEM.)

1 Cell wall
2 Pits containing plasmodesmata
3 Protein bodies
4 Mitochondrion

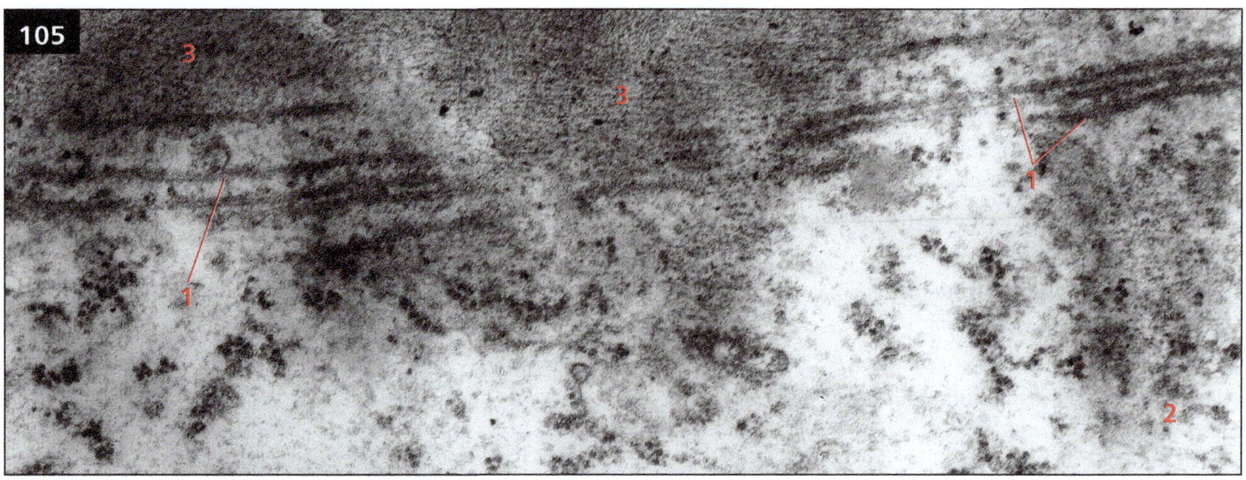

105 Interface between the cell wall and cytoplasm in *Pisum sativum* (pea) root. Note the group of longitudinally-sectioned, parallel microtubules (1) in the cytoplasm (2) and the similarly orientated microfibrils in the adjacent cell wall (3). (G-Os, TEM.)

1 Microtubules
2 Cytoplasm
3 Microfibrils in cell wall

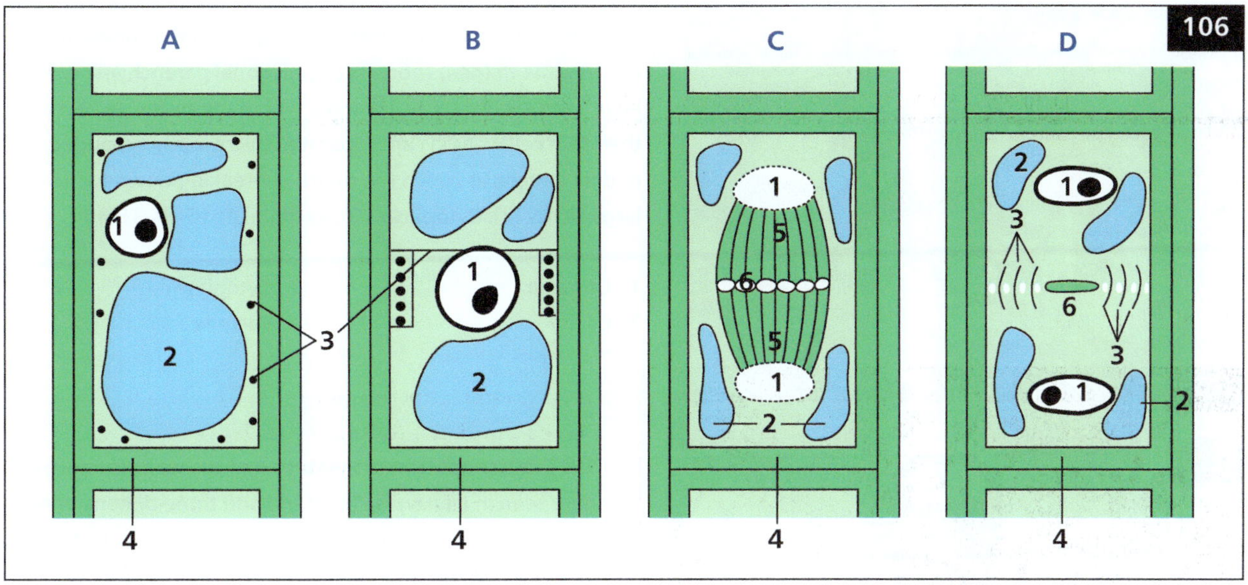

106 A–D: Diagrams showing the division of a parenchyma cell. A: Interphase nucleus (1) located in the peripheral cytoplasm among large vacuoles (2) while the microtubules (3) lie adjacent to the cell wall (4). B: The nucleus (1) migrates to the central cytoplasm and the microtubules (3) become concentrated into a pre-prophase equatorial band just within the cell wall (4). C: At the end of mitosis the envelopes (dotted) of the two progeny nuclei (1) are reconstituted at the poles of the mitotic spindle while at its equator dictyosome vesicles are fusing to form the cell plate (6). Peripheral microtubules are no longer present but the spindle fibres (5) are composed of bundles of microtubules. D: The progeny nuclei (1) are now fully formed, the spindle fibres have dispersed but a cell plate (6) has developed at the former equator. The plate spreads centrifugally and in the phragmoplast at its margins short phragmoplast microtubules (3) lie interspersed with anastomosing dictyosome vesicles. At a slightly later stage than illustrated, the cell plate joins to the mother cell wall and two daughter cells are formed.

1 Nucleus
2 Vacuoles
3 Microtubules
4 Cell wall
5 Spindle fibres
6 Cell plate

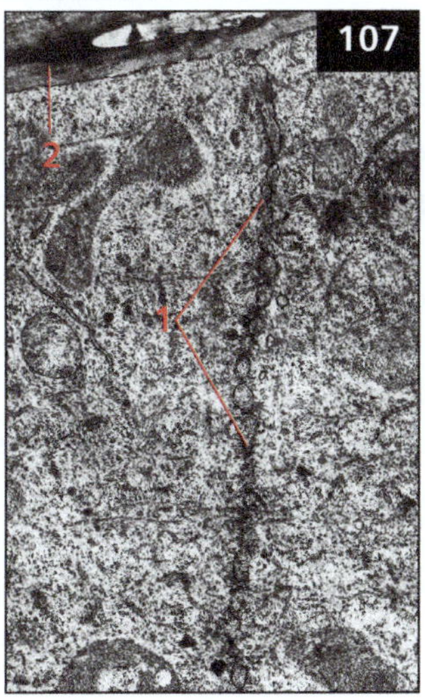

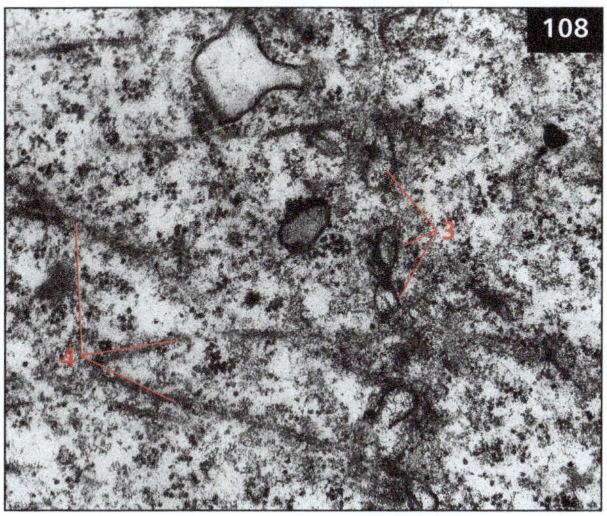

107, 108 LS of a dividing *Pisum sativum* (pea) root cell. 107 shows the centrifugally advancing cell plate (1) nearing the mother cell wall (2). 108 illustrates a detail of the beaded cell plate forming from the fusion of discrete dictyosome vesicles (3); note also the phragmoplast microtubules (4). (G-Os, TEM.)

1 Cell plate
2 Mother cell wall
3 Dictyosome vesicles
4 Phragmoplast microtubules

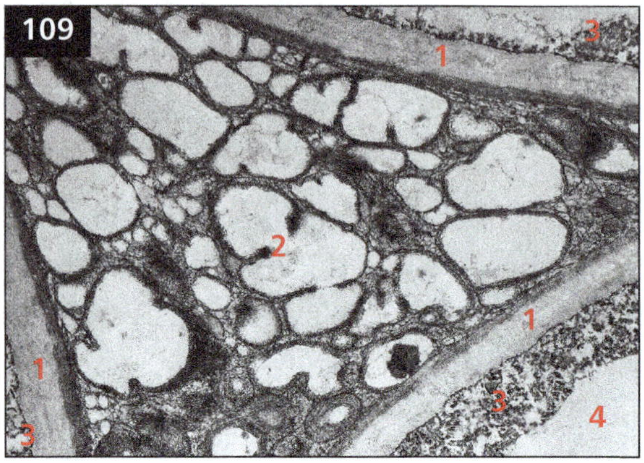

109 Development of an intercellular space in the cortex of *Pisum sativum* (pea) root. Three polygonal parenchyma cells have been sectioned at the angle where they interconnect and their thin cell walls (1) are separated by a greatly expanded middle lamella in which the matrix is breaking down to form a large cavity (2). Cytoplasm (3), vacuole (4). (G-Os, TEM.)

1 Cell wall
2 Cavity
3 Cytoplasm
4 Vacuole

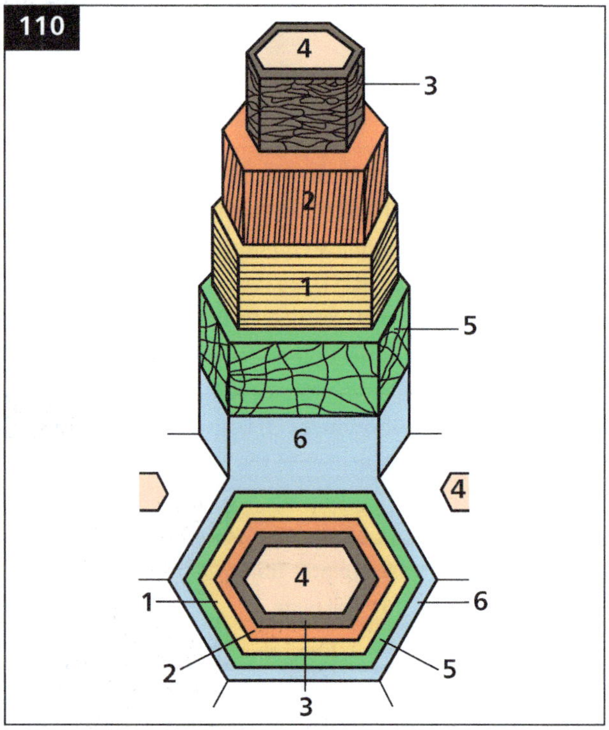

110 Diagrammatic representation of the wall structure of a fibre seen in transverse (bottom) and three-dimensional view (top). A thick secondary wall (1–3) surrounds the dead lumen (4). In the S1 and S3 layers of the secondary wall (1 and 3) the lamellae of parallel-orientated cellulose microfibrils are predominantly transversely orientated, but in the S2 layer (2) the microfibrils lie nearly vertical. In the primary wall (5) the cellulose microfibrils are less abundant than in the secondary wall and tend to lie haphazardly. The primary walls of adjacent fibres are separated by a thin middle lamella (6) from which cellulose is absent. No pits are indicated although a few simple pits normally connect adjacent fibres.

1–3 Secondary wall
  1 S1 layer
  2 S2 layer
  3 S3 layer
4 Lumen
5 Primary wall
6 Middle lamella

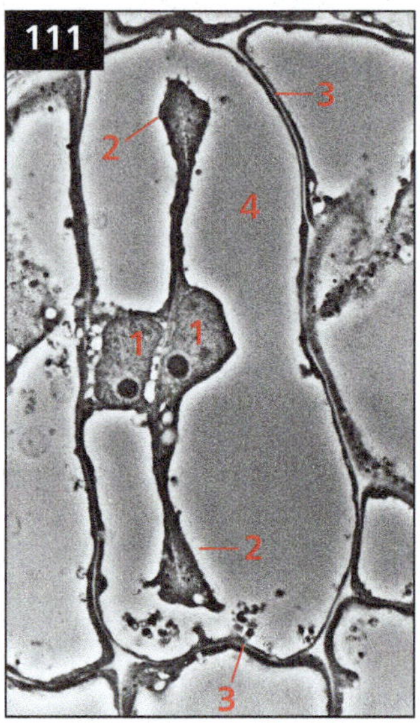

111 LS of a highly vacuolated parenchyma cell from the root of *Pisum sativum* (pea). The cell has recently undergone mitosis and the two progeny nuclei (1) are separated by a thin cell plate. This is covered on either side by a thin layer of cytoplasm which is distended at the margins of the plate to form a phragmosome (2). Mother cell wall (3), vacuole (4). (G-Os, Phase contrast LM.)

1 Nuclei
2 Phragmosome
3 Mother cell wall
4 Vacuole

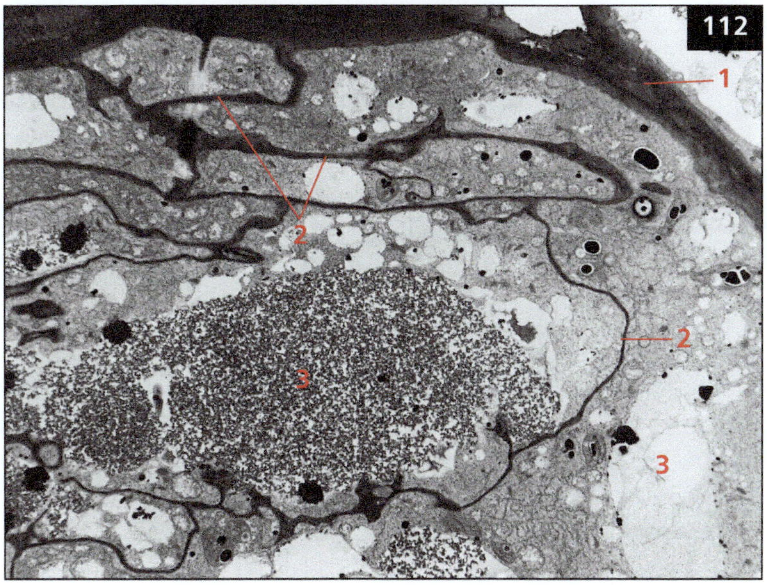

112 Storage cell from a callusing cotyledon of *Phaseolus vulgaris* (bean). The thick mother cell wall (1) is connected to the thinner, tortuous, freely-forming wall (2) which develops *in vitro*. Note also the large vacuoles (3) which develop after hydrolysis of the protein bodies. (G-Os, TEM.)

1 Mother cell wall
2 Freely-forming wall
3 Vacuoles

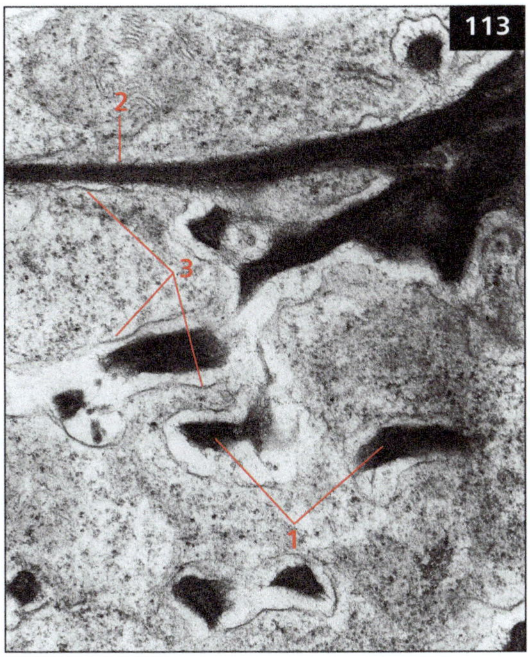

113 Transfer cells from the stem of *Linum usitatissimum* (flax). The lower cell shows numerous wall profiles (1) which represent blindly-ending ingrowths of the main wall (2) into the dense cytoplasm. The fibrillar darkly staining cores of these ingrowths are surrounded by relatively translucent areas (possibly callosic) separated from the cytoplasm by the plasmalemma (3). (G-Os, TEM.)

1 Wall ingrowths
2 Main wall
3 Plasmalemma

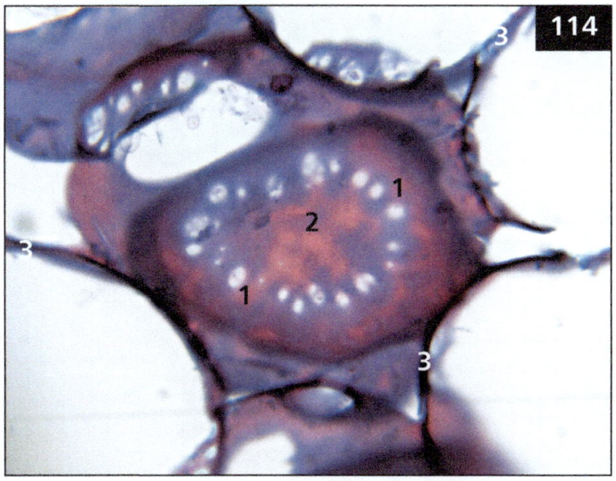

114 Face view of the region where two primary cell walls press against each other (contact face, 1) showing a circle of pit fields (2). The pit fields appear white, as if they were actually holes, but they are merely regions where the walls are especially delicate. Walls of the surrounding cells (3) were perpendicular to the plane of sectioning and thus reveal their thinness. (LM.)

1 Contact face
2 Primary pit fields
3 Cell walls cut in LS

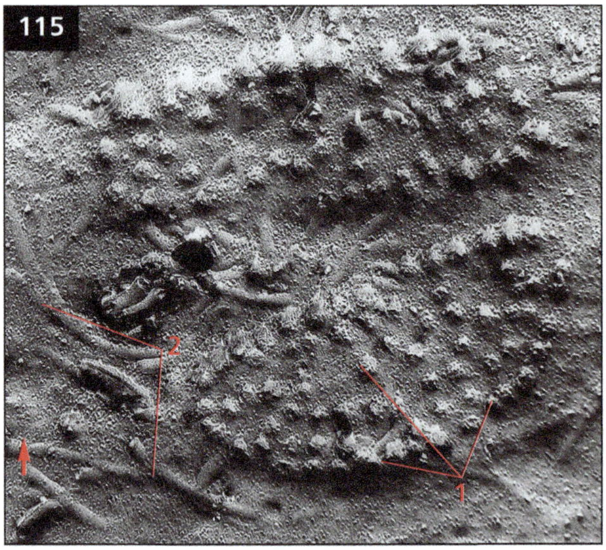

115 Pit field seen in face view of the plasmalemma of *Lycopersicon esculentum* (tomato). Note the numerous transversely fractured plasmodesmata (1) while the surface of the plasmalemma shows a number of small, probably proteinaceous, particles (cf. 64). The linear structures are thought to represent evaginations of the plasmalemma (2) caused by tubular endoplasmic reticulum adjacent to the under (cytoplasmic) surface of the plasmalemma. View from the direction of shadowing (arrow). (F-F, TEM.)

1 Plasmodesmata
2 Evaginations of plasmalemma

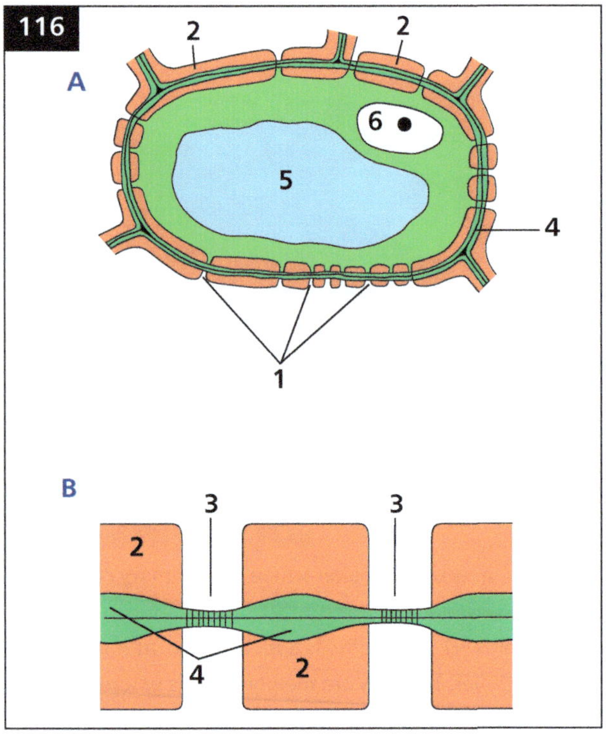

116 A, B: Diagrams of a xylem parenchyma cell. A illustrates numerous simple pits (1) in a thick secondary wall (2). B shows that the secondary wall (2) is not deposited at the pit fields (3, transverse lines represent plasmodesmata) and a simple pit with uniform diameter is formed. Primary wall (4), vacuole (5), nucleus (6).

1 Simple pits
2 Secondary wall
3 Pit fields
4 Primary wall
5 Vacuole
6 Nucleus

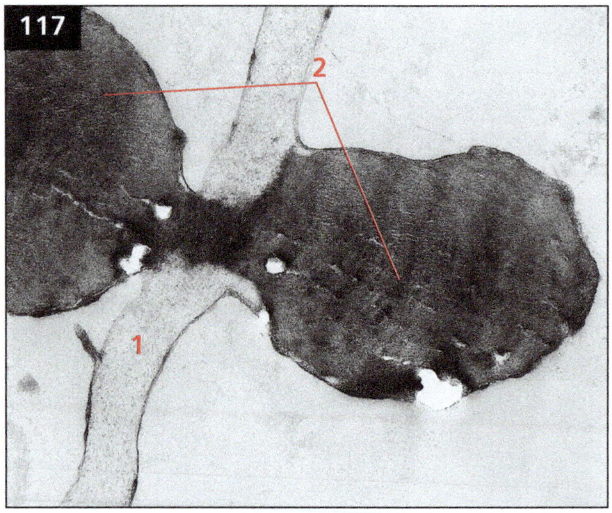

117 LS of an annular protoxylem element from the root of *Taraxacum officinale* (dandelion). Note the narrow 'holey' primary wall (1) and the densely staining, lignified secondary wall thickenings (2). (G-Os, TEM.)

1 Primary wall
2 Secondary wall thickenings

118 Development of a sieve tube and its companion cells in a flowering plant. A: Sieve element precursor cell; note its thin primary wall (1), large central vacuole (2), and two nuclei (3) with a developing cell plate (4) between them. B: Mature sieve element and companion cells. The largest cell formed from the precursor gives rise to the sieve element, which is enucleate but still retains its plasmalemma, modified plastids (5), mitochondria, endoplasmic reticulum. and proteinaceous fibrils (6). The end walls of the element now form the sieve plates and are perforated by pores which are often apparently occluded by fibrils. The cell wall is thickened at the sieve plate, and around the pores it contains amorphous callose. The two densely cytoplasmic, nucleated companion cells are formed after the further division of the smaller derivative of the precursor cell shown in A. Companion cell nucleus (7).

1 Primary wall
2 Central vacuole
3 Nuclei
4 Developing cell plate
5 Plastid
6 Proteinaceous fibrils
7 Companion cell nuclei

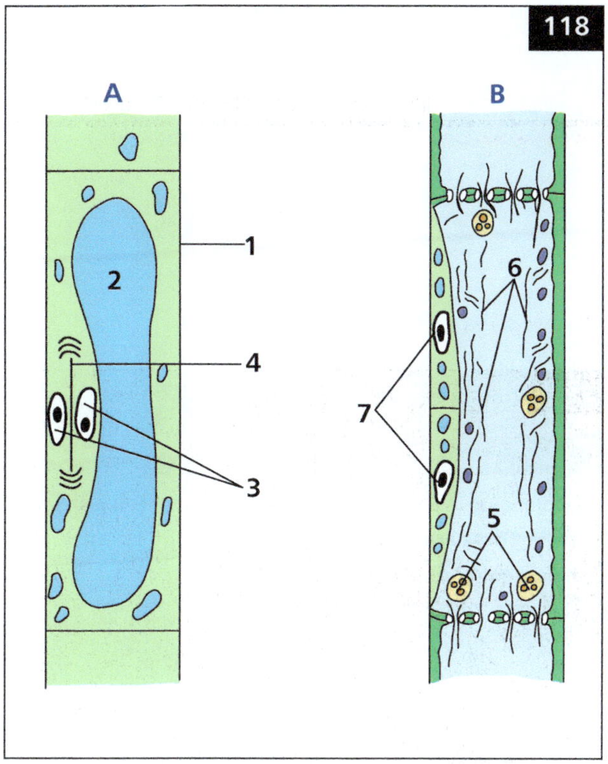

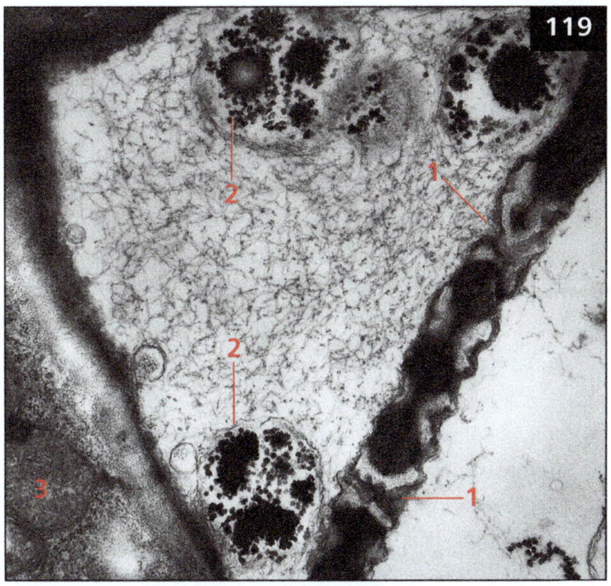

119 Sieve tube from the stem of *Linum usitatissimum* (flax). Note the obliquely inclined sieve plate with whitish regions of callose surrounding the sieve pores (1). Several plastids (2) containing starch are visible in the sieve tube and numerous fibrils of P-protein fill its lumen. Companion cell (3). (G-Os, TEM.)

1 Sieve pores
2 Plastids
3 Companion cell

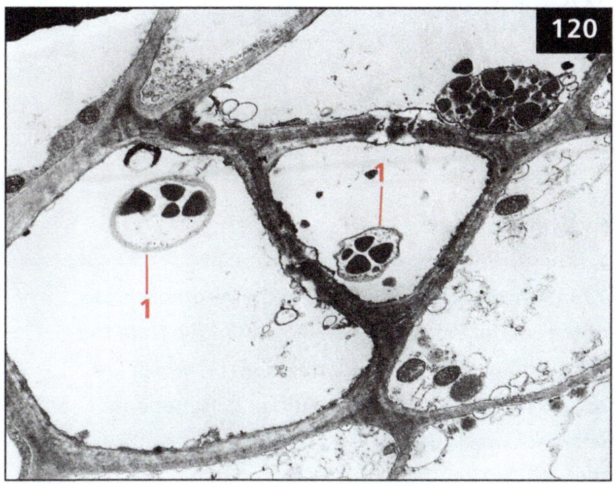

120 TS of secondary phloem (formed from a secondary thickening meristem) from the stem of the monocot, *Dracaena*. Note the oblique sieve plates with callose (white areas) around the pores. Plastids (1) with starch grains are present in the lumina but P-protein is absent. (G-Os, TEM.)

1 Plastids

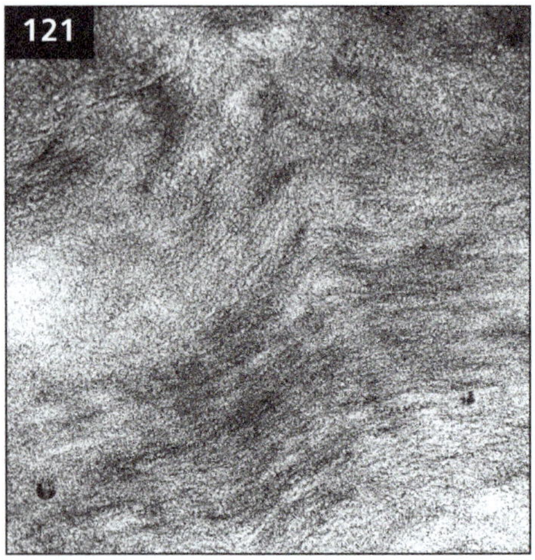

121 Glancing section of the face of a primary cell wall showing the parallel arrangement of its component cellulose microfibrils. (G-Os, TEM.)

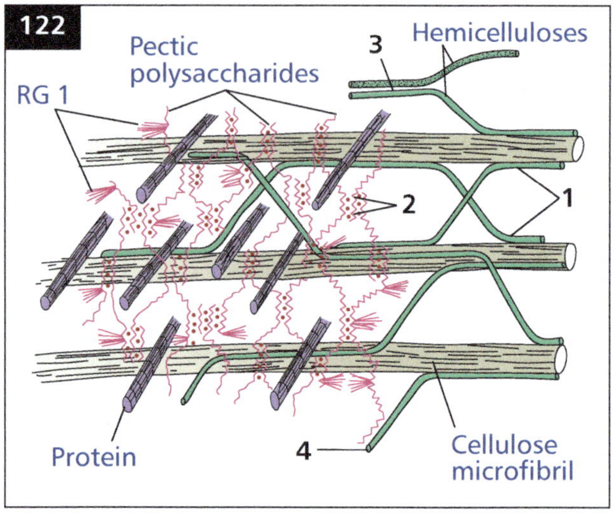

122 Model of the molecular structure of the primary plant cell wall. Cellulose consists of long chains of β-1,4-glucans, with 30 or more chains being aligned in parallel to form a microfibril held together by numerous hydrogen bonds, and resulting in a structure that is crystalline in its core, and partially crystalline at the exterior. Microfibrils do not bind laterally to one another, but matrix polymers (mainly hemicelluloses) hold them in place with hydrogen bonds. The cell wall matrix also contains a pectin network surrounding the cellulose-hemicellulose network, but with relatively few covalent or noncovalent bonds with it. Hemicelluloses bound to the surface of adjacent cellulose microfibrils (1), calcium ions cross-linking negatively-charged pectin molecules (2), bonding between hemicellulose molecules (3), covalent hemicellulose-pectin bonds (4). RG 1: Rhamnogalacturonan 1. (Diagram with permission of CT Brett and KW Waldron from *Physiology and Biochemistry of Plant Cell Walls,* 2nd edn, 1996, Chapman and Hall, Oxford.)

1 Hemicelluloses bound to the surface of adjacent cellulose microfibrils
2 Calcium ions cross-linking negatively-charged pectin molecules
3 Bonding between hemicellulose molecules
4 Covalent hemicellulose-pectin bonds RG 1: Rhamnogalacturonan 1

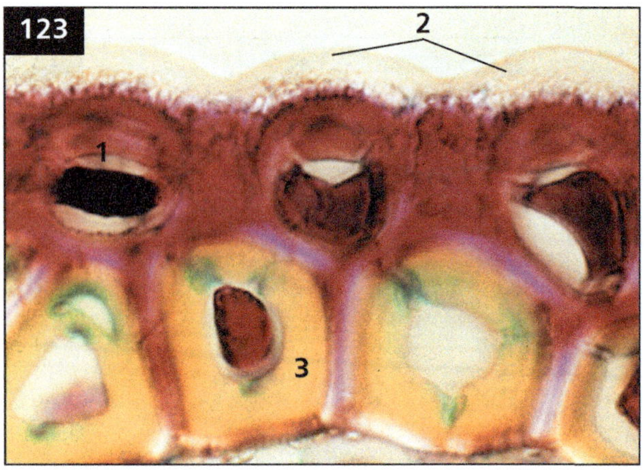

123 TS showing surface detail of the xeromorphic leaf of the conifer *Pinus monophylla* (nut pine). Note the thick-walled, lignified epidermal cells (1) which are coated externally by a thick cuticle (2). Hypodermal sclerenchyma (3). (LM.)

1 Epidermal cells
2 Cuticle
3 Hypodermal sclerenchyma

CHAPTER 3
Plant Histology

DISTRIBUTION OF CELLS AND TISSUES

The vascular plant is composed of dermal, ground, and vascular tissue systems (**124–127**). The phloem and xylem constitute the vascular system; both are complex tissues in which the conducting elements (**57, 119, 120**) are associated with other cell types (**10, 20, 128**). The ground system surrounds the vascular tissues (**124–127**) and comprises parenchyma, collenchyma, and sclerenchyma (**129–134**). The dermal system is initially represented by the epidermis (**125, 126**) containing various cell types (**126, 135**). Its structure is considered in Chapter 5.

In plants which undergo woody thickening, the epidermis typically becomes replaced by the cork and associated tissues (the periderm, **136, 137**) whose structure is discussed in Chapter 6. Secretory cells do not normally develop as distinct tissues (except, for example, in nectaries) but rather occur either as surface structures (**135**) or within other tissues (**127**).

PARENCHYMA

These cells generally have thin walls and large vacuoles (**71, 126, 129**). They form a continuous tissue in the cortex and pith (**40, 127**) and in the leaf mesophyll (**126**), while vertical strands and horizontal rays of parenchyma occur in the vascular system (**8, 138, 139**). The leaves and stem of seedlings and small herbs are largely held erect by the collective turgor of the parenchyma cells (**126**) and the shoot wilts if insufficient water is available from the root system to replace transpiration losses (**21**).

Parenchyma cells are often polyhedral or elongate (**126, 129**) but stellate and irregular forms occur (**140**). The mesophyll cell of the leaf is specialized for photosynthesis and contains numerous chloroplasts (**126**) while starch is frequently stored in the parenchyma of the root and stem (**23**). In plants growing in marshy conditions and in aquatics, the parenchyma of the shoot and root sometimes develops extensive intercellular spaces and is termed aerenchyma (**140**; see also Chapter 6, **369**).

Mature parenchyma cells of the stem and root may resume division (**141**) and undergo partial dedifferentiation to form the lateral meristems of the phellogen and vascular cambium (**128**) from which the cork and secondary vascular tissues originate (**136, 137, 142**). In some situations the parenchyma cells revert to an undifferentiated meristemoid state (**50**) and give rise to new lateral and adventitious roots (**143**) or adventitious shoots on various parts of the plant (**144, 145**).

The development of adventitious organs is often related to injury as in cuttings (**143**) and the plant's regenerative capacity is extensively exploited in various horticultural techniques (see Chapter 8). Parenchyma cells should therefore be regarded as potentially totipotent and single *in vitro* cultured cells derived from various species have given rise to completely new plants when grown on a suitable nutrient medium with exogenous growth substances.

The parenchyma cell wall is often thin, with its growth normally terminating at the end of cell vacuolation (**71, 126, 140**). However, in seed storage tissue the walls may be greatly thickened (as in *Coffea*, *Phoenix*, *Phaseolus*, **104**) by noncellulosic reserve carbohydrates, while parenchyma cells in the xylem and pith often develop thick lignified secondary walls (**8, 116**). In transfer cells labyrinthine ingrowths of the wall (**113**) vastly increase the plasmalemmal surface, while simpler vertical ingrowths of the wall also occur in the plicate mesophyll cells of several conifers (**146**).

COLLENCHYMA

These living cells have thick but relatively pliable walls (**130**) and are located in the outer ground tissue (**147, 148**). Collenchyma is of great value in the mechanical support of the young shoot but is rare in the root. The protoplasts of such cells are generally vacuolate (**130**) while their highly hydrated primary walls are unevenly thickened (**148**) and contain large amounts of pectins and hemicelluloses. Collenchyma cells are usually elongate and in transverse section often appear angular with thickening localized in the cell corners (**130, 148**) but the thickening may be confined to the tangential walls. Intercellular spaces sometimes occur in collenchyma, and in older cells the walls may become lignified.

SCLERENCHYMA

This tissue is characterized by its thick, normally lignified secondary walls and the protoplasts often die as part of maturation. Sclerenchyma is the principal mechanical tissue of plant organs (**149, 150**) and exists as either sclereids (**131, 132, 149**) or fibres (**10, 124, 134, 150–152**).

Sclereids frequently occur either singly (**131**) or in small aggregates (**132**) and may develop in the epidermis as well as internally in the plant. They vary considerably, from more or less isodiametric stone cells with prominent simple pits (**132**) to branched astrosclereids with tapering arms (**131, 133, 149**). Sclereids occur in leaves of some plants, in the hardened fruit walls of nuts and stone fruits and in the seed coats of many legumes.

Fibres commonly occur in groups, forming strands (**125, 134, 150, 151**). The individual cells are slender and highly elongate, with their tapering ends overlapping; simple pits link adjacent fibres (**134**). Their walls have extensive secondary deposition (**54, 110, 150**) and their lumina are generally much narrower than those of xylem tracheary elements (**151–153**). The walls of mature fibres are generally lignified, hard and incapable of extension so they mature only in regions of the plant in which elongation has ceased.

In some dicots the fibres are divided by several unthickened cross walls; these are designated as septate fibres and often retain their protoplasts (**154**). The thin-walled tips of fibres frequently continue to elongate after their mid-regions have formed extensive secondary walls (**10, 54**). In *Linum* (flax) the individual fibres are thickened but not lignified; their tips grow intrusively between neighbouring parenchyma cells (**155**) and may reach 6 cm in length.

Textile fibres are obtained from the extensive strands of primary phloem fibres that occur in the stems of several dicots; they are fairly flexible since they are usually only moderately lignified (for example *Boehmeria*, *Corchorus*) or nonlignified as in *Linum* (**10, 125**). However, in fibres from monocot leaves such as *Agave*, *Phormium*, *Sansevieria* (**150**) the lignification is greater and the extracted fibres are coarse and stiff.

SECRETORY TISSUES

Secretory trichomes (hairs) and glands (see Chapter 5) often develop in the epidermal and sub-epidermal tissues. In the insectivorous plant *Drosera* the leaves are covered by long and complex multicellular hairs (**135**). Their glandular heads are coated with a viscous secretion containing digestive enzymes in which insects become trapped. Nectaries are glands which secrete a sugary solution (nectar); they are located superficially and occur either on various parts of the shoot (extra-floral nectaries, **156**) or on the flower (floral nectaries). In the inflorescence of *Euphorbia* (**157**) a connate bract bears four oval nectaries. The several layers of secretory cells are densely cytoplasmic and the nectaries are supplied with vascular tissue. A shallow layer of nectar is secreted, so attracting various flies which effect pollination.

Hydathodes frequently occur on leaves and excrete water (**158**) from the leaf margins and tips. The hydathode consists of modified mesophyll tissue into which the water is discharged from tracheids. The enclosing epidermis bears stomata which remain permanently open and through which water is secreted (guttated). This guttation may be a mechanism which protects shoot tissues from becoming waterlogged in situations in which root pressure is excessive. However, many hydathodes contain transfer cells, indicating that minerals are unloaded within the tissue and become available for leaf growth at the same time that water is being discarded.

A large variety of secretory structures also occurs internally in the plant body (**127, 142**). In *Pinus* and other conifers, resin canals (**142, 146**) are formed schizogenously by the separation of adjacent cells from each other to form a central duct. Resin, which contains various terpenes, is secreted into the duct by the lining epithelial cells; felled conifer trunks sometimes weep copiously with resin exudations (**159**). Mucilage ducts (**127, 160**) are also formed schizogenously.

Laticifers produce a milky secretion termed latex (**161**) which probably represents a deposit of various metabolic byproducts. Laticifers occur in about 900 genera of angiosperms and are often branched and usually extend throughout the plant body in various tissues. In the fleshy, largely parenchymatous, root of *Taraxacum* the secondary phloem is especially well supplied with laticifers (**162**). These are closely associated with the strands of sieve tubes (**163**) which occur in concentric rings isolated by intervening phloem parenchyma. The articulated laticifers of *Hevaea* (the main source of commercial rubber), *Taraxacum*, and many other genera originate from the breakdown of the intervening walls between contiguous cells. However, nonarticulated laticifers form from a single cell which often becomes multinucleate. Both types of laticifers may branch and anastomose (**164**).

PHLOEM

This is a complex tissue composed of sieve elements, companion cells, parenchyma, and sclerenchyma (**10, 20, 165**), while laticifers sometimes also occur. The conductive sieve elements (**119, 120**) generally function for only a few months, but they conduct over several seasons in *Vitis* and *Tilia* (**165–167**) while in palms they apparently translocate for many years. In contrast, they are evanescent in the protophloem (**49**).

The angiosperm sieve tube is composed of sieve tube members joined end to end at their sieve plates (**118, 168**). The individual members are 50–150 μm long and up to 40 μm wide and have primary walls, although these may be thickened. The sieve plates separating sieve tube members are perforated by numerous pores (**53**). On the transversely situated sieve plates of *Cucurbita* (**169**) the pores may reach 15 μm in diameter. In species with obliquely-inclined end walls, the sieve plates are compound (**170**) and are composed of several sieve areas with small pores only one to several micrometres wide.

The sieve tube member is a living but highly modified, enucleate cell (**118**). During its maturation the tonoplast and most organelles are lost (**118–120**). The plasmodesmata develop into the vastly enlarged sieve pores (**53, 169**), but the plasmalemma remains intact and lines the margins of the pores and wall (**53**). Sieve areas may occur on the longitudinal walls (**170**) but these are less well-defined than the sieve plates and their pores are smaller.

In actively translocating sieve tubes the sieve pores are generally considered to be open (**171**). However, in most sections prepared from dicotyledonous material the pores are blocked by plugs of P-protein (**53, 119, 168**) while callose deposition at the margins of the pores (**169**) greatly reduces their diameter. Similar sealing of sieve tubes is thought to occur in damaged tissue on the intact plant. In monocots (**120**) P-protein is rare and it is apparently absent in gymnosperms. In nonflowering vascular plants the translocating elements are discrete sieve cells; these are commonly very elongated with small sieve areas distributed over both the lateral walls and tapering ends.

Sieve tubes are typically associated with both parenchyma cells and more densely cytoplasmic companion cells (**170**). Sieve tube members and companion cells arise from common precursor cells (**118**) and plasmodesmatal connections between them are abundant, with those in the companion cell wall usually branched (**171**). Companion cells apparently supply ATP to the sieve tubes; in the minor veins of the leaf they function, along with parenchyma, as intermediary cells in the accumulation and loading of photosynthates into the sieve tubes.

The pressure-flow hypothesis of translocation (**171**) suggests that sugars and other nutrients are loaded by molecular pumps into the sieve tubes of the leaf. This generates a negative osmotic potential in the tubes and consequently water is absorbed. The increased turgor pressure causes flow from one element to another via the open sieve pores. In growing regions of the plant (**49**), and in storage regions (**23**), the sugars are pumped out from the sieve tubes into the adjacent tissues and water follows, so that a mass flow of nutrients is established from source to sink (**171**). Translocation in the phloem allows fast movement of nutrients (usually about 1 m/hr) and the sap contains up to 250 mg/l of sugar plus other nutrients and plant hormones.

In plant organs with secondary tissues the older phloem has additional roles. It is a principal component of the protective bark of trees (**136, 137, 167**) and in small shoots or twigs the phloem fibres provide considerable mechanical support (**165**). Secondary phloem parenchyma (**165**) provides an important storage tissue. When the buds break in deciduous woody species, large quantities of carbohydrates and nitrogenous substances are mobilized in the phloem parenchyma. The soluble products are then transported to the expanding new leaves, sustaining them before they are fully photosynthetic. Secretory tissues often occur within the phloem and in *Hevaea* (rubber tree) the laticifers in the phloem secrete various polyterpenes; the milky liquid which exudes when the tree is tapped is refined to provide rubber.

Stretched by the increased circumference of the growing root or stem, the outermost secondary phloem expands laterally by the resumed division and growth of its parenchyma. This expansion is called dilatation growth and is commonly obvious in the ray tissue which may flare outwards from the vascular cambium towards the nonfunctional outer phloem (**165**). In most woody species, after the first cork cambium (phellogen) becomes nonfunctional, new cambia arise progressively more deeply internally and eventually from the parenchyma of secondary phloem, where they often appear in cross-section as discontinuous but overlapping layers (**167**). This older phloem becomes sloughed off in the successively formed layers of the peeling bark (see Periderm in Chapter 6).

XYLEM

Xylem is complex tissue with two principal roles: the transport of large quantities of water from the root to the shoot in the tracheary elements (**21**) and the mechanical support of the plant body (**172**). This support is provided both by the tracheary elements and the associated (often thicker-walled) nonconducting fibres (**8, 153**). Additionally, the axial and ray parenchyma in the secondary xylem store food and water (**8, 138, 173**).

Mature tracheary elements (**20, 142**) are dead and have lost their protoplasts (**56, 57**). Their secondary walls are thickened relative to the primary walls (**55, 57**) and, due to lignification, are impermeable except at the pits where only primary wall is present (**57**). The tracheary elements are elongated and water moves along their lumina from the root to the shoot in the transpiration stream (**21**). The absence of a plasmalemma allows the water to pass fairly freely from one element to another via the numerous pits (**174**).

In nearly all angiosperms the tracheary elements comprise both tracheids and vessels (**175**) but generally only tracheids occur in gymnosperms and lower vascular plants (**142, 151, 176**). A tracheid is derived from a single cell and has no perforations; it is elongated, has tapering ends (**175**), and in conifers the tracheids (**177, 178**) may reach more than a centimetre in length. By contrast, vessels are composed of a tube-like series of many vessel elements aligned end to end. They are directly linked through their perforation plates (**175, 179–181**) which represent the remnants of their original end walls (**182**). Vessel elements tend to be shorter but wider than tracheids (**175**).

The perforation plate shows either a single large perforation (**175, 180**) or in compound plates a number of elongated perforations which are commonly scalariform (**181**). Because of these open perforations, vessels generally show a lower resistance to water movement than tracheids, where the closed pits impede water flow. In ring porous wood (**201**), the wide vessels apparently extend many metres along the tree trunk. The tips of vessels are imperforate but the numerous pits allow water to move into adjacent tracheary elements.

The extent and type of pitting in tracheary elements are variable (**175–184**). The protoxylem in the shoot shows secondary wall deposition of an annular or helical pattern (**184**). The primary wall between thickenings becomes greatly extended after the protoxylem element dies while its noncellulosic components are digested. The stretched wall often appears 'holey' under TEM (**117**) and may rupture leaving a protoxylem cavity (**20**). In non-elongating regions of the plant, metaxylem elements show much more extensive secondary wall deposition of various patterns (**183**).

In a scalariform element at least half of the primary wall is covered by secondary wall and the pits are horizontally elongate and usually bordered (**183, 184**). In reticulate elements the thickening is more irregular, while in pitted elements (**175, 176, 183**) a greater proportion of the wall is secondary. Their pits occur in horizontal rows (opposite pitting, **183**) or diagonally (alternate pitting, **177**).

In these various modes of secondary wall thickenings the intervening pits are bordered (**57**). In conifer tracheids the centre of the pit membrane (torus, **176**) is thickened and lignified, but the periphery (margo) has only a loose cellulose network and is permeable. The pits between adjacent tracheary elements are abundant and bordered (**56, 57, 174**), but there are few connections to fibres. The pits which link with parenchyma cells are either simple or half-bordered on the tracheary element side (**174**).

STRUCTURE OF WOOD

In woody plants the formation of secondary vascular tissue is generally periodic, since the vascular cambium becomes dormant in unfavourable environmental conditions. This typically results in the formation of growth rings in the secondary xylem of the tree (**172, 173**). In nontropical species the cambial activity is limited by temperature and the rings usually represent annual increments (**142, 167, 172, 173**). Generally in temperate species the last-formed layers of xylem in a growth ring are composed of narrower cells with thicker walls than the earlier wood (**185**), so that growth rings are often visible to the naked eye (**172, 173**). In many tropical trees (**186**) and desert succulents, growth rings are not obvious.

Water movement through a tree occurs though the sapwood and passes from the root to shoot. It is powered by evaporation of water, via the stomata, from the leaf surfaces (**21**) and the consequent pull exerted on the water columns in the individual tracheary elements. Water moves from tracheid to tracheid via their common pits, where only the unlignified and permeable primary wall is present. Vessels, although generally longer than tracheids, are not of infinite length and water must pass from vessel to vessel via the pits in their side walls. Water movement is much more rapid along wide vessels (some vessels are up to 360 μm in diameter) than in narrower vessels/tracheids.

In species of *Quercus* (oak) and *Fraxinus* (ash) individual vessels may extend for a metre or more along the tree's axis, but in *Acer saccharum* (sugar maple) they do not exceed about 30 cm in length. In some species of *Eucalyptus* and temperate conifers such as *Sequoia sempervirens* (coastal redwood) water columns are pulled up by leaves in a canopy which may be up to 100 m above ground level. In certain deciduous trees such as species of *Acer*, *Betula* (birch), and *Juglans* (walnut) a positive pressure develops in the wood some weeks prior to the buds bursting. This is due to the secretion of sugars and minerals into the xylem elements; water then enters the xylem by osmosis and the sugary sap is forced up the tree. In northeast America the sap of *Acer saccharum* is tapped and boiled down to produce maple syrup.

The newly-formed tracheary elements of the sapwood conduct water for a relatively short time. In stressful environments (for example nutrient deficient conditions) they often cease to function by the end of the first year, but in tropical trees they may remain active longer. The cessation of water transport in a tracheary element results from cavitation, in which a bubble of water forms an embolism, due to excess tension on the water column (**174**). The tension on the water columns within tracheary lumina is greatest in the wide elements characteristic of many broadleaf trees and may become very severe. The cohesion of the water molecules in the transpiration columns is reinforced by the adhesion of the molecules to the walls of the tracheary elements, but this effect is negligible at the centre of a wide vessel. It follows that when a tree's water supply is restricted or it is buffeted by a severe gale, cavitation in vessels in the branches and twigs of an oak is much more likely to occur than in the much narrower tracheids of a pine tree. Cavitation blocks the entire length of an affected vessel or tracheid to water movement. However, passage of the embolism to adjacent tracheary elements is blocked by their common pit membranes.

In some temperate trees the sapwood becomes cavitated and ceases to conduct water by the end of its first growing season. But in others, such as *Robinia pseudoacacia* (false acacia, black locust), the sapwood functions for 2–3 years while in *Juglans nigra* (black walnut) it may be active for up to 20 years. In tropical broadleaf trees, the tracheary elements remain uncavitated and conduct water for many years. In palm trees, no secondary xylem is formed and yet the primary xylem continues to transport water during the life of the tree for decades and sometimes even centuries.

The nonconducting tracheary elements in the heartwood of a tree (**172**) constitute the great bulk of the wood and provide the tree's principal support. Vessels in the heartwood frequently become filled by parenchymatous tyloses (**187, 188**), which intrude via the pits from adjacent parenchyma cells. Gums and various polyphenols are also often deposited and provide at least a partial barrier against the vertical spread of infectious micro-organisms through the wood. It also seems that resins in tracheary elements located at the boundaries between growth rings and in the medullary rays help make resistant barriers to the spread of infection. In heartwood the walls of parenchyma cells may become lignified after the reserve food and water are withdrawn from the living ray and axial parenchyma cells, so that overall the wood becomes drier. Some trees lose their heartwood as a result of termite assault (this is estimated to occur in up to half of Australian eucalypts) or fungal and microbial infections. But such trees often survive for many years, as is strikingly demonstrated in various veteran trees. In such ancient specimens, abundant adventitious roots frequently develop in the upper trunk and grow down its hollow interior thus reclaiming some of the nutrients released from its decayed heartwood (**189**).

The dicots are termed hardwoods because their secondary xylem generally contains a high proportion of thick-walled fibres (**138, 153, 190**) although balsa wood (*Ochroma*) is very soft because of its thin-walled fibres and extensive rays (**191**). Conifers are designated softwoods because they lack fibres (**142, 185, 192**). In conifers axial parenchyma is rare and food reserves are stored in the rays (**192**). Radial movement of water in conifer wood is largely restricted to these rays since pits normally occur only on the radial walls of tracheids (**176, 185**). The wood of dicots is usually more complex than that of conifers and contains vessels of several diameters, tracheids, and fibres while axial and ray parenchyma are normally abundant (**138, 192–196, 370**).

Rays generally consist of procumbent (somewhat radially elongated) parenchyma cells. However, in angiosperms the top, bottom and margins of the ray may also contain specialized upright (axially elongated) parenchyma. In some gymnosperms (**176, 178**) the top and bottom of the ray is composed of dead and lignified ray tracheids, while the ray parenchyma may also be lignified. Rays in gymnosperm wood are almost always one cell wide (uniseriate, **185, 192**) but in dicots both uniseriate and multiseriate rays (**196–199**) may occur.

Carbohydrates are often stored in large quantities in rays. In deciduous trees these are mobilized at the onset of the growing season and transported to the expanding buds. Sometimes, as in sugar maple (*Acer saccharum*), these solutes are conducted in the xylem elements and pass from the ray to the tracheary elements via the numerous pit connections in the upright ray parenchyma cells.

Both vessels and tracheids are present in the wood of many broadleaf tree species, but in *Drimys winteri*, *Tetracentron sinense*, and *Trochodendron aralioides* (all members of basal angiosperm clades) and in conifers only tracheids occur (**199, 200**). In ring porous trees (**201**), the new season's early growth contains wide vessels but relatively few tracheids and fibres (*Catalpa*, *Fraxinus*, *Quercus*, *Robinia*). Later in the year the vascular cambium produces narrower vessels and a greater proportion of fibres and tracheids. Diffuse porous species (**202**) show narrower vessels ('pores') and are more common than ring porous species. Wide, isolated vessels sometimes occur in the wood and are associated with numerous tracheids (**201**), but more commonly the vessels are clustered in various patterns (**196**).

The main trunk of a tree usually initially develops a more or less upright habit but its branches typically grow horizontally or often obliquely outwards from the trunk (**203**) and these branches form compression or tension (reaction) wood, each with its distinctive anatomy. In a wind-blown tree, the subsequent straightening of its trunk or branches into an upright position (**204**) involves the formation of reaction wood. Compression wood is formed on the lower side of a conifer branch (**205**) or leaning trunk; this involves an increased production of xylem, which shows as wider growth rings, containing a high proportion of shorter, thick-walled and heavily lignified tracheids.

By contrast, tension wood is said to characterize the upper side of a leaning trunk or branch in broadleaf trees. Nevertheless, in many such species the growth rings clearly show greater development on the lower side of the branch or trunk (**206**, cf. **205**). In tension wood, fewer and narrower vessels occur than in normal xylem but numerous unlignified and cellulose-rich gelatinous fibres are present. In felled trees, reaction wood is best avoided for use as timber: tension wood splits easily on drying whereas compression wood is hard but brittle to work. On windblown specimens of *Pinus sylvestris* (Scots pine) 20–50% of the bulk is represented by compression wood.

TECHNIQUES OF PLANT ANATOMY

Simple observation with a hand lens (often supplemented with a binocular dissecting microscope) reveals tremendous morphological detail of a plant (**144, 158**) (*Table 3.1*). For light microscopic examination many nonwoody specimens can be cut free-hand with a razor blade and mounted in water or an aqueous stain (**150**) but the sections are sometimes too thick for critical examination. For thinner permanent preparations, the material is first dissected, then fixed (chemically preserved), dehydrated, and finally embedded in paraffin wax. As thin sections (slices) are cut with a microtome knife (**207**), the paraffin holds tissues and structures in place and prevents the knife from pushing crystals or fibres against softer parts of the specimen. The sections are then floated on warm water, allowing the tissues to hydrate and re-expand to life-size. Once the sections are glued to glass slides, the wax is dissolved and the sections are stained (**151, 208**); finally, a film of transparent mounting medium is applied with a thin glass coverslip on top. For tough woody specimens a sledge microtome, utilizing a much heavier and stronger knife, is used to cut unembedded but fixed material; the individual sections are then further processed. Various processing artefacts sometimes occur: the tissues shrink slightly; natural pigments are replaced with stains; the knife may displace crystals; mucilage may swell out of position; and starch grains may fall out as paraffin dissolves (**146, 209–211**).

The section thickness affects the view: at 1.0 µm thick only a fraction of any cell is visible but the view is unobscured because material behind or in front has been cut away (**126, 128**). Sections 20–40 µm thick have more information but the view is cluttered so most sections are cut 8–15 µm thick. Views of tissues must be interpreted: in a section sequence through a cell, one may contain the front wall, another contains the nucleus, and the last shows the back wall. An undulating tubular cell may appear round in transverse section but as unconnected ovals in longitudinal section (**212**).

Table 3.1. Types of microscopes

Light microscopy

1. Dissecting (stereo) microscopes use separate left and right light paths (e.g. left objective, left ocular, left eye); the brain constructs a three-dimensional image (135).
2. Compound microscopes have one light path. Brightfield microscopes use a broad beam of light transmitted through the specimen, which is stained if tissues are naturally pale or colourless (139, 208). Phase microscopes construct images of density variations in unstained living cells (141, 169).
3. Confocal microscopes scan a narrow laser beam through a thick specimen, then a computer constructs an image. By marking different structures with specific stains, a three-dimensional map is constructed.

Electron microscopy

1. Transmission electron microscopes (TEMs) use magnetic lenses to form a high-resolution image of a specimen illuminated by a broad beam of electrons (48, 163).
2. Scanning electron microscopes (SEMs) scan samples with a beam of electrons and a computer constructs a three-dimensional image with all parts in focus, even if the specimen is thick (237, 529).

124 TS of the petrified stem of the fossil seed fern *Lyginipters*. In this Carboniferous plant the same tissue systems occur as in present-day flowering plants. The ground tissue is represented by a prominent pith (1) and cortex which shows an outer network of fibres (2). The dermal system and leaf bases have, however, not been preserved. The ring of secondary xylem (3), composed of radially seriated tracheids, and the poorly preserved phloem (4) represent the vascular system. A leaf trace (5) is also apparent.

1 Pith
2 Network of fibres
3 Secondary xylem
4 Phloem
5 Leaf trace

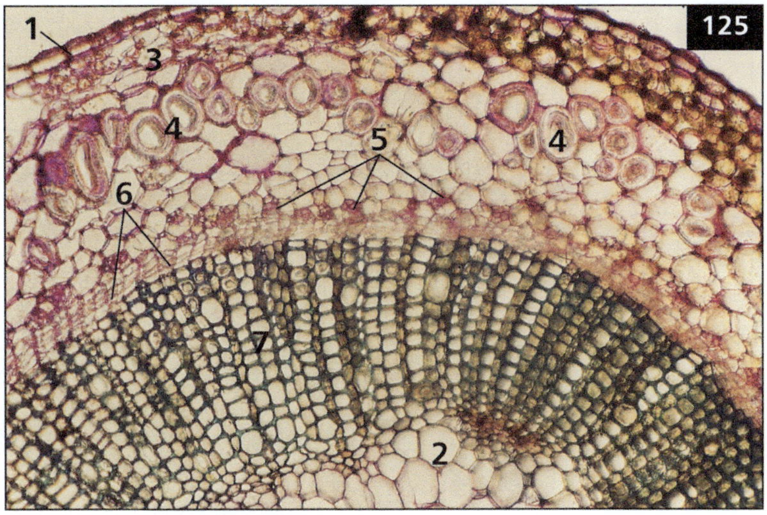

125 *Linum usitatissimum* (flax) showing the distribution of vascular, ground, and dermal systems in the stem of a dicot. The latter is composed of a single-layered epidermis (1) but the ground system comprises both the parenchymatous pith (2) and the cortex (3). The outer limit of the vascular system is marked by wide phloem fibres (4) while internally groups of conducting elements (5) lie in the phloem parenchyma. A conspicuous vascular cambium (6) separates the phloem from the secondary xylem (7). (LM.)

1 Epidermis
2 Parenchymatous pith
3 Cortex
4 Phloem fibres
5 Phloem conducting elements
6 Vascular cambium
7 Secondary xylem

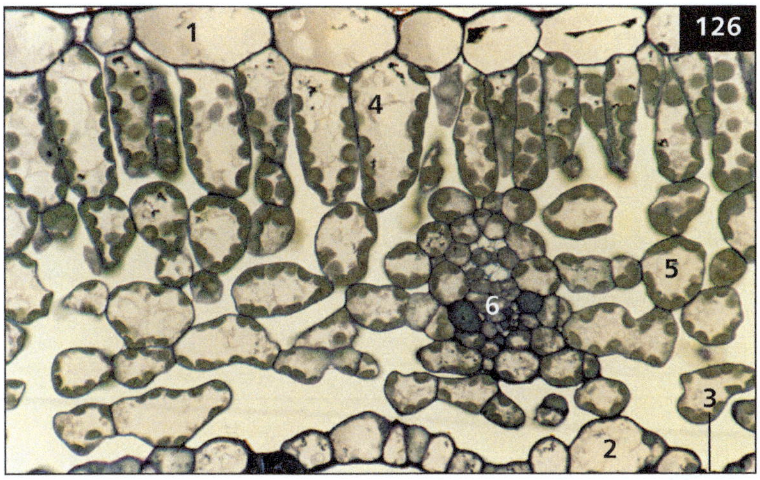

126 TS of the lamina of a bifacial leaf of the dicot *Glechoma hederacea* (ground ivy). The dermal system comprises the ad- and abaxial epidermis (1 and 2) with the stomata (3) confined to the abaxial surface. The chlorenchymatous ground tissue consists of a single palisade layer (4) and a thicker layer of spongy mesophyll (5). Numerous large chloroplasts are visible in the mesophyll but are absent from the epidermis. The vascular system is represented by the veinlet (6). (G-Os, LM.)

1 Adaxial epidermis
2 Abaxial epidermis
3 Stoma
4 Palisade layer
5 Spongy mesophyll
6 Veinlet

127 TS of the aerial root of *Monstera* (a monocot). The central polyarch vascular system shows prominent wide vessels (1), while the ground tissue consists of a thick-walled, lignified pith (2) and a wide parenchymatous cortex (3). These are demarcated by the endodermis (4). Mucilage ducts (5). (LM.)

1 Vessels
2 Lignified pith
3 Parenchymatous cortex
4 Endodermis
5 Mucilage ducts

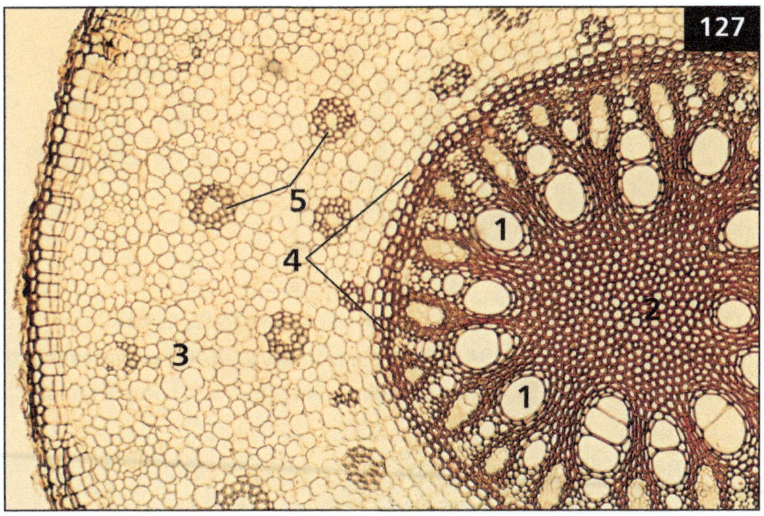

128

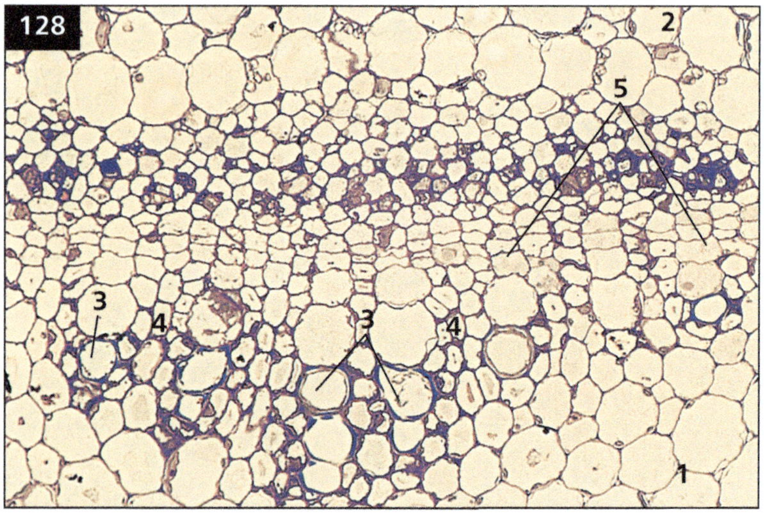

128 TS of the herbaceous stem of the dicot *Zinnia* showing detail of a vascular bundle. This is enclosed by the parenchymatous ground tissue of the pith (1) and cortex (2). The xylem contains several mature tracheary elements (3) with wide lumens and thickened walls but these are interspersed by axial parenchyma (4). The vascular cambium (5) forms xylem centripetally and phloem centrifugally. The latter consists of axial parenchyma in which are interspersed densely staining companion cells and apparently empty sieve tubes. (G-Os, LM.)

1 Pith
2 Cortex
3 Tracheary elements
4 Axial parenchyma
5 Vascular cambium

129

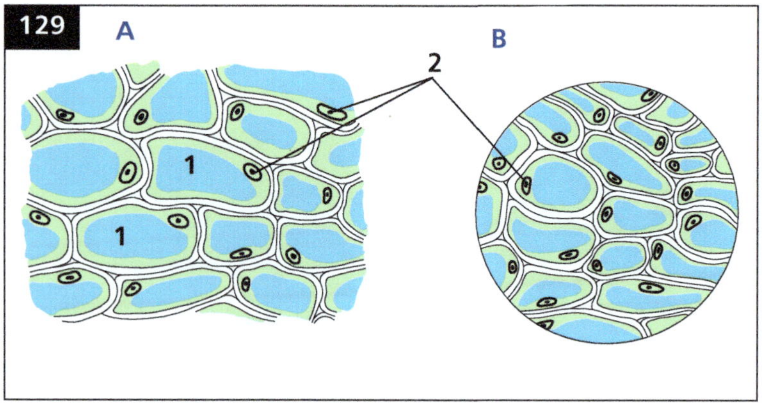

129 Diagrams of parenchyma cells viewed in LS (A) and TS (B) of a stem. Note the large vacuoles (1) and the peripheral cytoplasm containing the nuclei (2) and other organelles. Thin cellulosic primary walls enclose the protoplasts and small intercellular spaces occur at the angles of the cells where the middle lamellae are breaking down.

1 Vacuoles
2 Nuclei

130

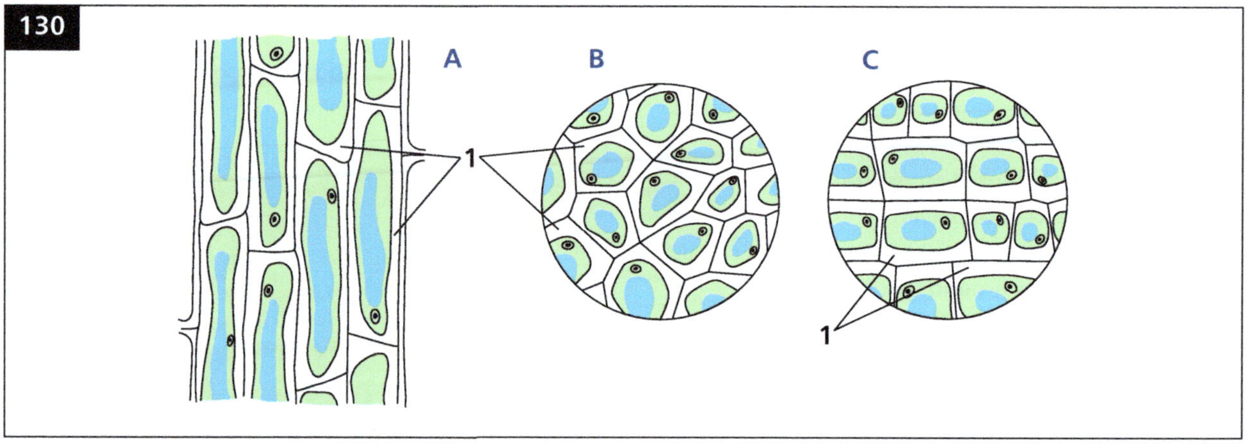

130 Diagrams of collenchyma cells in LS (A) and TS (B, C) of a stem. Their protoplasts are similar to those of parenchyma cells (cf. 129A, B) but the primary walls (1) are unevenly thickened. Two common variants are illustrated: angular (A, B) and lamellar collenchyma (C) and intercellular spaces are absent in both.

1 Primary walls

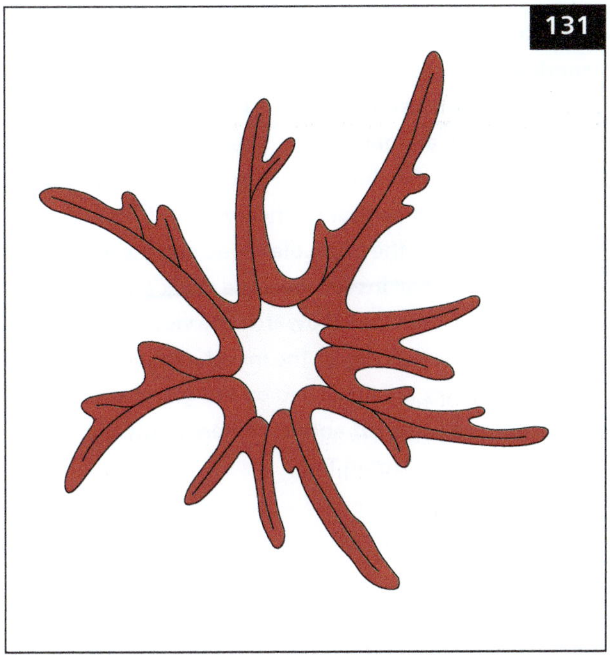

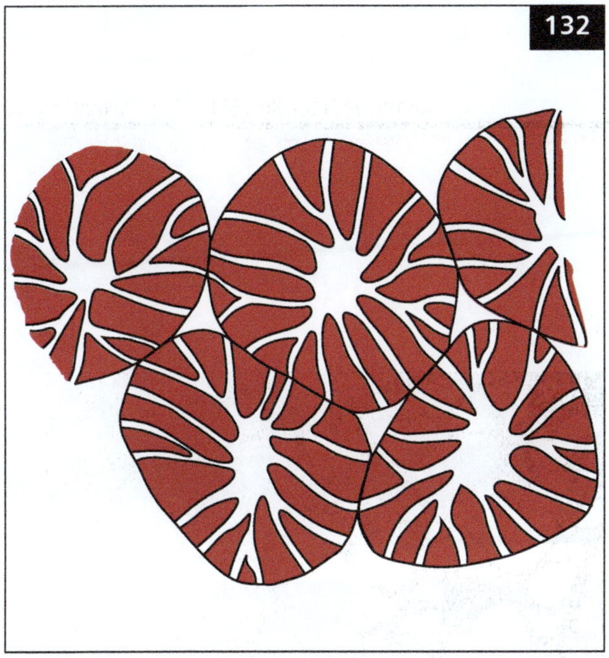

131, 132 Diagrams of sclerenchymatous elements with thick, lignified secondary walls. Mature cells are generally dead and degradation of their protoplasts leaves nothing but prominent lumina. 131 Much-branched astrosclereid. 132 Group of isodiametric stone cells with branched simple pits.

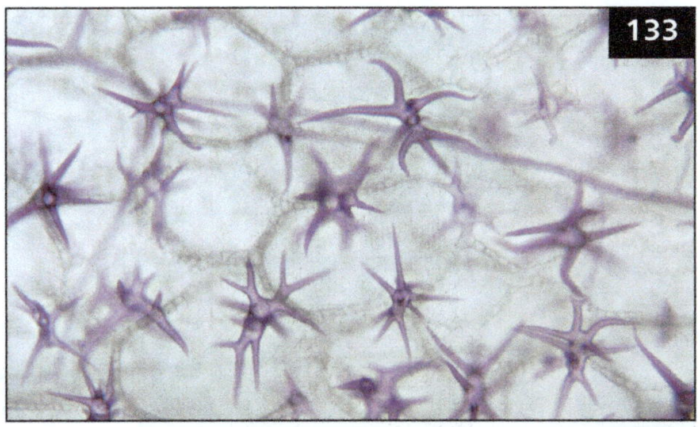

133 Clearing of a water lily leaf (*Nymphaea*, a basal angiosperm); the clearing technique made most cells transparent, but stained the numerous astrosclereids. Each astrosclereid is a single cell with many arms. (LM.)

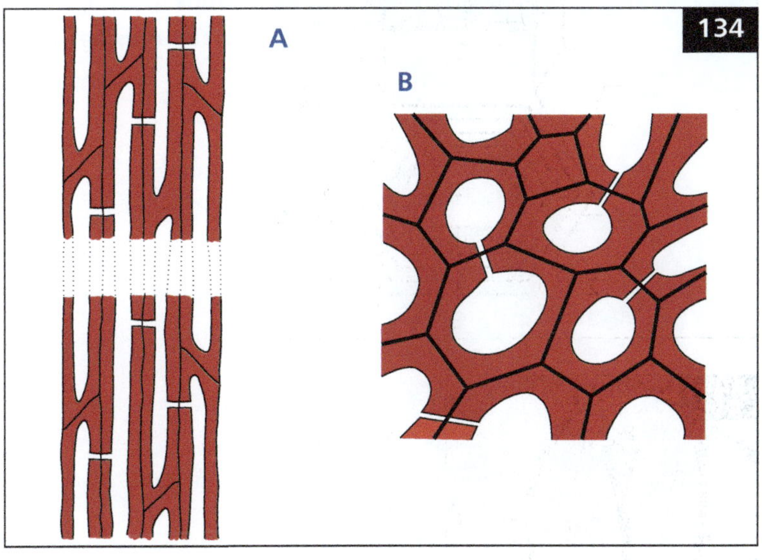

134 Diagrams of fibrous sclerenchymatous elements (with thick, lignified secondary walls) seen in (A) LS and (B) TS. These highly elongate cells have tapering (often branched) tips and simple pits are often frequent in their walls. Fibres constitute a major mechanical support system in the shoot.

135 Leaf of the insectivorous dicot *Drosera rotundifolia* (sundew). The crowded long hairs, with glandular heads sticky with secretions, are especially prominent on the adaxial laminal surface. These epidermal glands are complex multicellular structures with the cylindrical stalk containing a central tracheary strand. The epidermal cells of the glandular head secrete a viscous fluid in which small insects become stuck and the adjacent hairs then bend towards the victim. Enzymes within the secretion digest the insect's tissues and the soluble products are absorbed by the gland and translocated to the leaf and elsewhere in the plant. (Copyright of T. Norman Tait.)

136 Stack of cork harvested from the dicot *Quercus suber* (cork oak). This cork replaces the epidermis of the young stem and represents the dermal system of the tree.

137 Tree of the dicot *Quercus suber* (cork oak) being harvested for cork. At intervals of about 10 years a layer several centimetres thick (consisting of cork formed by successive cork cambia) is removed from the tree but a thin layer of newly produced cork is left on the trunk to protect the phloem within. Commercial cork is highly water-resistant and provides excellent thermal insulation.

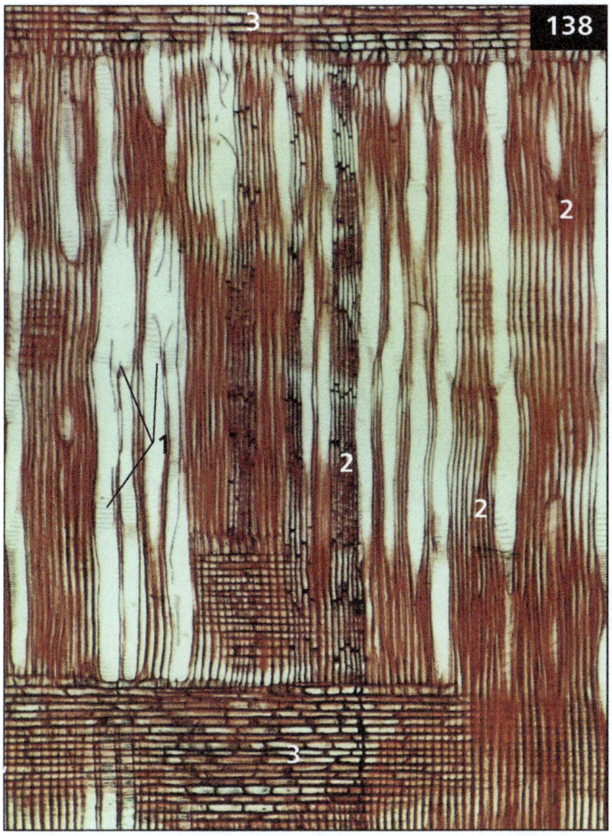

138 RLS through the diffuse porous wood of the magnoliid *Magnolia grandiflora*. Note the wide vessels with scalariform perforation plates (1) and the narrow fibres (2). Wide parenchymatous rays (3) are present and axial parenchyma is also evident. (LM.)

1 Scalariform perforation plates
2 Narrow fibres
3 Wide parenchymatous rays

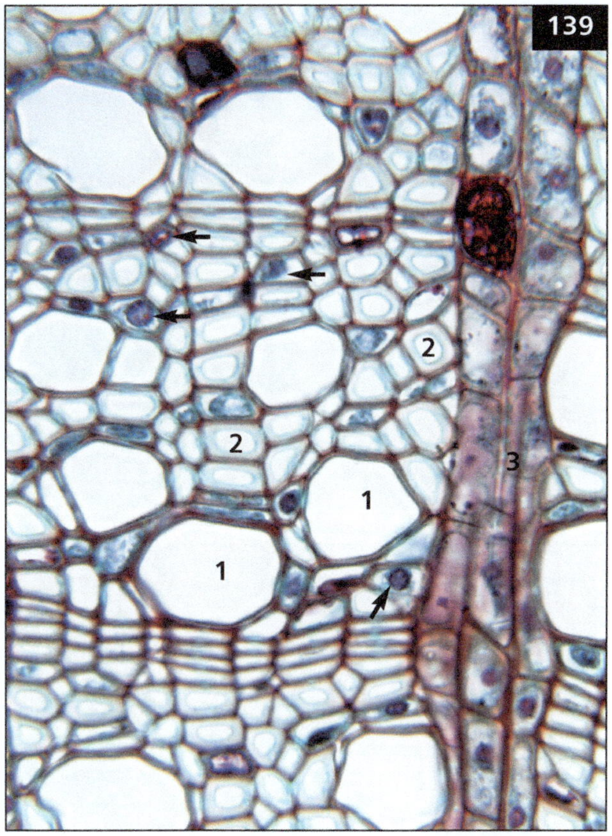

139 TS of wood of the dicot *Platanus* (plane tree) showing large amounts of living, nucleate parenchyma cells (arrows) in the wood, being especially abundant near vessels (1) but also occurring among wood fibres (2). Rays (3) also consist of living parenchyma cells. (LM.)

1 Vessels
2 Wood fibres
3 Ray

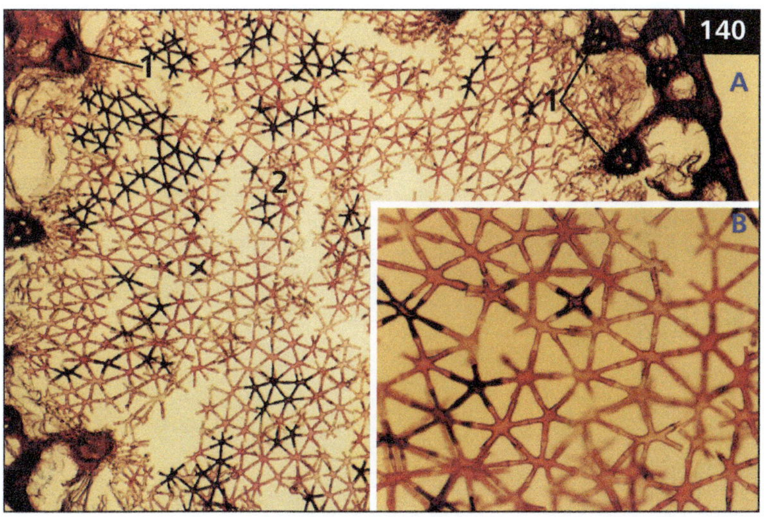

140 TS of the stem of the hydrophyte monocot *Juncus communis* (rush). A: Although this is a monocot, the vascular bundles (1) are peripherally distributed and most of the stem is occupied by an aerenchymatous ground tissue (2). B: Detail of the aerenchyma cells which connect, via stellate arms, with adjacent cells; note the enormous apoplastic system represented by the intercellular spaces and cell walls. (LM.)

1 Vascular bundles
2 Aerenchymatous ground tissue

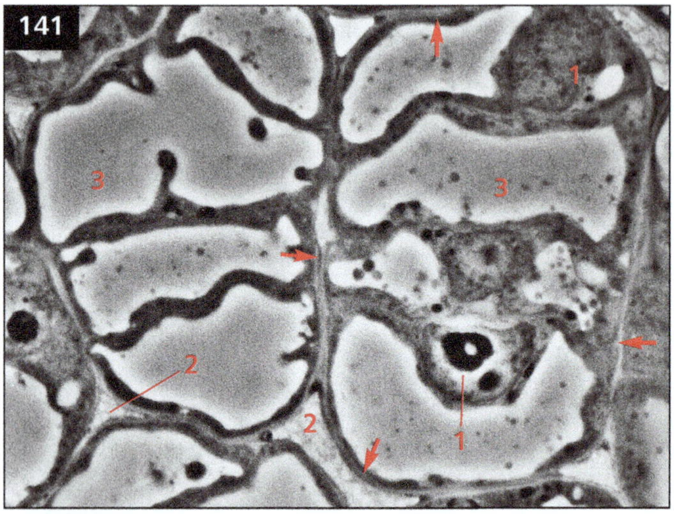

141 Reactivated cortical parenchyma cells from an *in vitro* cultured *Pisum sativum* (pea) root. The parental parenchyma cell (with its wall demarcated by arrows) has divided into four smaller cells and three derivative nuclei (1) are visible. The other cell has similarly divided but no nuclei are visible. Intercellular space (2), vacuole (3). (G-Os, Phase contrast LM.)

1 Nucleus
2 Intercellular space
3 Vacuole

142 TS of a young stem of the gymnosperm *Pinus* (pine). Note the origin of the secondary vascular tissues and cork from the vascular and cork cambia respectively. The secondary xylem (1) shows several growth rings whose conducting elements consist of tracheids only, while numerous rays (2) traverse it. Cortex (3), pith (4), resin duct (5), secondary phloem (6), vascular cambium (7), cork cambium (8). (LM.)

1 Secondary xylem
2 Rays
3 Cortex
4 Pith
5 Resin duct
6 Secondary phloem
7 Vascular cambium
8 Cork cambium

143 Stem tip cuttings of the dicot *Kalanchoe* with numerous basal adventitious roots formed after growing for 6 weeks in compost. These roots originated endogenously from vascular parenchyma cells which as a result of division and dedifferentiation (cf. 141) eventually gave rise to new roots.

144 Leafy adventitious bud (1) arising *de novo* at the base of a parent leaf (2) of *Begonia rex* (a dicot). This leaf was still attached to the parent plant although normally in this genus adventitious buds only develop on detached leaves.

1 Adventitious bud 2 Parent leaf

145 Nonsterile excised root segments of the dicot *Armoracia rusticana* (horseradish) cultured for several weeks *in vitro*. Note the numerous leafy adventitious buds which arise from the cork cambium of the root; adventitious roots also arise from this tissue.

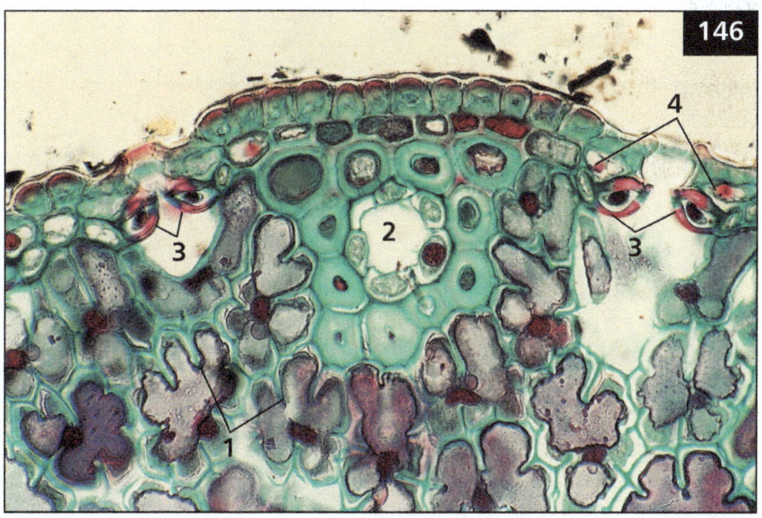

146 TS of a xeromorphic leaf of the conifer *Pinus* (pine) showing the epidermis and mesophyll. The thin mesophyll cell walls possess numerous vertical ingrowths (1), but the protoplasts of these cells have become plasmolysed so that a gap separates them from the walls. Note the resin duct (2) in the mesophyll and also the guard cells (3) which are sunken beneath the subsidiary cells (4) in the epidermis. (LM.)

1 Wall ingrowths
2 Resin duct
3 Guard cells
4 Subsidiary cells

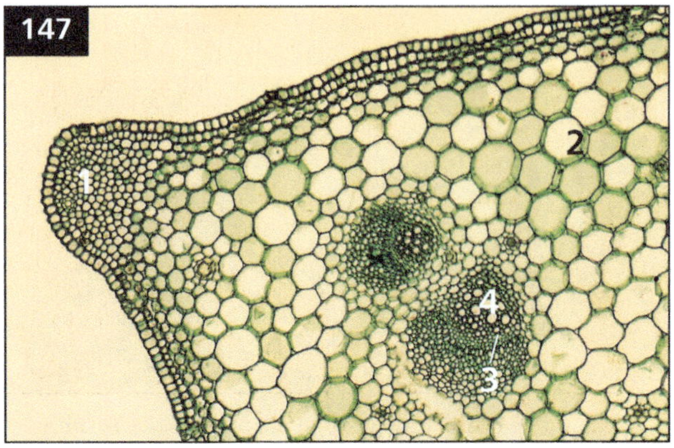

147 TS of the petiole of the dicot *Sanicula europea* (sanicle) showing the peripheral location of collenchyma (1). This living tissue has thickened walls and its location helps to support the young leaf. Parenchyma (2), primary phloem (3), primary xylem (4). (LM.)

1 Collenchyma
2 Parenchyma
3 Primary phloem
4 Primary xylem

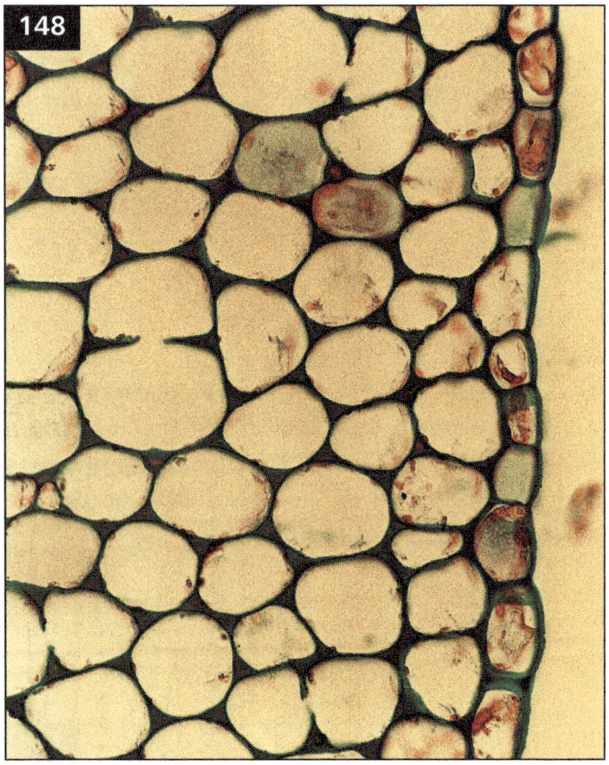

148 TS of the stem of *Coleus* (a dicot) showing detail of the collenchyma. This peripheral tissue (cf. 147) is of the angular form with additional cellulose and pectin deposited at the angles of these cells (cf. 130). (LM.)

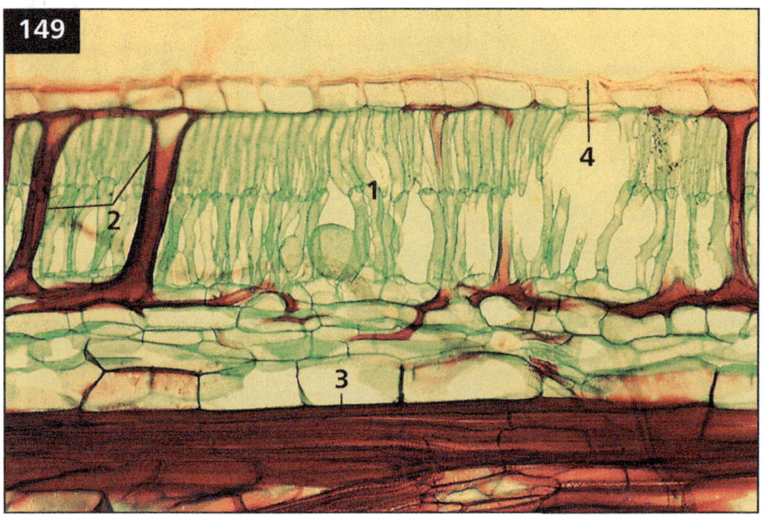

149 LS of the xeromorphic leaf of *Hakea* (a dicot) showing columnar sclereids. The palisade mesophyll (1) of this leaf is strengthened by red-stained sclereids (2) which have branched ends terminating beneath the epidermis and at the sheath of the vascular strand (3). Note the thick epidermal cuticle (4). (LM.)

1 Palisade mesophyll
2 Sclereids
3 Vascular sheath
4 Cuticle

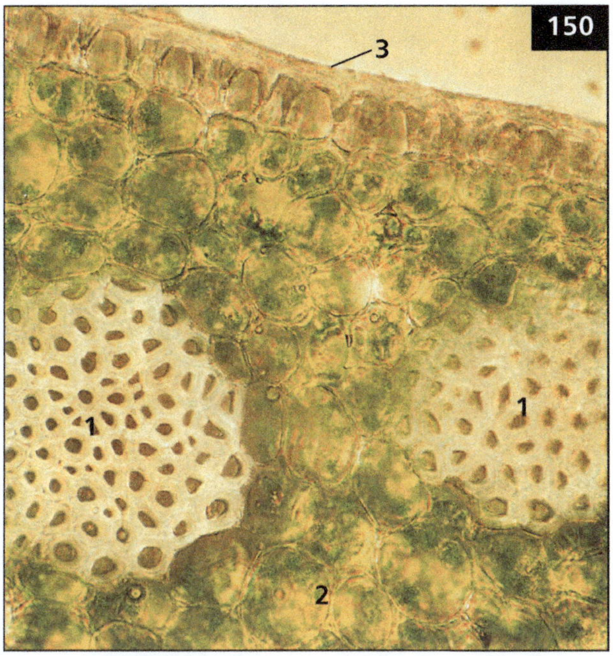

150 TS of the xeromorphic leaf of the monocot *Sansevieria trifasciata* (bowstring hemp). Note the longitudinal strands of thick-walled, unstained lignified fibres (1) in the mesophyll (2); these fibres are used commercially for cordage. A thick cuticle (3) covers the epidermal cells. (Fresh unstained section, LM.)

1 Lignified fibres
2 Mesophyll
3 Cuticle

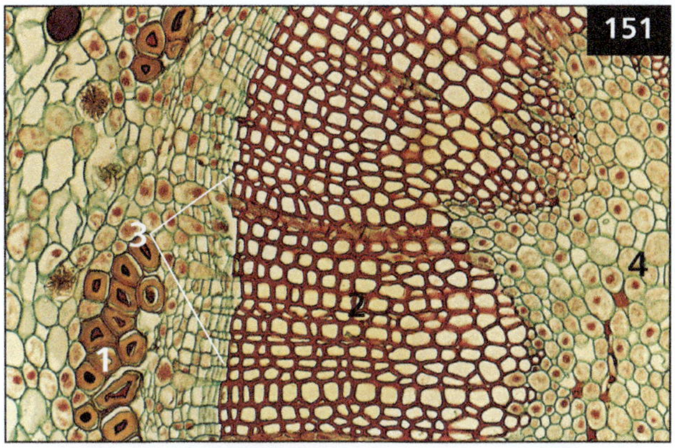

151 TS of a young twig of the gymnosperm *Ginkgo biloba* (maiden hair tree). Note the contrast between the thick-walled phloem fibres (1) and the thinner-walled tracheids of the secondary xylem (2). Vascular cambium (3), pith (4). (LM.)

1 Phloem fibres
2 Secondary xylem
3 Vascular cambium
4 Pith

152 Macerated wood of the dicot tree *Homalium* showing abundant long, slender fibres (1), one large vessel element (2) with one of its two perforations visible (arrow), and several small wood parenchyma cells (3), some still attached to the vessel element, others floating freely attached to each other.(LM.)

1 Fibres
2 Vessel element
3 Wood parenchyma cells

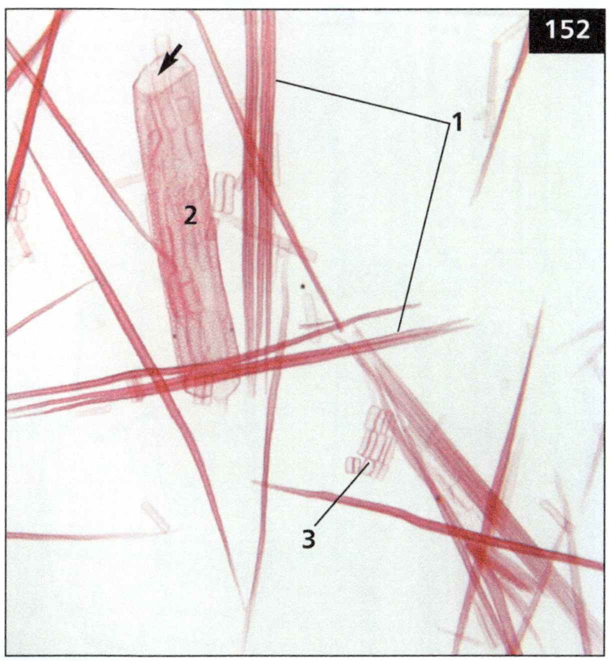

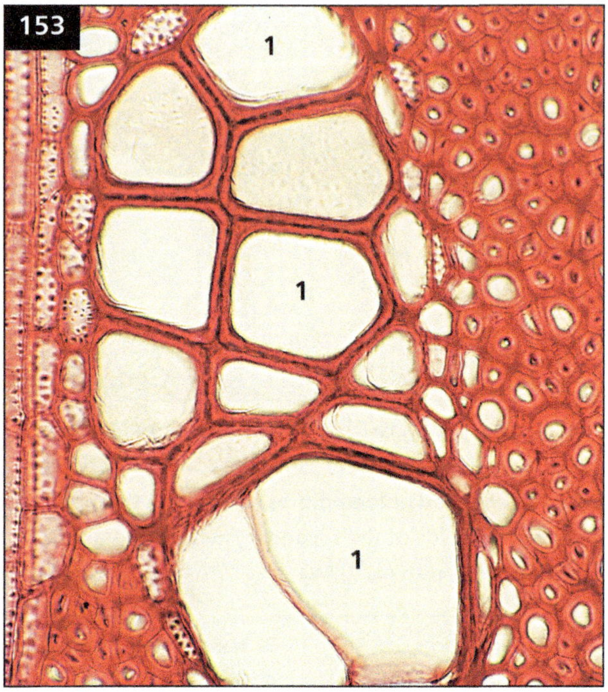

153 TS of the wood of the dicot *Robinia pseudoacacia*. Note the cluster of wide vessels (1) surrounded by much narrower fibres with very slender lumina. (LM.)

1 Wide vessels

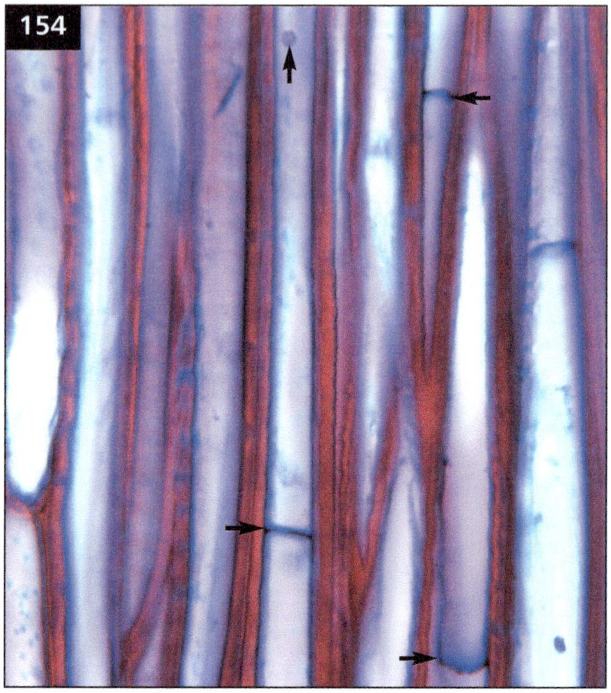

154 TLS of wood of the cactus *Acanthocereus tetragonus* (a dicot); after wood fibres deposit and lignify their secondary wall, they undergo one round of nuclear and cell division resulting in a living fibre with two nuclei (but only one is visible, vertical arrow) and a septum (horizontal arrows) consisting of a primary wall. (LM.)

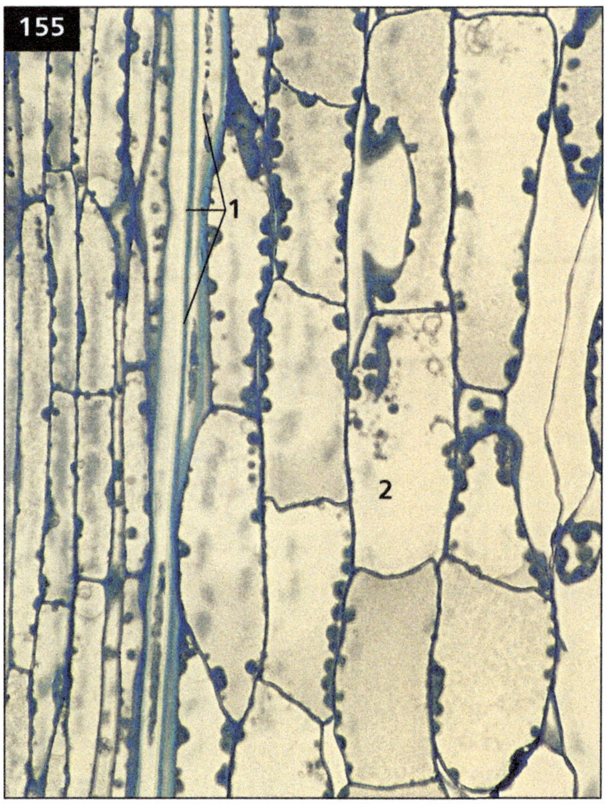

155 LS of the young stem of the dicot *Linum usitatissimum* (flax) showing intrusive fibres. The thin-walled fibres (1) are growing between adjacent chlorenchyma cells (2). Flax fibres are generally unlignified and the multinucleate protoplasts are persistent. At maturity the fibres are extremely thick-walled (cf. 54) and may be extracted to produce linen. (G-Os, LM.)

1 Fibre
2 Chlorenchyma cells

156 Close up of a young spine cluster in *Ferocactus* (barrel cactus). Three spines of each cluster develop as ordinary spines, consisting of just hard, sclerified, dead fibres, but other spine primordia (arrows) develop as extrafloral nectaries: their cells remain alive, parenchymatic and they secrete sugary nectar. (Plant cultivated by Dr. B. Barth.)

157 Clustered inflorescences of the dicot *Euphorbia cyparissias*. Each 'flower' is actually a small inflorescence with a pair of bracts (1) at its base. A single cup-shaped bract bears four yellow nectaries (2) and in the centre lies a prominent ovary (3) terminated by three styles. This represents the solitary female flower which lacks a perianth. At the base of the pedicel (4), which bears the female flower, lie the male flowers (5) with each represented by a single stamen.

1 Bracts
2 Nectaries
3 Ovary
4 Pedicel
5 Male flowers

158 Leaf of the dicot *Alchemilla* showing guttation. Water is excreted from numerous hydathodes (modified stomata) located at the leaf margins. Guttation may be a mechanism to protect the intercellular space system of the leaf from becoming waterlogged in conditions of excessive root pressure. (Copyright of T. Norman Tait.)

159 Felled trunk of the conifer *Araucaria araucana* showing copious exudation of resin (now congealed) from the severed resin canals mainly situated in the secondary phloem.

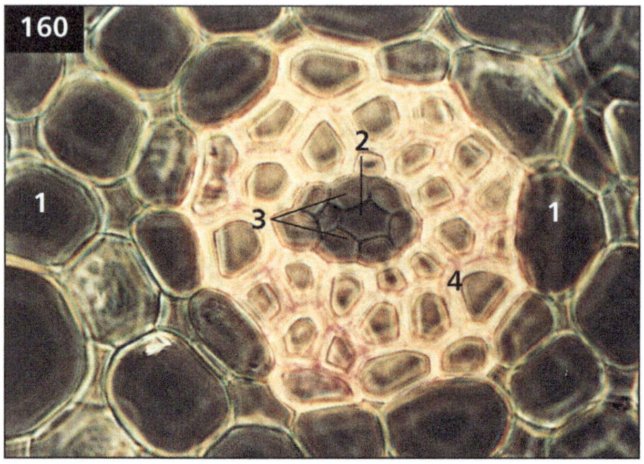

160 TS of a cortical mucilage duct in the aerial root of the monocot *Philodendron sagittifolium*. These ducts occur in the parenchymatous cortex (1, cf. 127) and each consists of a narrow central duct (2) surrounded by a single layer of secretory cells (3). The mucilage duct is itself enclosed within several layers of fibres (4). (LM.)

1 Parenchymatous cortex
2 Central duct
3 Secretory cells
4 Fibres

161 Xeromorphic shoot system of the dicot *Euphorbia canariensis* with its leaves reduced to spines. The stem contains an elaborate system of laticifers and the wounded stem has exuded latex (1) which is congealing on to the volcanic rock on which the bush is growing.

1 Latex

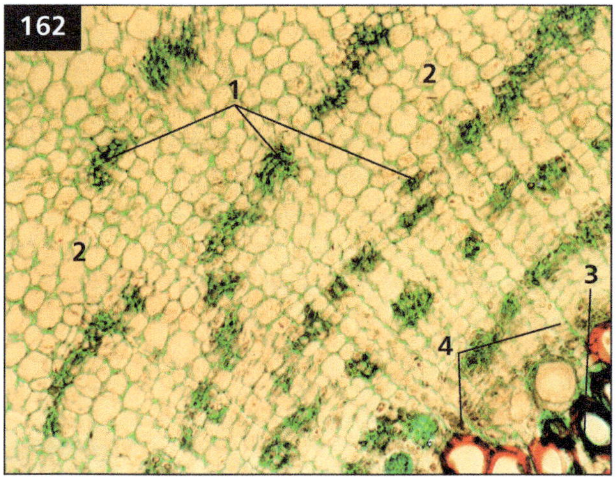

162 TS of the secondary phloem in the root of the dicot *Taraxacum officinale* (dandelion). The phloem contains numerous laticifers associated with the strands of conducting elements (1, cf. 163). Copious latex is contained within the plant body and another species of *Taraxacum* was cultivated as an alternative source of rubber to the rubber tree in the Second World War. Phloem parenchyma (2), secondary xylem (3), vascular cambium (4). (LM.)

1 Phloem conducting elements
2 Phloem parenchyma
3 Secondary xylem
4 Vascular cambium

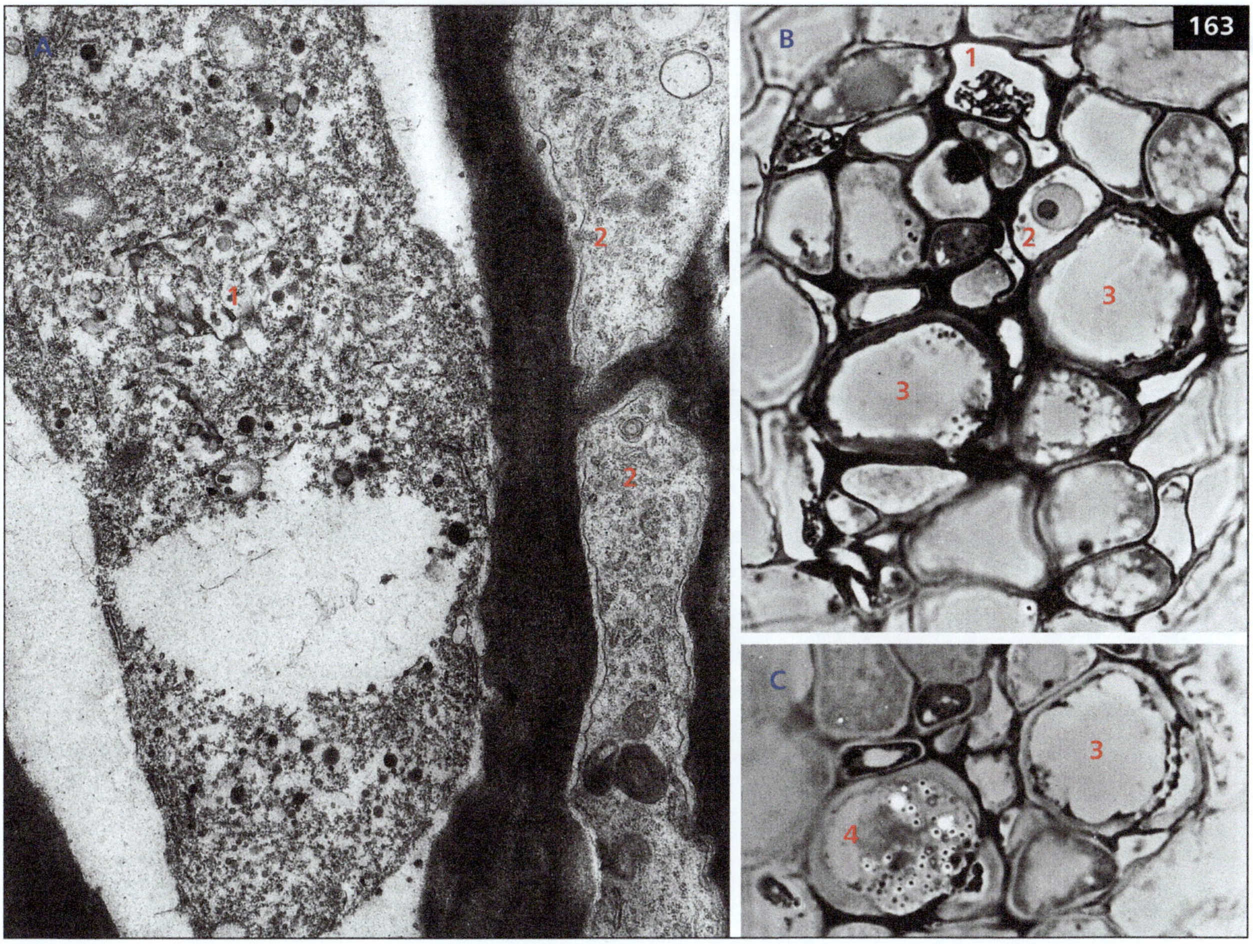

163 Laticifers in the root of the dicot *Taraxacum officinale* (dandelion). A: Detail of the latex in a laticifer (1) and a pair of companion cells (2). B, C: Adjacent sections of the same conducting strand in TS of the secondary phloem. Two sieve tubes (3) are evident and in one a sieve plate is sectioned (4); these conducting elements are associated with laticifers (1) and companion cells (2). (A: G-Os, TEM; B: G-Os, Phase contrast LM.)

1 Laticifer
2 Companion cells
3 Sieve tubes
4 Sieve plate

164 TS of cortex of the dicot *Euphorbia* (spurge) with several long tubular latex cells (1) that ramify and extend among cortex parenchyma cells (2). Each latex cell has many nuclei (not visible here). Several cortex cells have filled themselves with tannins (3), which have been stained red. (LM.)

1 Tubular latex cells
2 Ordinary parenchyma cells of cortex
3 Tannin cells of cortex

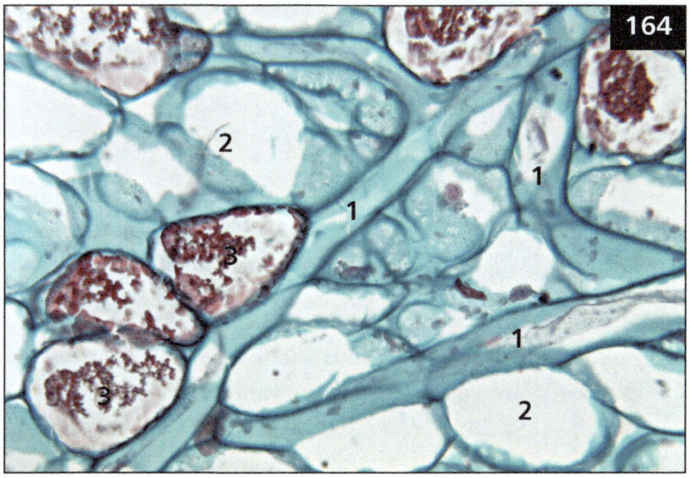

165

165 TS of a young twig of the dicot *Tilia cordata* (lime) showing the secondary phloem. This is a complex tissue with wide flares of ray parenchyma cells (1) which divide tangentially and accommodate the increasing circumference of the stem as secondary thickening progresses. The conductive phloem elements (2) function over several seasons and they are interspersed with tangential bands of thick-walled fibres (3). Vascular cambium (4), secondary xylem (5). (LM.)

1 Ray parenchyma
2 Phloem elements
3 Fibres
4 Vascular cambium
5 Secondary xylem

166

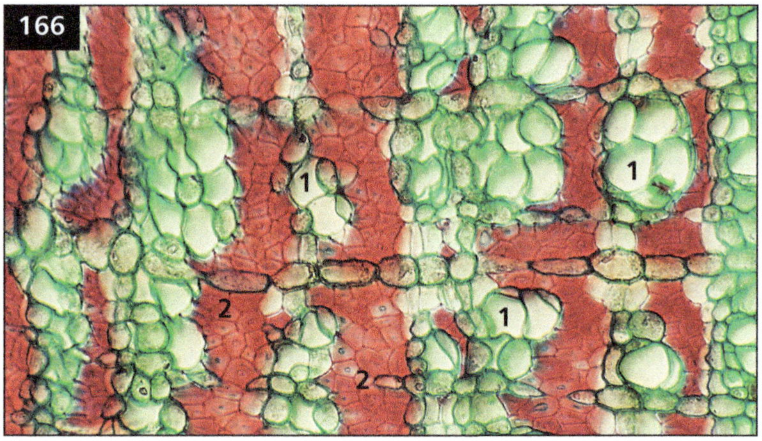

166 TS of the dicot *Tilia cordata* (lime) stem showing detail of the phloem (cf. 165). The wide sieve tubes (1) are sandwiched between tangential bands of very thick-walled fibres (2). (LM.)

1 Sieve tubes
2 Fibres

167

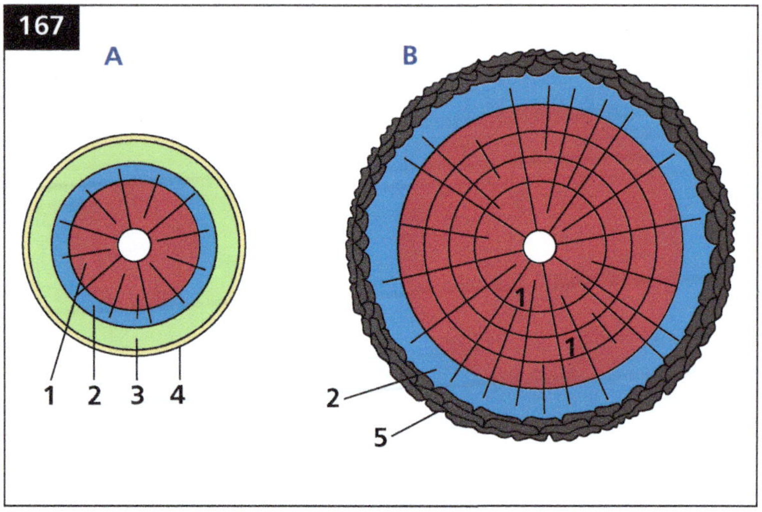

167 Diagrammatic representation of bark formation in a woody dicot stem. A: Secondary xylem (1) and phloem (2) have formed but the cortex (3) and epidermis (4) are still intact. (Primary vascular tissues not shown.) B: The epidermis has been replaced by a thick layer of bark (5) which protects the adjacent cylinder of secondary phloem (2). The first-formed cork cambium generally arises hypodermally but is short lived; successive cambia arise internally from any remaining cortex and later the secondary phloem. These cork cambia are usually discontinuous and form overlapping concave shells. The outermost, nonfunctional secondary phloem becomes sloughed off in the various layers of the peeling bark.

1 Secondary xylem
2 Secondary phloem
3 Cortex
4 Epidermis
5 Bark

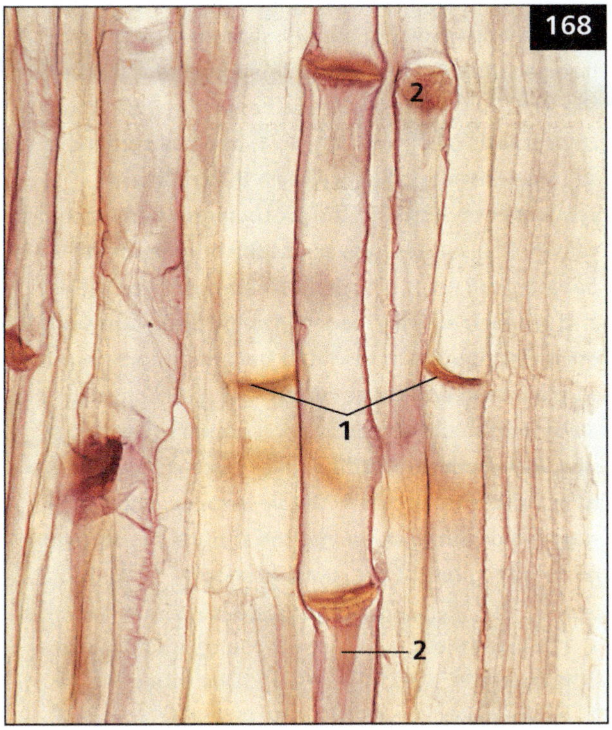

168 LS of the phloem of the dicot *Cucurbita* showing numerous wide sieve tubes in the stem. These are interrupted periodically by transverse or slightly oblique sieve plates (1). The P-protein fibrils coagulated when the specimen was excised and now form plugs (2) at the sieve plates. (LM.)

1 Sieve plates
2 P-protein plugs

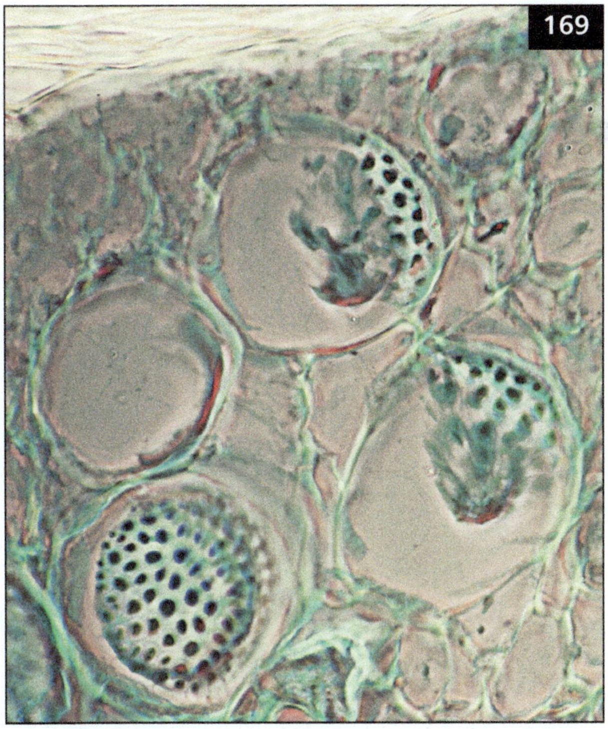

169 TS of the phloem of the dicot *Cucurbita* showing several broad sieve plates with wide sieve pores which are bordered by deposits of callose. (Phase contrast LM.)

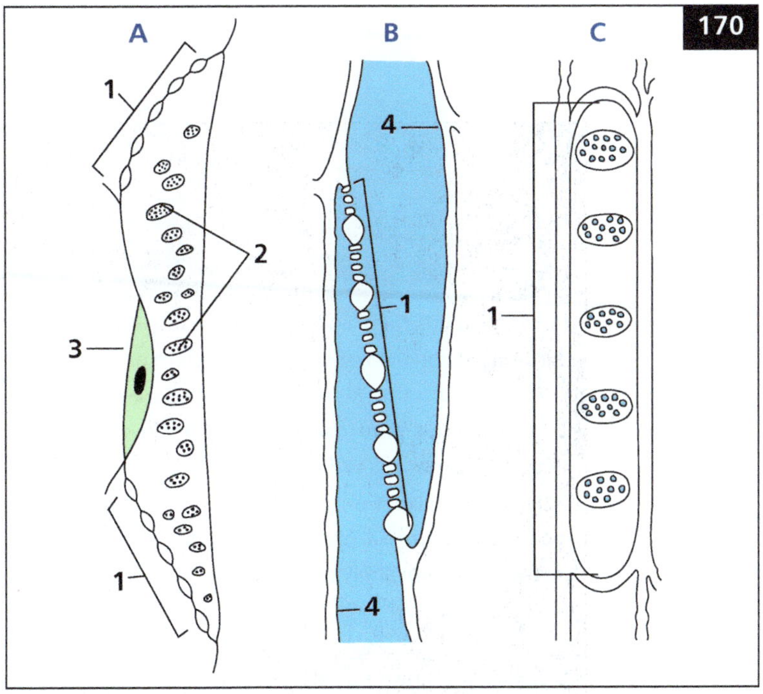

170 Diagrams of a sieve element with compound sieve plates. A: LS showing oblique sieve plates (1) on the end walls and sieve areas (2, seen in face view) on the side walls; a nucleated companion cell (3) is also evident. B: Detailed view of the compound sieve plate shown in A. Note the beaded appearance of the compound plate; the sieve element nucleus has degenerated and the few remaining organelles are greatly modified, but the plasmalemma (4) remains intact. C: Detailed view of a compound sieve plate seen in face view.

1 Oblique sieve plates
2 Sieve areas (face view)
3 Companion cell
4 Plasmalemma

171

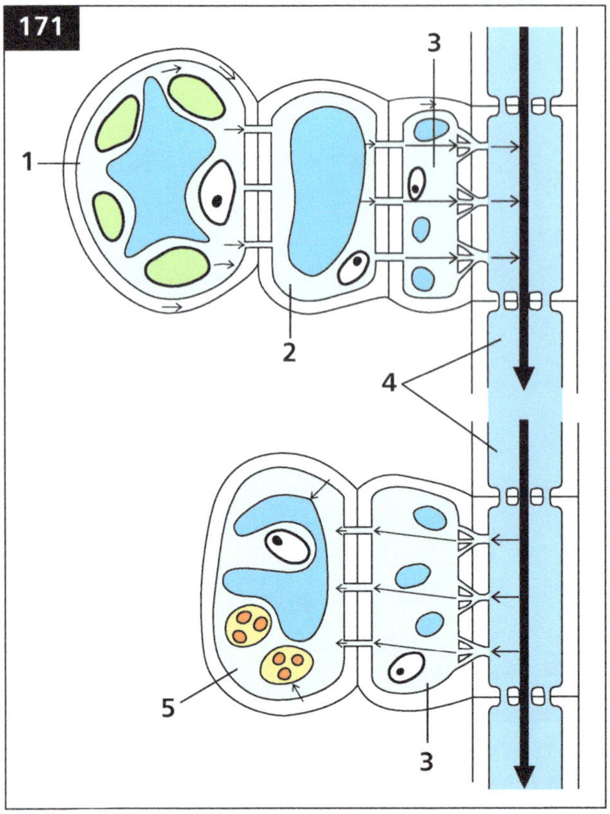

171 Diagram showing the pressure-flow model of translocation through the phloem from the source to the sink. Sugars are photosynthesized in the chloroplasts of the mesophyll (1) and transported in solution (both apoplastically via the walls and symplastically through the protoplasts) until reaching a veinlet. The solutes pass across the bundle sheath parenchyma (2) and nucleated companion cells (3) and are then actively loaded into the sieve tubes (4). The osmotic potential of the latter becomes more negative so that water enters the system and a bulk flow of water and solutes (wide arrows) occurs towards the sink. Here active unloading takes place so that a lower turgor pressure occurs in the sieve tubes, and the sugars move in solution into the storage tissue (5) or growing regions of the root and shoot.

1 Photosynthetic mesophyll
2 Parenchyma
3 Companion cells
4 Sieve tubes
5 Storage tissue

172

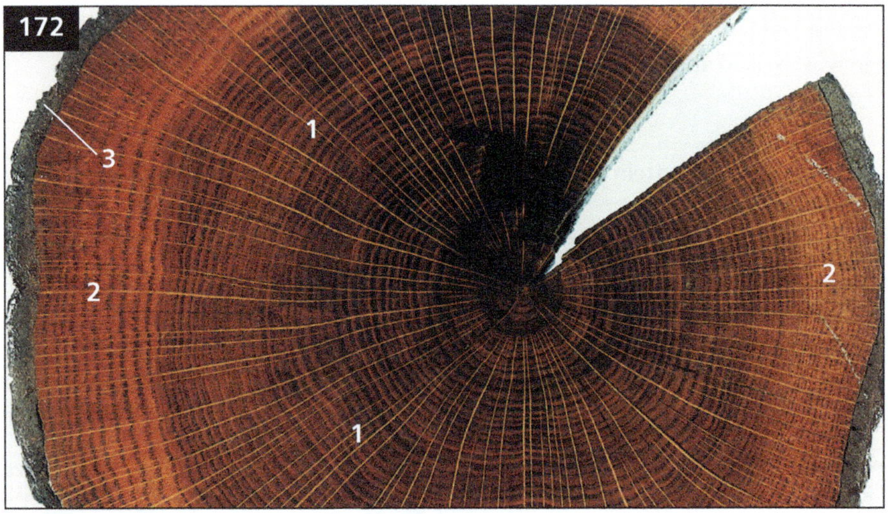

172 TS of a lateral branch of the dicot *Laburnum anagyroides*. The great bulk of the branch consists of secondary xylem and about 50 growth rings are present; the nonconducting heartwood (1) is darker than the surrounding sapwood (2). On the outside a narrow, darker layer (3) of bark and secondary phloem is evident. Numerous radial rays transverse the wood and the branch has split down one of these as the wood dried out. Note the eccentric appearance of the wood; in life the narrower (split) side of the branch corresponded to the lower surface of the branch since the production of reaction (tension) wood thickened the upper side.

1 Nonconducting heartwood
2 Sapwood
3 Bark and secondary phloem

173 TS of a young branch of the dicot *Tilia cordata* (lime tree). This shows eight clearly defined growth rings but the twig is unevenly thickened due to the formation of tension wood on its upper side.

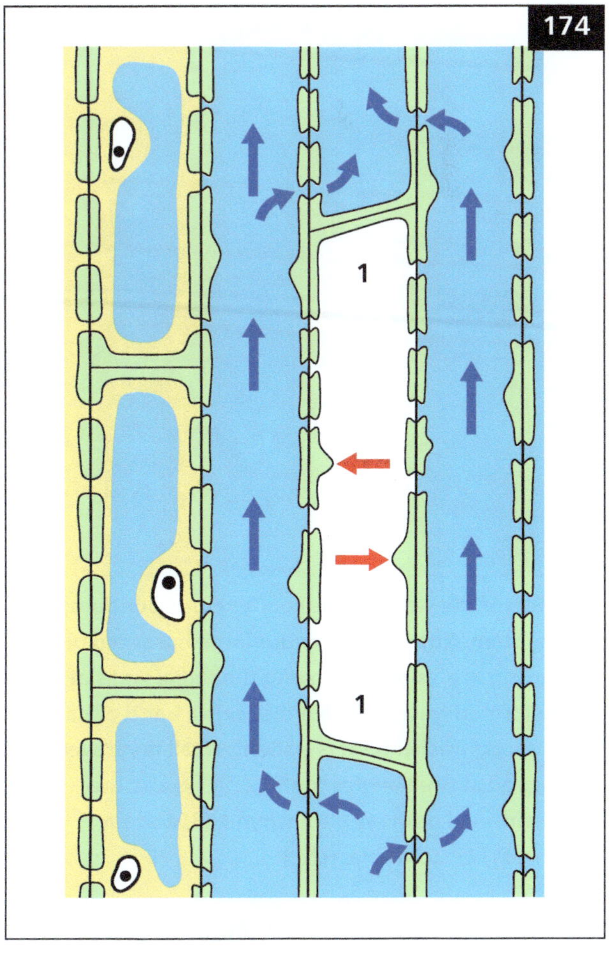

174 Diagrammatic representation of cavitation in the xylem of an angiosperm. The air bubble (white, 1) has blocked movement along a short vessel (most would be much longer) consisting of two members interconnected by a steeply tilted perforation plate with its rim indicated by red arrows. The embolism was able to spread through the perforation but cannot easily penetrate the pit membranes. The interrupted flow of water (blue arrows) is diverted laterally, via the pit-pairs in the vessel's lateral walls, to adjacent vessels.

1 Air bubble

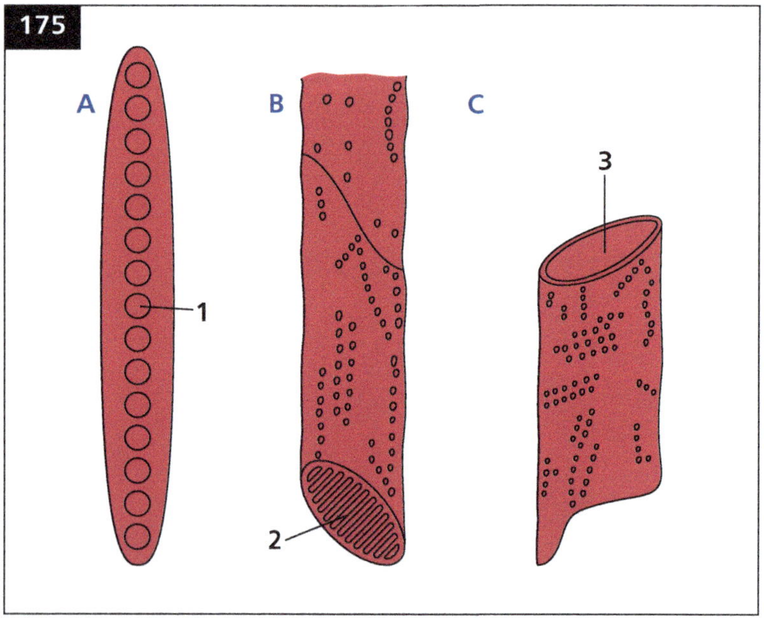

175 Diagrammatic representation of isolated tracheary elements as seen in macerated wood. Numerous pits (1) are present on their longitudinal walls and these are generally bordered where they interconnect with other tracheary elements. A: Tracheid consisting of a single thick-walled elongate cell with tapering ends. B: Part of a pitted vessel with an oblique, scalariform perforation plate (2) visible in face view. This vessel element is joined to another but the intervening perforation plate shows as an oblique line. C: Isolated vessel element showing a simple, approximately transverse, perforation plate (3) at one end.

1 Pits
2 Scalariform perforation plate
3 Simple perforation plate

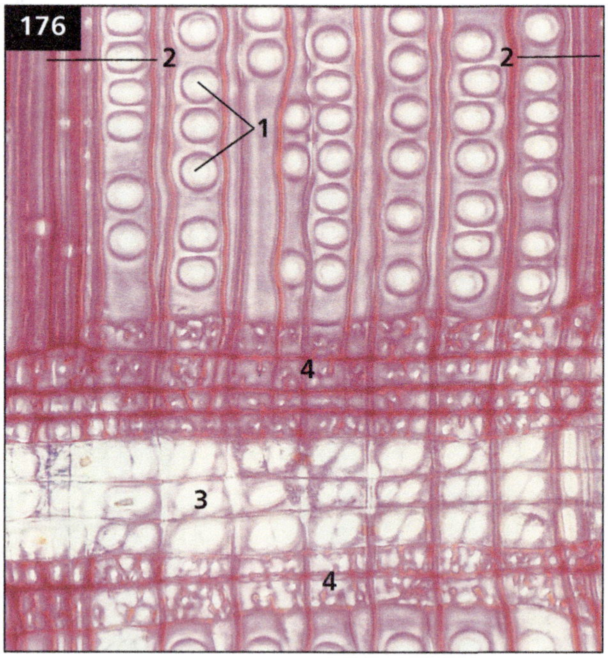

176 RLS of the wood in the conifer *Pinus ponderosa* (pine) stem. Note the numerous large bordered pits (1) occurring in a single row, along the radial walls of the early formed tracheids. To either side a band of narrow later wood tracheids (2) is evident. A tall ray traverses the wood at right angles to the long axes of the tracheids and both ray parenchyma (3) and ray tracheids (4) are visible. (LM.)

1 Pits
2 Late wood tracheids
3 Ray parenchyma
4 Ray tracheids

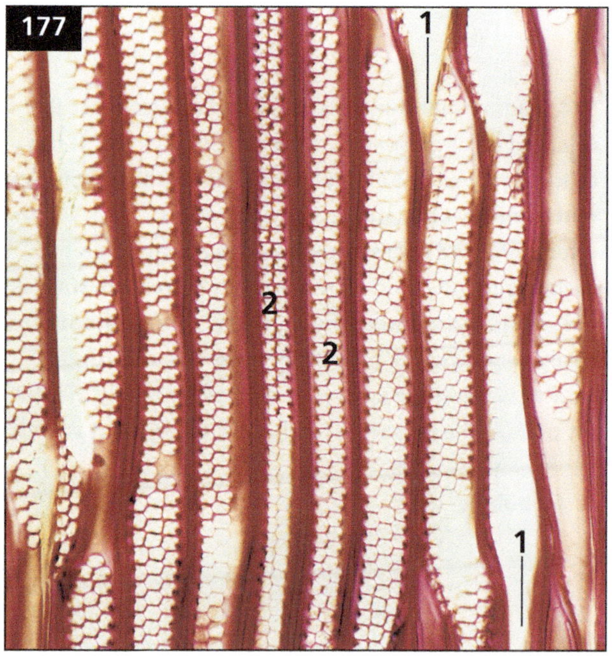

177 RLS of the secondary xylem of the conifer *Araucaria angustifolia*. The elongate tracheids have tapered ends (1) which interdigitate but are not perforated; however, water moves from one element to another via the numerous bordered pits (2) which occur in an alternate arrangement in their radial walls. (LM.)

1 Tracheid tapered ends
2 Bordered pits

178 RLS of wood of *Pinus strobus* (white pine), showing a ray oriented horizontally across the top and axial tracheids aligned vertically. The large, round pale areas where the ray parenchyma cells contact the axial tracheids (1) are cross-field pits. Pines also have horizontally aligned ray tracheids (2) along the bottom margin of the rays. Ray tracheids interconnect with each other by circular bordered pits (arrows) in their end walls, and they connect to the axial tracheids by circular bordered pits in their side walls. (LM.)

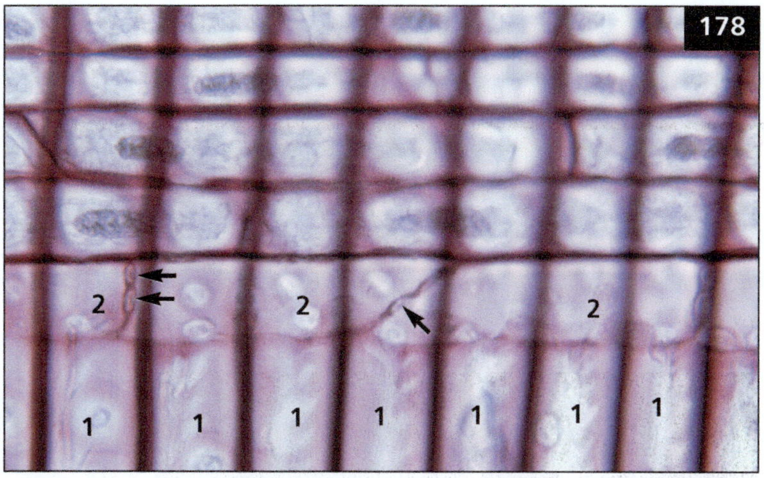

1 Axial tracheids 2 Ray tracheids

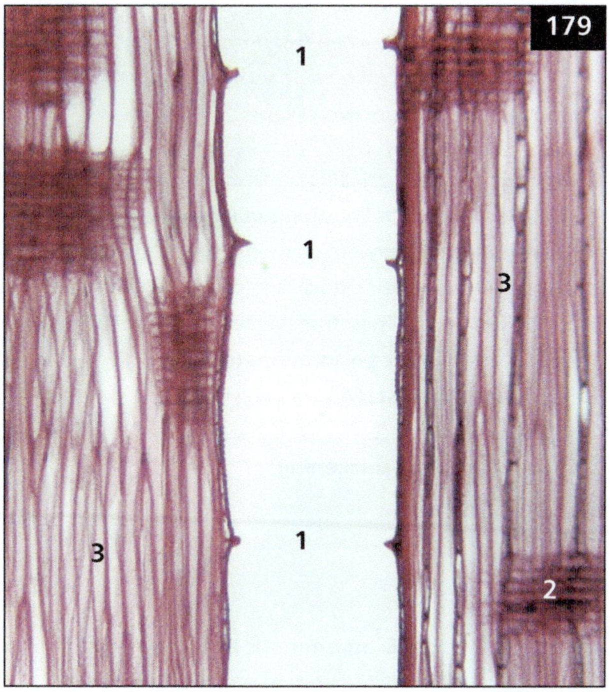

179 RLS of wood of the dicot tree *Juglans cinerea* (butternut) with a wide vessel aligned vertically in the centre. Four of its vessel elements are visible here, interconnected by perforations (1). Because the cell did not digest away all of each end wall while creating the perforations, a narrow rim of perforation plate remains around each perforation. The front and back walls of these vessel elements have been cut away. Rays (2), fibres (3). (LM.)

1 Perforations with rim-like remnants of perforation plates
2 Rays
3 Fibres

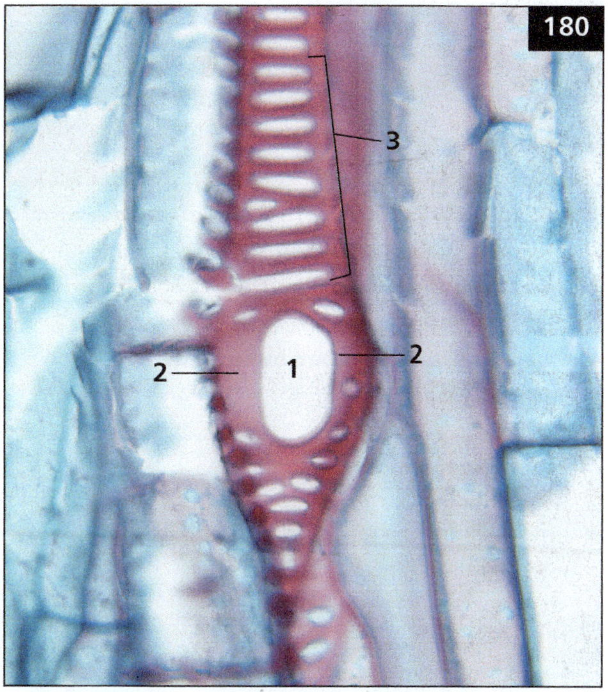

180 RLS of wood of the dicot *Euphorbia horrida* (a spurge) showing the tilted, tapered end (perforation plate) of a vessel element with a single oval perforation (1); undigested portions of the perforation plate constitute a rim-like margin (2). Scalariform pitting occurs on the sides of the element (3). (LM.)

1 Simple perforation
2 Perforation plate margin
3 Scalariform pitting

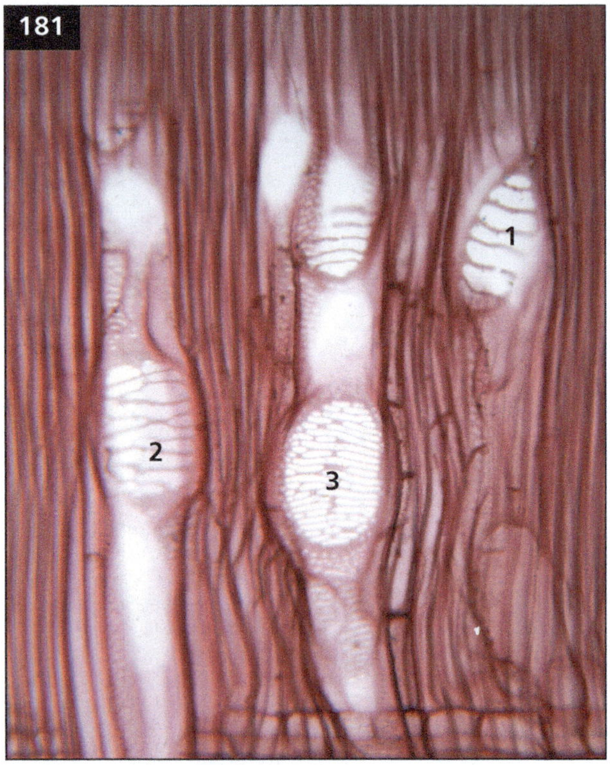

181 RLS of wood of the dicot *Schefflera digitata* with three vertically aligned vessels, showing several perforation plates in face view. One is a scalariform perforation plate (1), another is reticulate (2), and another is mostly scalariform but with slight reticulation (3). (LM.)

1 Scalariform perforation plate
2 Reticulate perforation plate
3 Intermediate plate, mostly scalariform, slightly reticulate

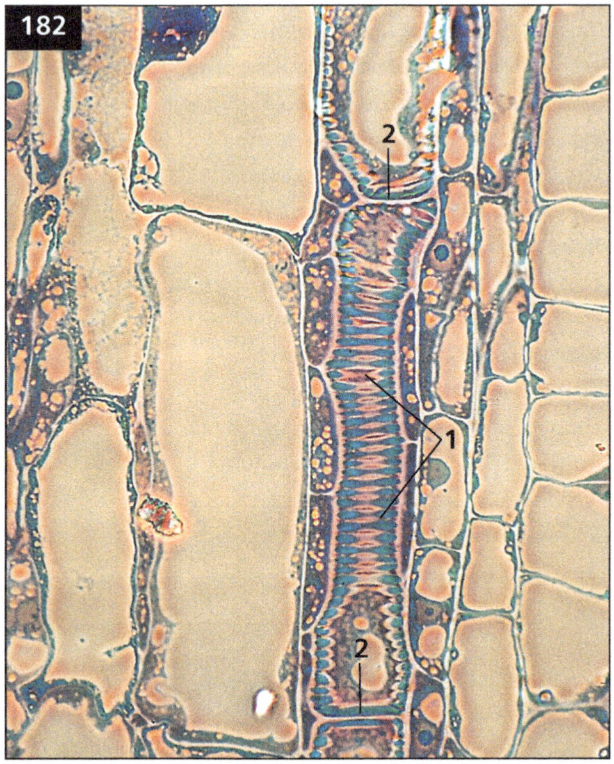

182 Stem of the dicot *Phaseolus vulgaris* (bean) showing differentiating metaxylem elements. The narrow elements (which still retain their cytoplasm, cf. 55) show scalariform wall thickening (1). Their transverse walls (2) later break down to form simple perforations (cf. 180). (G-Os, Phase contrast LM.)

1 Scalariform wall thickening
2 Transverse walls

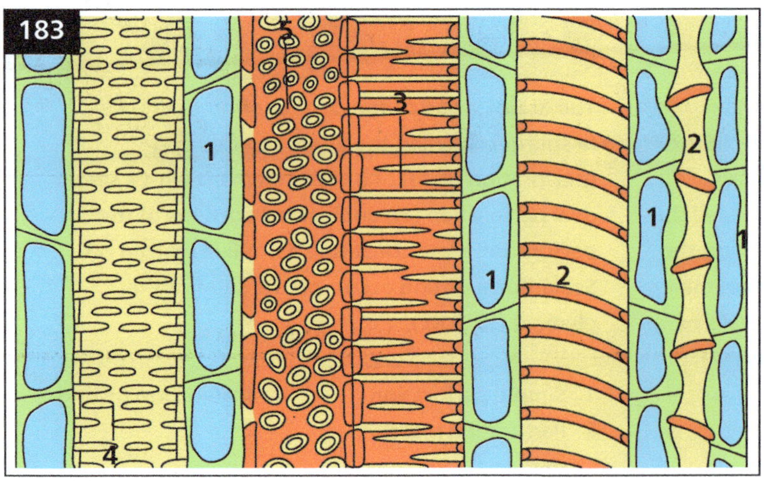

183 Diagrammatic LS through the primary xylem of a dicot. Xylem is a complex tissue (cf. 128) and living axial parenchyma cells (1) are interspersed with dead tracheary elements. The latter consists of: I: protoxylem in which lignified secondary wall is deposited internal to the cellulosic primary wall in an annular (2) or spiral pattern; II: metaxylem in which the lignified secondary wall is much more extensive and is deposited in a scalariform (3), reticulate (4) or pitted (5) pattern.

1 Axial parenchyma cells
2 Annular protoxylem
3 Scalariform metaxylem
4 Reticulate metaxylem
5 Pitted metaxylem

184 LS of primary xylem of the monocot *Dracaena fragrans*. Tracheary elements on the left (1) have helical secondary walls; their primary walls are so thin they seem transparent and invisible here. Element in the centre (2) has scalariform pitting. Phloem (3). (LM.)

1 Helical secondary wall
2 Scalariform secondary wall
3 Phloem

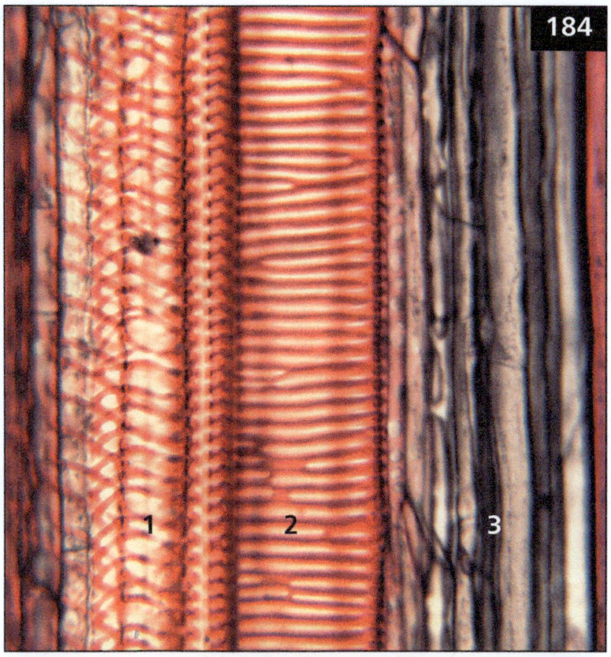

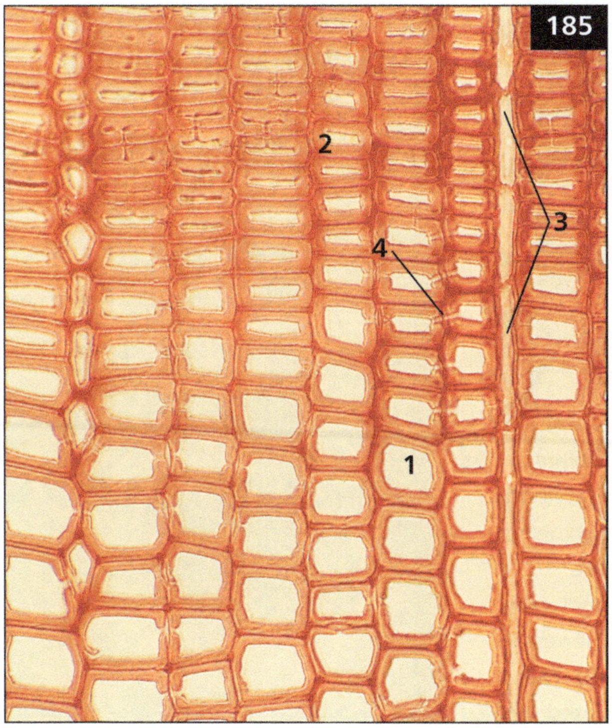

185 TS of the wood of the conifer *Thuja* showing radially aligned tracheids. The wide lumina of the early tracheids (1) contrast with their narrow lumina in the late wood (2). Note the narrow ray (3) traversing the wood, and the pits (4) on the radial walls of the tracheids. (LM.)

1 Early tracheids
2 Late wood
3 Ray
4 Pits

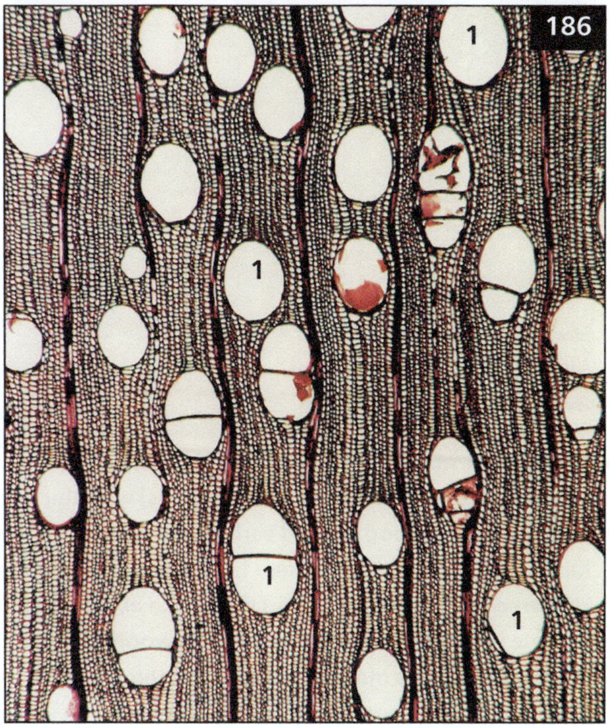

186 TS of the secondary xylem of the dicot *Aucomea klaineana*. This is native to west Africa and, as in many tropical trees, growth rings are not obvious since the wide vessels (1) are evenly distributed throughout the wood. (LM.)

1 Wide vessels

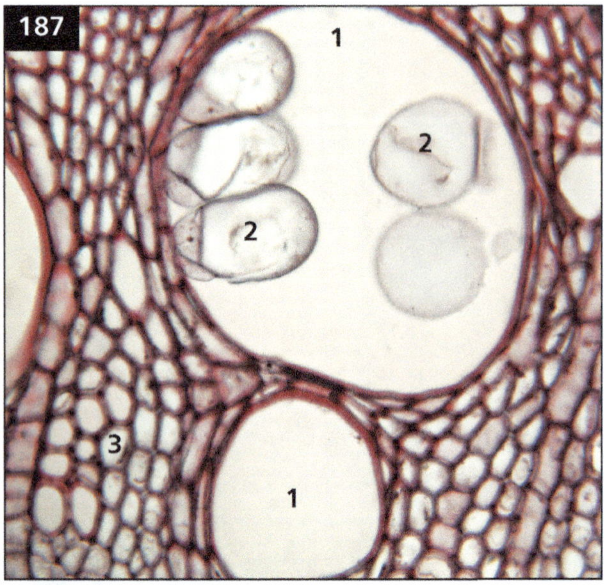

187 TS of wood of the dicot *Machaerium purpurascens* (a legume), showing three vessels (1); five tyloses (2) have begun to push into the lumen of one. They would continue to enlarge until the vessel had become completely occluded. Wood matrix consists of fibres (3). (LM.)

1 Vessels
2 Tyloses
3 Wood fibres

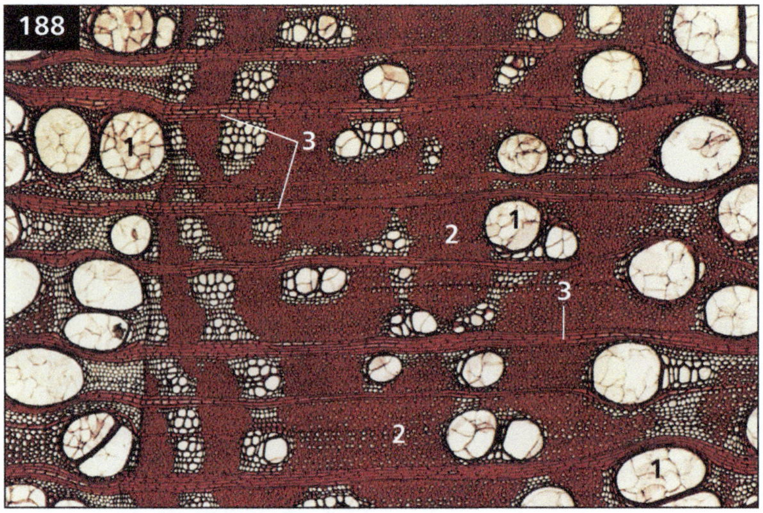

188 TS illustrating ring porous wood (secondary xylem) of *Robinia pseudoacacia*, a dicot. Wide pores (vessels) with thick secondary walls are confined to wood formed early in the growing season and their lumina are packed with tyloses (1). Numerous small vessels are clustered in the later wood and there is an abrupt boundary between the late wood of one year and the early wood of next year. The vessels are embedded in a mass of narrow, very thick-walled fibres (2). The axial parenchyma consists of narrow, thin-walled cells associated with the vessels. Numerous parenchymatous rays (3), several cells wide and with thickened walls, traverse the wood radially. (LM.)

1 Tyloses
2 Thick-walled fibres
3 Parenchymatous rays

189 Trunk of an ancient pollarded specimen of *Fagus sylvatica* (beech) which has split open to reveal its hollow centre in which the heartwood has decayed; note the prominent adventitious roots which have grown down the rotted centre and are recycling some of the nutrients released by the decay micro-organisms.

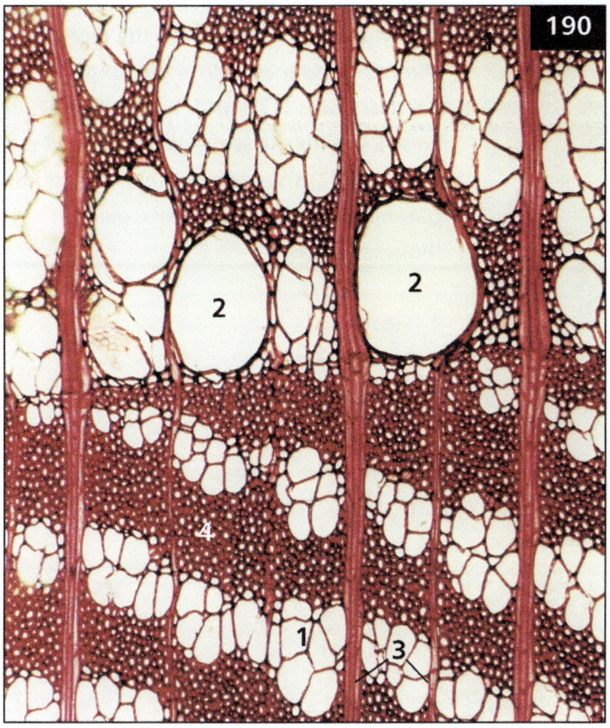

190 TS of the secondary xylem of the dicot *Ulmus americana* (American elm). This is a ring porous wood and clearly shows the abrupt transition between the narrower vessels (1) of the previous year's later growth and the wide vessels (2) formed at the beginning of the new season. Ray parenchyma (3), fibres (4). (LM.)

1 Narrower last season's vessels
2 Wider new season's vessels
3 Ray parenchyma
4 Fibres

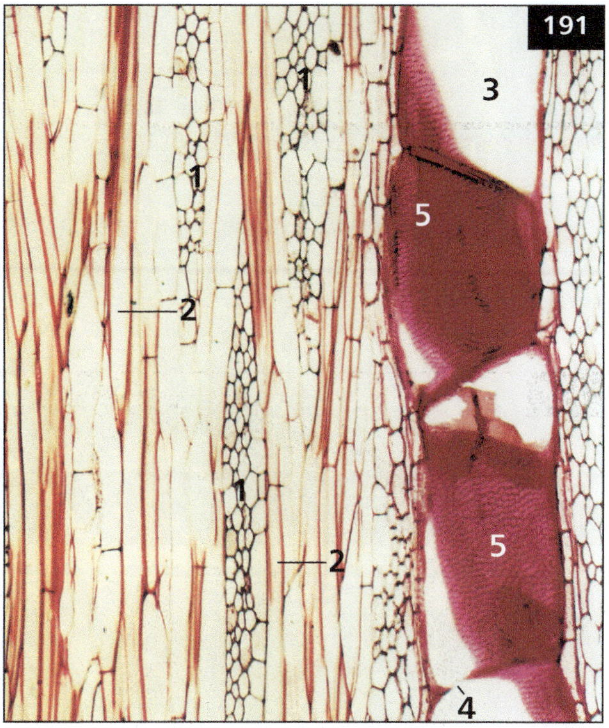

191 TLS of the secondary xylem of the dicot *Ochroma lagopus* (balsa). The abundant and large parenchymatous rays (1), the thin-walled fibres (2), and the relatively few and thin-walled vessels (3) make this wood very light (specific gravity 0.1–0.16). Perforation plate (4), alternately arranged pits in vessel (5). (LM.)

1 Parenchymatous rays
2 Thin-walled fibres
3 Thin-walled vessels
4 Perforation plate
5 Pits

192 TLS of the conifer wood of *Pinus* (pine). Note that the bordered pits (1) occur in the radial walls of the tracheids but are absent from the tangential walls (2). The numerous nonstoried rays (3) are only a single cell wide but extend a number of cells in height (cf. 176). (LM.)

1 Bordered pits in radial walls
2 Tangential walls (no pits)
3 Nonstoried rays

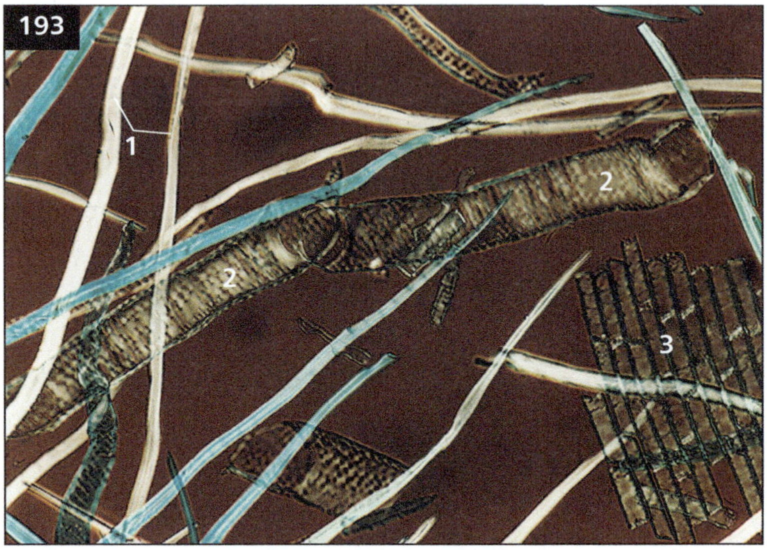

193 Macerated wood from the dicot *Quercus alba* (oak). Note the long, narrow fibres (1) with few pits; these contrast with the wide-diametered vessel elements (2) showing numerous pits. Ray parenchyma cells (3) are also evident. (Polarized LM.)

1 Fibres
2 Vessel elements
3 Ray parenchyma cells

194 Macerated wood from the dicot *Fraxinus americana* (ash). Note the disassociated vessel elements with abundant pitting on the lateral walls and the simple perforations which terminate the vessel elements. (Polarized LM.)

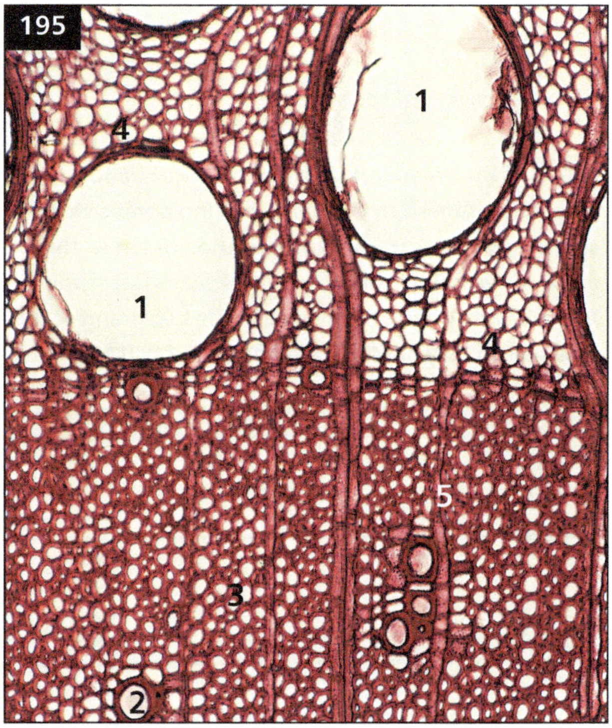

195 TS of the wood of the ring porous stem of the dicot *Fraxinus americana* (ash). Note the very wide vessels (1) in the early wood and the few narrow, single or aggregated (often paired) vessels (2) in the later wood. The latter lie in a ground mass of narrow, thick-walled fibres (3), but the fibre tracheids (4) of the early wood have thinner walls. Narrow rays (5) traverse the wood. (LM.)

1 Wide vessels
2 Narrow vessels
3 Thick-walled fibres
4 Fibre tracheids
5 Rays

196 TS of the secondary xylem of the dicot *Pterygota kamerumensis.* The wood shows broad transverse bands of parenchyma (1) and wide and tall multiseriate rays (2). Vessel (3). (LM.)

1 Parenchyma
2 Multiseriate rays
3 Vessel

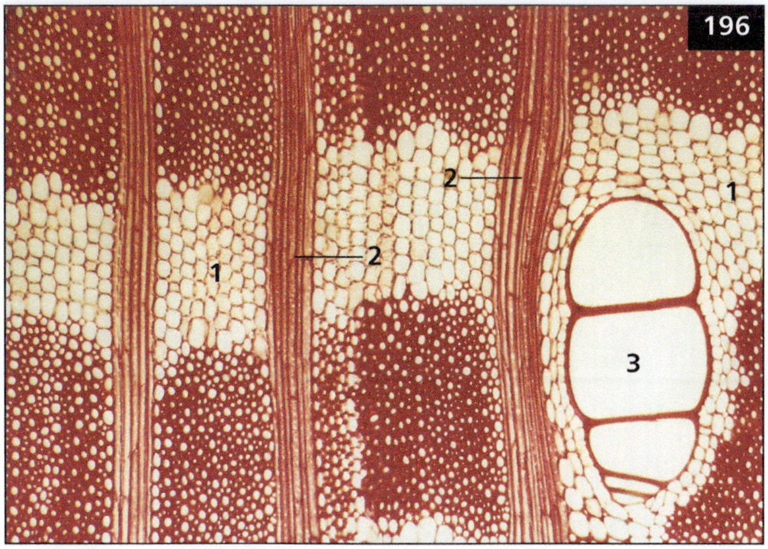

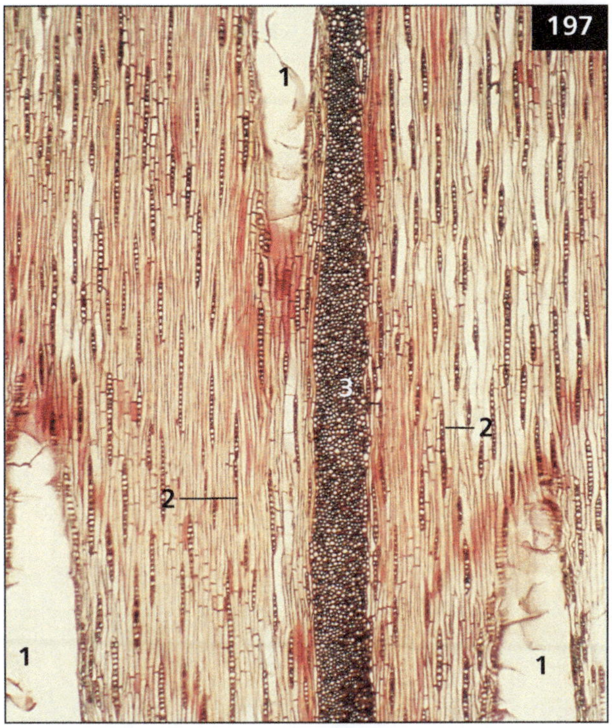

197 TLS of the secondary xylem of the dicot *Quercus alba* (oak). Note the several wide vessels (1) containing tyloses and the numerous short uniseriate rays (2). These contrast greatly with the very wide multiseriate rays (3). Axial parenchyma bands are also present amongst the narrow tracheary elements and fibres. (LM.)

1 Vessels
2 Uniseriate rays
3 Wide multiseriate ray

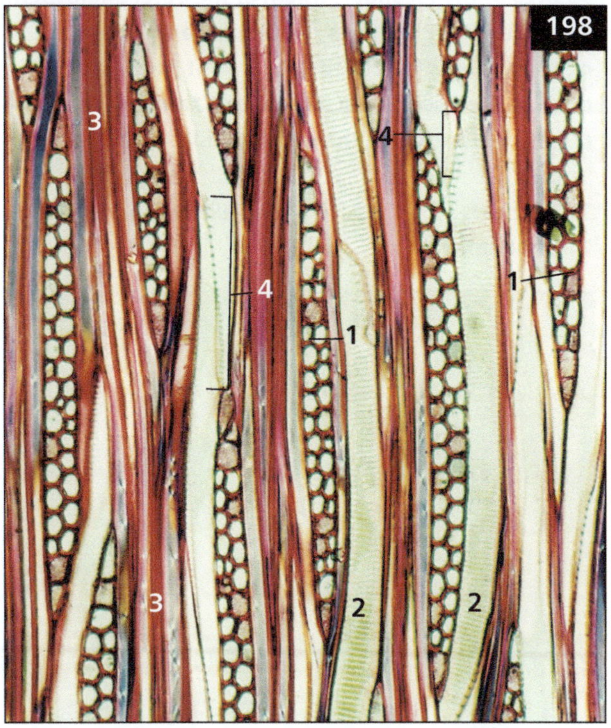

198 TLS of the secondary xylem of the magnoliid *Magnolia grandiflora*. Abundant multiseriate rays (1) occur between the uniformly diametered vessels (2) and narrower fibres (3). Scalariform perforation plate (4). (LM.)

1 Multiseriate rays
2 Vessels
3 Fibres
4 Scalariform perforation plate

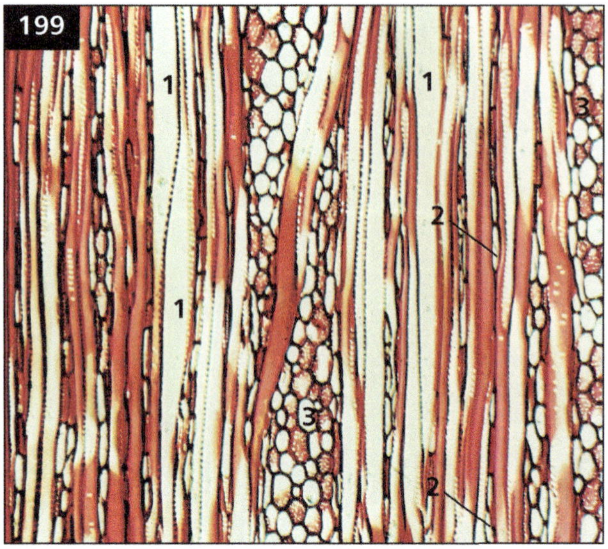

199 TLS of the secondary xylem of the basal angiosperm *Drimys winteri*. This primitive angiosperm does not develop vessels (cf. 200) but abundant tracheids (1) occur in the xylem. Both uniseriate (2) and multiseriate (3) rays are present in the wood. (LM.)

1 Tracheids
2 Uniseriate ray
3 Multiseriate ray

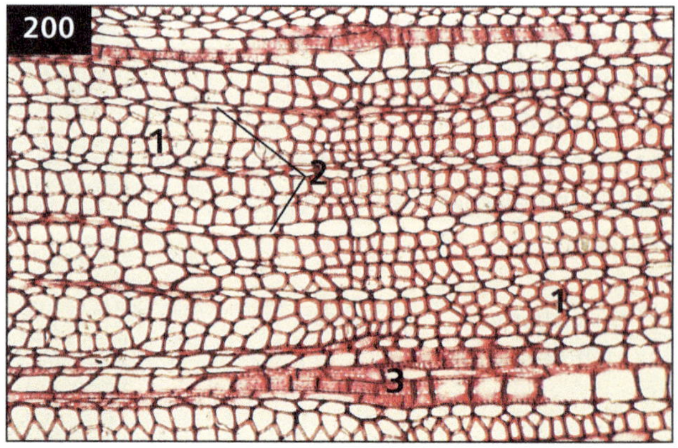

200 TS of the wood of the primitive basal angiosperm *Drimys winteri*. This does not contain vessels and the tracheids (1) are of rather uniform diameter. Both uniseriate (2) and multiseriate rays (3) are present. (LM.)

1 Tracheids
2 Uniseriate ray
3 Multiseriate ray

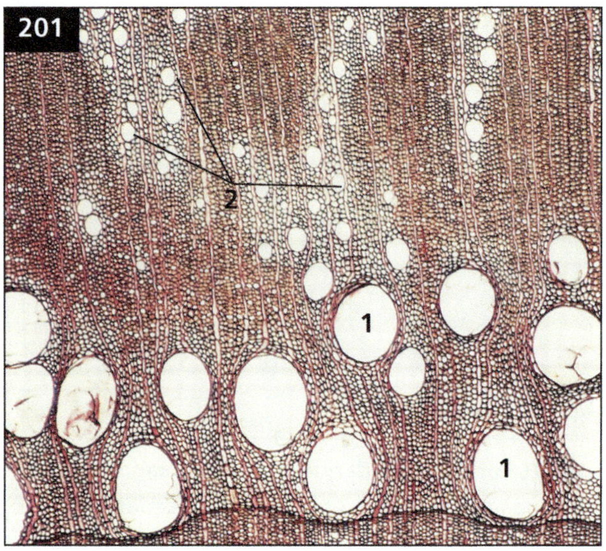

201 TS of the ring porous wood of the dicot *Quercus alba* (oak). The wide early wood vessels (1) contrast with the smaller-diametered vessels of the late wood. The latter are distributed in more or less radial bands of vessels (2) alternating with areas of very narrow tracheary elements. Numerous rays are also evident. (LM.)

1 Early wood vessels
2 Late wood vessels

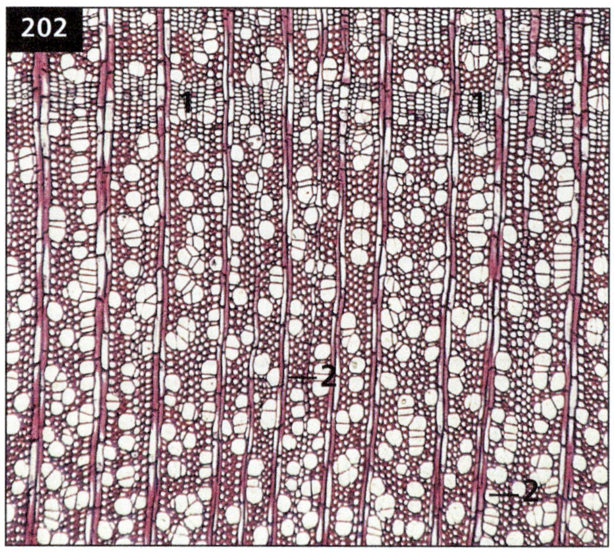

202 TS of the diffuse porous wood of the magnoliid *Magnolia grandiflora*. The narrow and fairly uniform vessels occur evenly throughout the growth ring except for a narrow band of late wood (1). Numerous, narrow multiseriate rays are present (2). (LM.)

1 Late wood vessels
2 Multiseriate rays

203 *Tilia* x *europea* (common lime) showing the principal semi-oblique branches arising from its vertical trunk and numerous adventitious brushwood shoots formed at its base.

204 Prone trunk of a wind-blown specimen of *Populus* from which several branches have developed into large upright trunks.

205 Trunk of a large specimen of *Cedrus deodora* (cedar) showing the scar of a large side branch. In gymnosperms, the reaction (compression) wood develops on the underside of the branch. Original position of pith (1). Note the callusing at the margins of the scar which is beginning to cover the wound with cork (2).

1 Original position of pith 2 Wound cork

206 Cross-cut trunk of *Castanea sativa* (a dicot); this is one of several very large branches which developed from an ancient coppice stool and shows clearly that wood has developed predominantly on the lower side of the branch. Original position of pith (1).

1 Original position of pith

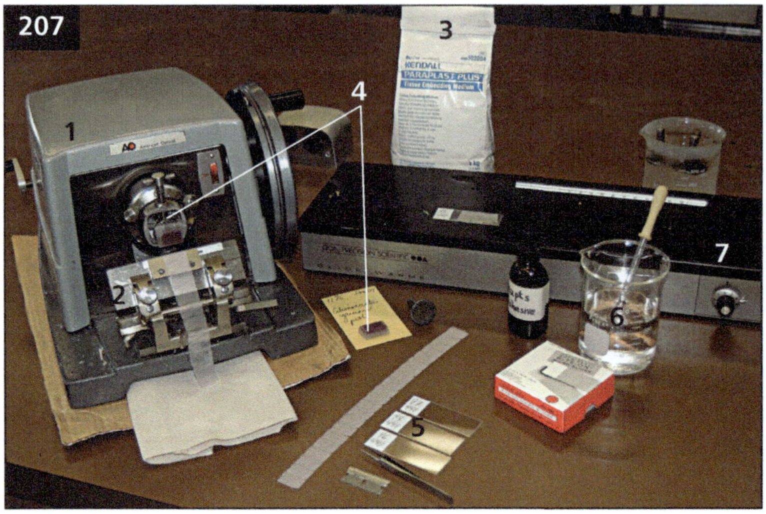

207 Materials for cutting paraffin sections of plant material include a rotary microtome (1), knife (2), paraffin (3), paraffin-embedded sample (4), adhesive-coated slides (5), distilled water (6), and a slide warmer (7).

1 Microtome
2 Knife (disposable razor blade)
3 Paraplast embedding medium
4 Sample embedded in block of paraffin
5 Microscope slides
6 Distilled water
7 Slide warmer

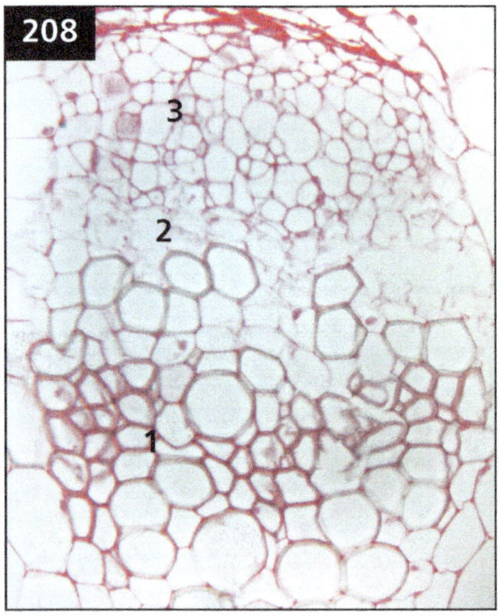

208 Unstained TS of stem of *Corryocactus*: wood (1), vascular cambium (2), and secondary phloem (3) after being sectioned, attached to a slide and deparaffinized. Without artificial stains, almost no detail is distinguishable. (LM.)

1 Wood
2 Vascular cambium
3 Secondary phloem

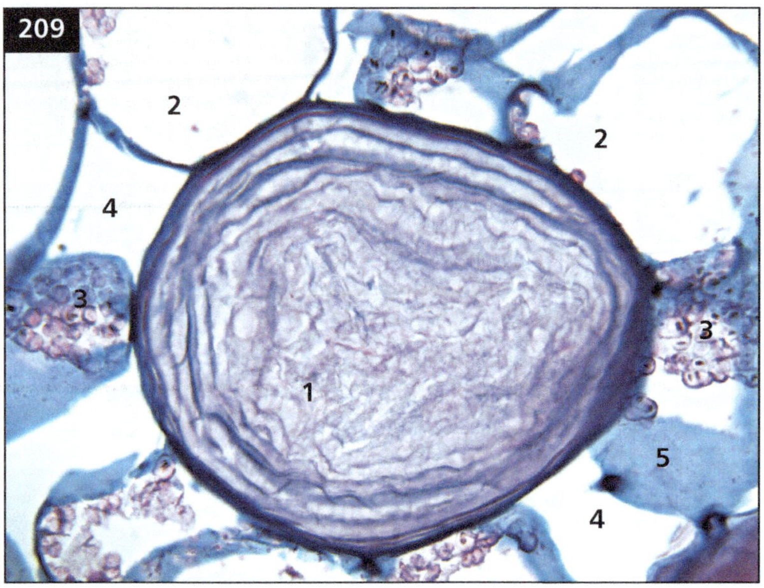

209 TS stem of *Corryocactus*. Some cortical cells secrete mucilage between the plasmalemma and cell wall; as mucilage accumulates (1), the protoplast shrinks then dies and degenerates. Fixation and dehydration stabilized this mucilage and held it in place during processing. Cortex parenchyma cell (2), starch grains (3), intercellular space (4), primary cell wall in face view (5). (LM.)

1 Mucilage
2 Cortex parenchyma cell
3 Starch grains
4 Intercellular space
5 Primary cell wall in face view

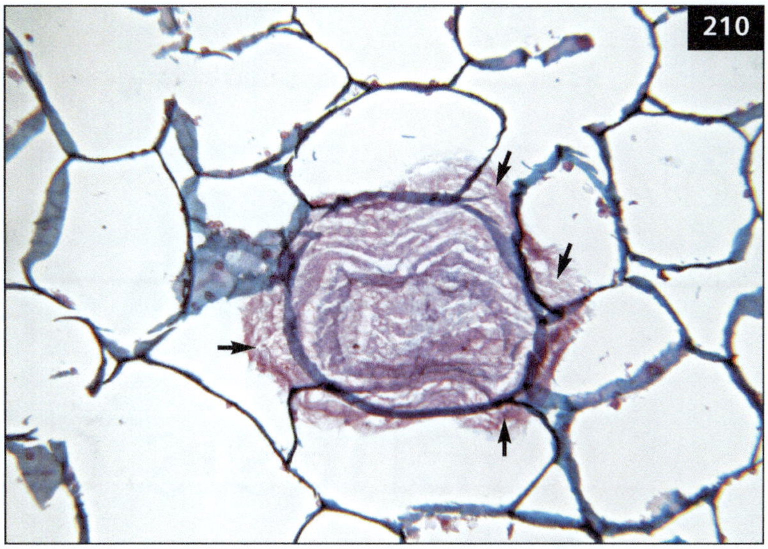

210 TS stem of *Corryocactus* showing an artefact of incomplete fixation: the mucilage in this central cell was not completely stabilized by the fixative. When the section was floated on warm water on the slide, the mucilage cell contents (stained purple) swelled upward and outward, then settled outside its wall (arrows) as the slide dried. In life, mucilage would not have been present in intercellular spaces or parts of surrounding cells. (LM.)

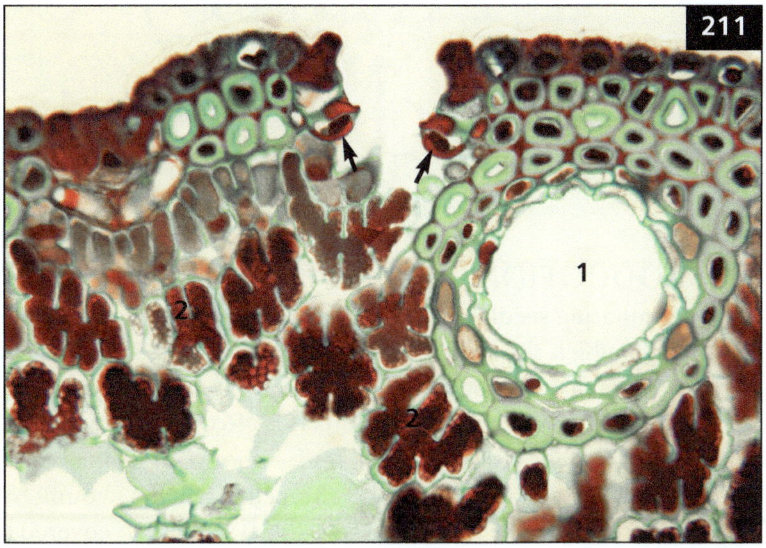

211 TS of *Pinus* needle showing an artefact caused by section movement during processing. The stomatal pore is much too wide, and the guard cells (arrows) too far apart. The microtoming removed the epidermis behind and in front of the cells visible here, thus there was nothing to stabilize these guard cells except for the paraffin, which must have torn during processing. Resin canal (1), mesophyll chlorenchyma (2). (LM.)

1 Resin canal
2 Mesophyll chlorenchyma

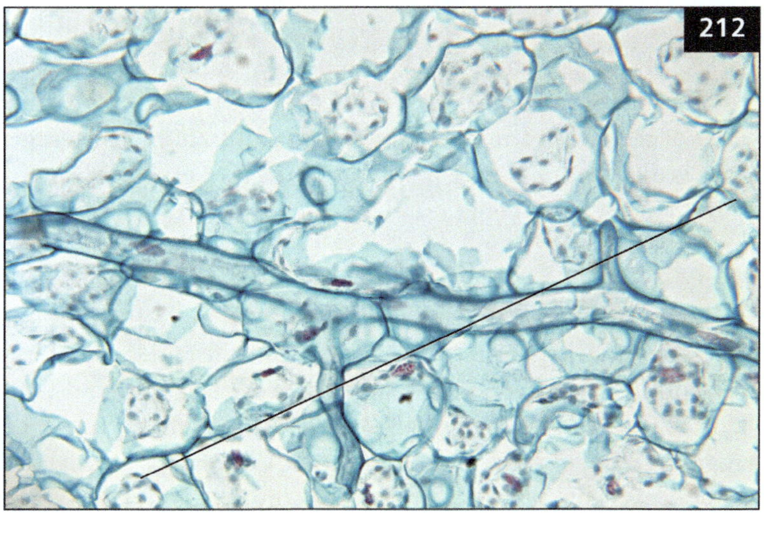

212 TS of stem of *Euphorbia fortuita*. Microscope views must be interpreted: this tubular structure in the pith is a latex duct with at least two branches within the field of view but which might also have other branches (which would not be visible) coming up toward the observer or back and away. The two visible side branches might be short or much longer but turned out of the plane of section. If this sample had been cut as indicated by the line, the duct would have appeared as three small circles. (LM.)

CHAPTER 4

Apical Meristems: Genesis of Primary Shoot and Root

INTRODUCTION

The germinating seedling originates from a bipolar embryo in which the radicle is situated at one pole and the plumule at the other (**46, 213**). The plumule is flanked by one or more cotyledons (**32, 46**). The meristematic cells of the root and shoot tips (**47, 213–217**) are the source of the cells which differentiate within the apical and sub-apical regions (**215–220**) into the primary tissues of the stem and root. The leaves and buds originate from surface layers at the margins of the shoot apex (**215, 216, 218, 221**) but lateral roots develop deep within the root, some distance from the root tip (**222**). The rates of cell division, differentiation, and maturation vary greatly within different regions of the root and shoot apices. Meristematic activity usually extends for some distance behind the apex.

VEGETATIVE SHOOT APEX

The shoot apex is usually domed or conical and is typically invested with leaf primordia (**218, 223–229**). In radial longitudinal section (RLS) of the angiosperm apex there is an outer tunica of one to several layers of cells. This surrounds a corpus of less regularly arranged cells (**214, 215, 227**). In the tunica the cells divide anticlinally so that the newly formed division walls always lie perpendicular to the outer surface of the apex. Thus the tunica remains discrete and its outermost cell layer differentiates into the epidermal covering of the leaves, buds and young stem (**215, 227**). The shoot epidermis is covered by cuticle and even at the apex a thin cuticle occurs (**229**). In gymnosperms the multicellular apex lacks a distinct tunica layer (**230**).

In addition to zonation based upon planes of cell division (tunica-corpus), the cells within the shoot apices of some species appear cytologically heterogeneous with a core of somewhat vacuolated rib meristem cells enclosed by the more densely-staining flank meristem cells (**215, 221**). In some cases the cells at the extreme tip of the apex stain less densely and form a zone of central mother cells (**221**; but not evident in **215, 227**) which apparently divide much more slowly than those of the rib and flank meristems. The rib meristem consists of vertical rows of vacuolating cells (**215, 221**) which differentiate into the central ground tissue (pith) in the young stem of dicots (**218, 219**), while the flank meristem (**215**) gives rise to the epidermis, cortex, and procambium (**216, 218, 219, 221**). However, no rigid boundary occurs between the rib and flank meristems, so that the cortical and procambial tissues are somewhat variable in their derivation from the apex.

Shoot apices are usually microscopically small, between 90–300 µm wide and containing a few hundred cells (**215, 223, 227, 228**). Cycads and many cacti, however, have giant shoot apical meristems up to 2500 µm across (easily visible to the naked eye; **221**) with each tunica layer and each corpus zone containing thousands of cells. Plants with narrow apical meristems produce thin stems with a slender pith and cortex; such shoots only broaden if they later produce wood. In contrast, shoots produced by giant apical meristems are exceptionally broad, having a wide pith, cortex, or sometimes both (**231, 232**). Slender shoots weigh (relatively) so little that a tree trunk can support numerous branches and twigs which are produced by the activity of thousands of small shoot apices; broad shoots are so heavy that their trunks can support only a few branches, so each plant is produced by just a few, or single, giant apical meristems (**231, 232**).

EARLY LEAF AND BUD DEVELOPMENT

Leaves are initiated from the margins of the shoot apex (**221, 227**) by anticlinal divisions in the outermost tunica layer and variously oriented divisions in the internal tissue. The arrangement of leaves relative to the shoot apex and stem (phyllotaxy) is usually characteristic of the species (**224, 226**), and sometimes of the family, as in grasses and labiates (**35, 218, 223**). In monocots the leaf primordia are initiated singly from the shoot apex, and the leaf base extends around a broad arc of the apex (**35, 36, 228**). By contrast, in dicots a leaf primordium usually develops from a narrower sector of the apex and initially forms a peg-like protuberance (**224, 225, 229**).

In dicots the leaf primordia may be initiated singly from the shoot apex giving rise to a spiral phyllotaxy (**224, 226**). This actually consists of two sets of spirals, one set running in a clockwise direction, the other running counterclockwise, with leaf primordia located at every point where the two sets intersect (**224, 226**). Giant shoot apical meristems have so much room available for leaf primordia that they may have as many as 55 spirals in one set, 34 in the other. In many succulent stems, leaf primordia are aligned vertically and the shoot forms ribs (**231–233**). Leaf primordia can also be arranged in opposite pairs (**215, 223**) or in whorls of three or more primordia (**225**). In grasses, and commonly in other monocots, the leaf arrangement is distichous (**34**) and leaves lie in two rows which are usually 180 degrees apart. However, spiral and other leaf arrangements also occur in monocots.

Axillary bud primordia are frequently delimited as lateral meristematic swellings close to the shoot apex, at the adaxial junction of the young leaf and stem (**215, 223**), and are linked by procambium to the vascular supply of the young shoot (**234**). However, dichotomous branching, which results from division of the shoot apex into two equal parts (**235**), occurs in several angiosperm families and is common in many nonseed-bearing vascular plants. In some angiosperms axillary buds may be absent or develop some distance behind the shoot apex from axillary parenchyma. In many species one or more accessory buds develop in addition to the axillary buds (**236**).

Buds may also develop adventitiously on the shoot: when the plumule of *Linum usitassimum* (flax, linseed) is damaged the hypocotyl forms adventitious replacement buds (**237**) from dedifferentiated epidermal and cortical tissue (**238**). Adventitious buds are common on other stems and may also develop on roots and leaves (**144, 145**). Shoots frequently develop from dormant buds located on the trunk or main branches of trees (**239**); these buds are commonly adventitious and arise endogenously from vascular parenchyma or cambial tissue. A number of tropical trees such as *Artocarpus altilis* (breadfruit), *Ficus* (figs), and *Theobroma cacao* (cocoa) are cauliflorous, developing their flowers from persistent bud complexes on the mature trunk (**240**).

In many tropical species, the axillary buds develop into lateral shoots just beneath the terminal bud (sylleptic growth). However, in other plants the terminal bud exerts dominance over the axillaries; these are commonly invested by bud scales and undergo a period of dormancy before sprouting (proleptic growth, **241**).

REPRODUCTIVE SHOOT APEX

In a number of flowering plant species the vegetative shoot apex becomes considerably modified, and ultimately completely depleted, as it converts to reproductive growth and is transformed into a floral or inflorescence apex. However, in some plants the vegetative apex does not appear to undergo any structural changes when forming axillary floral primordia instead of vegetative buds. For instance, *Glechoma hederacea* (ground ivy) perennates by runners as a soil-hugging herb, but in late spring its horizontal shoots grow vertically upwards, forming many small purple-blue flowers at its nodes (**242**). Nevertheless, there are no obvious differences between the vegetative apex (**215**) and the apex giving rise to floral primordia (**243**). At the end of its reproductive phase, the shoot apex resumes vegetative activity and continues growth as the terminal bud of a runner. In a number of other species, the shoot apices appear to continue monopodial growth, despite forming lateral floral primordia (**244, 245**) in their reproductive phase.

TISSUE DIFFERENTIATION IN THE YOUNG STEM

In the terminal bud the procambium (incipient vascular tissue) develops acropetally into the apex (**216, 243**) from the older procambial tissue at its base. Within the apex the procambium becomes differentiated from the inner flank meristem, with each leaf linked from its inception to the procambium (**215, 216**). In the procambial strand the first vascular tissues begin to differentiate close to the apex (**49, 219**). The protoxylem normally develops at the inner margin of the strand while protophloem forms on the outer margin nearest to the epidermis (**20**).

Protoxylem usually first differentiates within the procambial strand at the base of a leaf primordium (**228**), forming a short longitudinal file of tracheary elements which then differentiates bidirectionally: both upwards into the leaf and downwards into the young internode where it links with older and larger xylem strands (**246**). The longitudinal pathway of protophloem differentiation in the procambium is normally acropetal into the young axis and leaf primordia and is in continuity with the phloem elements in the older bud. The metaphloem differentiates somewhat later and is located inwards (centripetally) to the protophloem, while the metaxylem develops centrifugally to the protoxylem (**247**). In many dicots and gymnosperms a narrow strip of procambium remains undifferentiated between the xylem and phloem and constitutes the fascicular vascular cambium (**128**), but in monocots this is absent (**20**).

In most dicots a large parenchymatous pith occupies the centre of the primary stem and is surrounded by a ring of discrete vascular bundles, with a narrow cortex situated externally (**41, 221, 231, 232**). In monocots a distinct pith is uncommon, and the vascular bundles normally occur as a complex three-dimensional network throughout the ground tissue (**40, 228**).

ROOT APEX

In the great majority of species the root apex is subterminal since it is covered by a protective root cap (**248**), although in some aquatic plants this is absent. Due to massive dictyosome activity in the outer cap cells (**66**), a large quantity of mucigel is secreted into the soil rhizosphere (**249**). More mucigel is contributed by the root hairs (**250**) which develop behind the root apex. In some plants such as *Zea mays*, the root cap has its own distinct initials (calyptrogen, **248**). The incipient epidermis (protoderm) and cortex can be traced to a single tier of cells adjacent to the calyptrogen, while the central procambial cylinder (**217**) apparently originates from a third tier of initials immediately within those of the protoderm-cortex (**248**). However, in some taxa the protoderm and root cap originate from a common tier of initials, while in others the cap and other regions all converge to a common group of initials (**251**).

In actively growing and elongating root tips the tip of the apex represents a quiescent centre (**252**), so that the patterns of apical initials described previously reflect the architecture of the apex before active growth of the root primordium had commenced. In *Zea mays* the 'initials' of the protoderm, cortex, and procambium (**248**) all lie within an extensive quiescent centre whose cells divide on average once every 174 hours in contrast to every 12 hours in the calyptrogen. On the surface of the quiescent centre, remote from the calyptrogen, the cells also divide rapidly and are the real initials of the protoderm, cortex, and procambium. If the dividing cells of the root tip are damaged by ionizing radiation, the resistant quiescent centre cells become reactivated and regenerate a new apex.

The root apex does not normally give rise to the lateral roots; instead these normally develop from the maturing root several millimetres from its tip (**222**), basal to the root hair zone (**250**).

TISSUE DIFFERENTIATION IN THE YOUNG ROOT

The longitudinal differentiation of both xylem and phloem within the procambium is acropetal. These vascular tissues occur nearer to the apex in slow growing or dormant roots than in actively elongating roots. The first mature protophloem elements appear at the circumference of the procambial cylinder and differentiate closer to the root apex than the first mature protoxylem (**220**). By contrast the prospective metaxylem becomes demarcated, by its prominent vacuolation growth (**217**), nearer the apex than the protophloem. Some distance basal to the protophloem, mature protoxylem elements develop between the protophloem files. Maturation of the metaphloem and metaxylem proceeds centripetally (**253**) on the radii already demarcated by the protophloem and protoxylem (**220**), and the vascular system of the mature primary root is formed (**254**). The outer layer of the procambium gives rise to the parenchymatous pericycle (**220, 254**).

In dicot roots the tissue between the xylem and phloem often forms a vascular cambium (**254**), which later spreads laterally over the protoxylem poles to form a continuous meristem (**38**). The vascular cylinder is invested by the parenchymatous cortex (**42, 253**) which frequently contains conspicuous intercellular air spaces. The pericycle is bounded by the single layered endodermis (**42, 254, 255**) whose radial and transverse walls are impregnated with lignin and suberin to form Casparian bands (**256**). These laterally continuous bands are impermeable, so that all water and solute movement across this layer is confined to the symplast (**256**). In the roots of many species, particularly monocots, the lignification and suberization of the endodermis later extends to all walls and additional cellulose thickening may be deposited (**255**). Such cells, however, still allow symplastic transport from the cortex to the stele via their plasmodesmata.

The endodermis is an important selective barrier but it allows the active transport into the vascular cylinder of certain beneficial ions (potassium, phosphate) absorbed from the soil. Calcium, however, apparently moves apoplastically and cannot enter the vascular cylinder through the endodermis, but instead passes into the vascular cylinder in the very immature root where Casparian bands have not developed. The active transport of ions into the vascular cylinder accounts for the phenomenon of root pressure which sometimes plays an accessory role to transpiration in the movement of water to the shoot.

In some roots the outer cortex differentiates as a one- to several-layered exodermis (**42**) but the Casparian bands in the radial walls are usually masked by deposition of suberized lamellae adjacent to the protoplast (**257**). A short distance behind the apex a zone of absorptive root hairs develops from the epidermis (**250**); water absorption also occurs over the rest of the epidermal surface and in some species root hairs are absent. In the aerial roots of some epiphytic orchids and aroids the multilayered epidermis develops into a dead velamen (**257, 258**), whose walls are thickened by bands of cellulose. This is probably an adaptation to absorb water from the humid atmosphere of tropical forests.

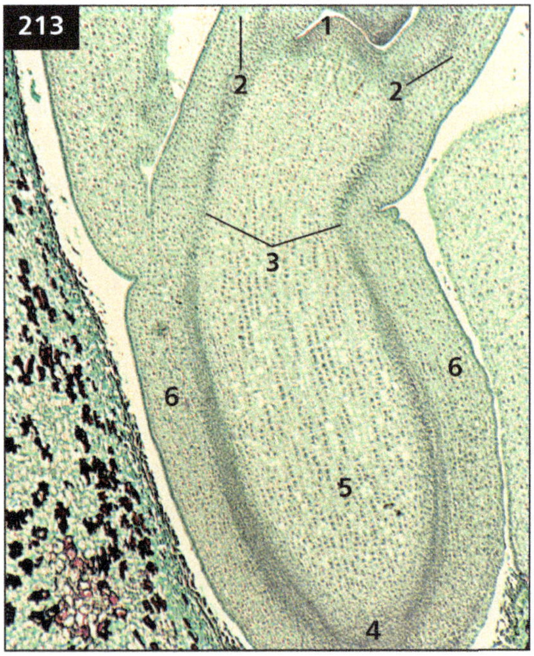

213 RLS of the bipolar embryo of the dicot *Phaseolus vulgaris* (French bean). The hemispherical, densely-staining and small-celled shoot apex (1) bears a pair of foliage leaf primordia (2) from which prominent strands of procambium (3) extend down into the radicle (4). Pith (5), cortex (6). (LM.)

1 Shoot apex	4 Radicle
2 Leaf primordium	5 Pith
3 Procambium	6 Cortex

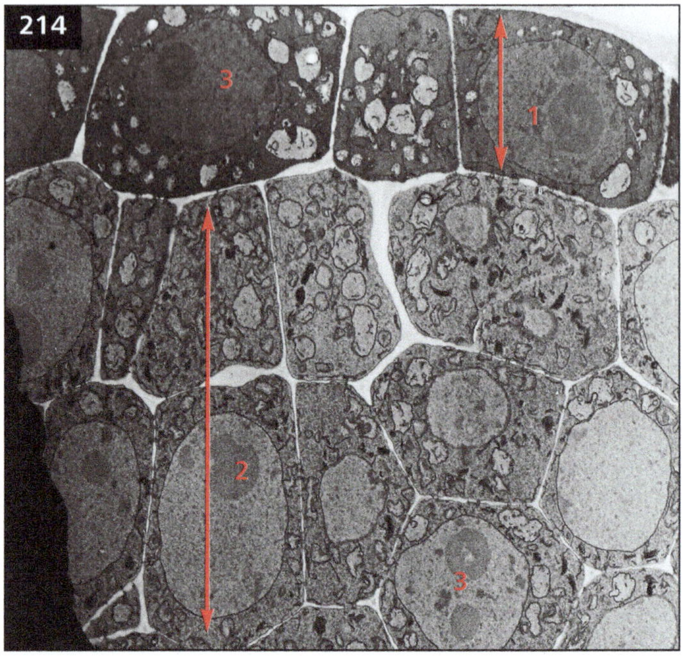

214 RLS of the extreme tip of the vegetative shoot apex of the dicot *Glechoma hederacea* (ground ivy). Note the single tunica layer (1) where only anticlinal divisions occur and the underlying corpus (2) in which cells divide in various planes. The small thin-walled cells possess large nuclei (3) whereas the cytoplasm is relatively unvacuolated (cf. 215). (KMn, TEM.)

1 Tunica layer	3 Nucleus
2 Corpus	

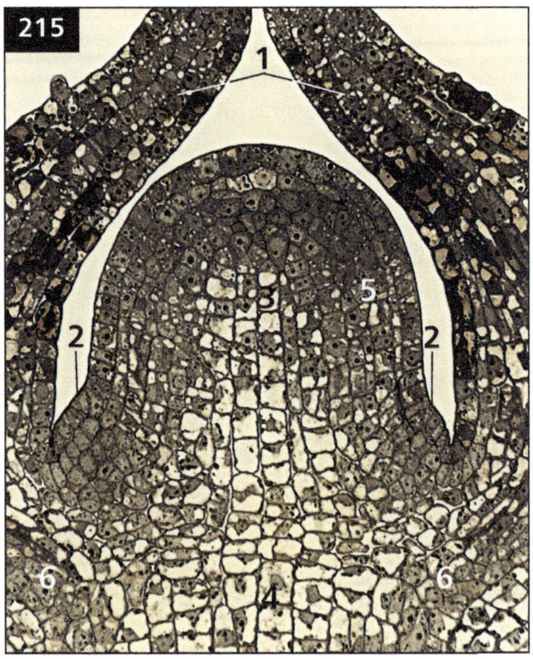

215 RLS of the vegetative shoot apex of the dicot *Glechoma hederacea* (ground ivy). Note the pair of leaf primordia (1) at its base and a pair of bud primordia (2) in their axils. The extreme tip of the apex (cf. the corresponding area in 214) consists of small densely cytoplasmic cells, but the rib meristem cells (3) below are vacuolating and become the pith (4) of the young stem. The margins of the apex consist of densely-staining flank meristem cells (5). Procambium (6). (G-Os, LM.)

1 Leaf primordium	4 Pith
2 Bud primordium	5 Flank meristem
3 Rib meristem	6 Procambium

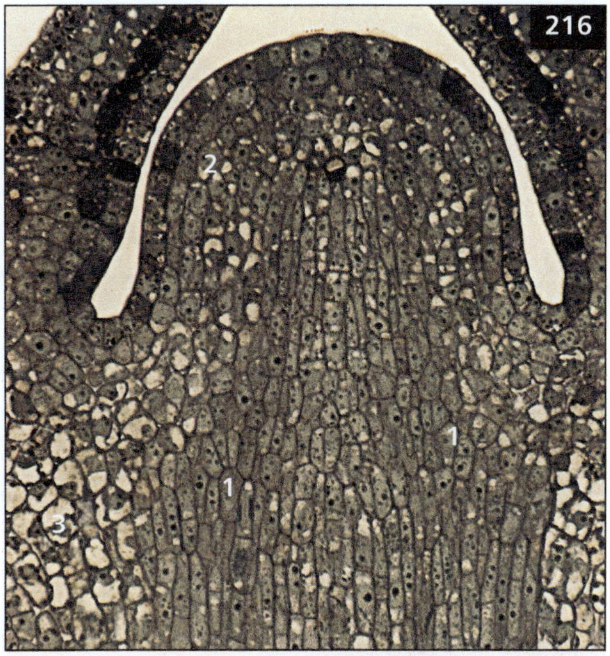

216 LS of the vegetative shoot apex of the dicot *Glechoma hederacea* (ground ivy). This section is slightly tangential to the same apex shown previously (cf. 215). Note how the narrow, densely cytoplasmic procambial cells (1) merge with the flank meristem (2) within the shoot apex. Cortex (3). (G-Os, LM.)

1 Procambium
2 Flank meristem
3 Cortex

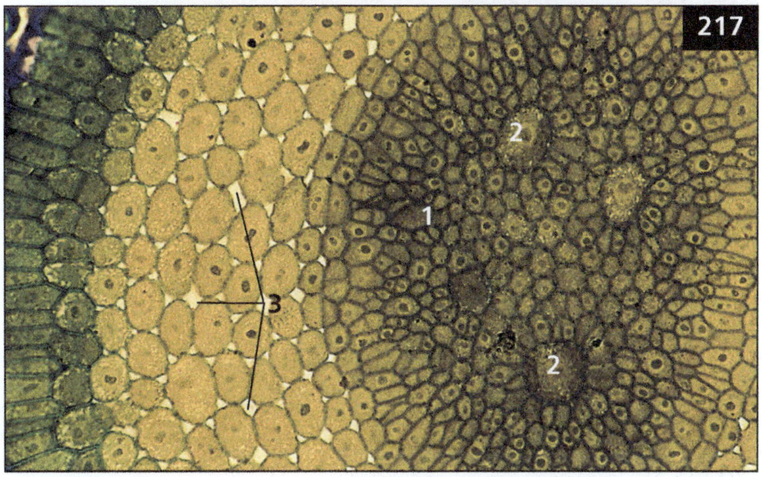

217 TS just behind the root apex of the monocot *Zea mays* (maize). Note the core of small-diametered, densely cytoplasmic procambial cells (1) in which several larger cells (2) are differentiating into large metaxylem elements. The procambium is bounded by a wide cortex of larger cells, with intercellular spaces (3) already evident. (G-Os, LM.)

1 Procambial cells
2 Prospective metaxylem
3 Intercellular spaces

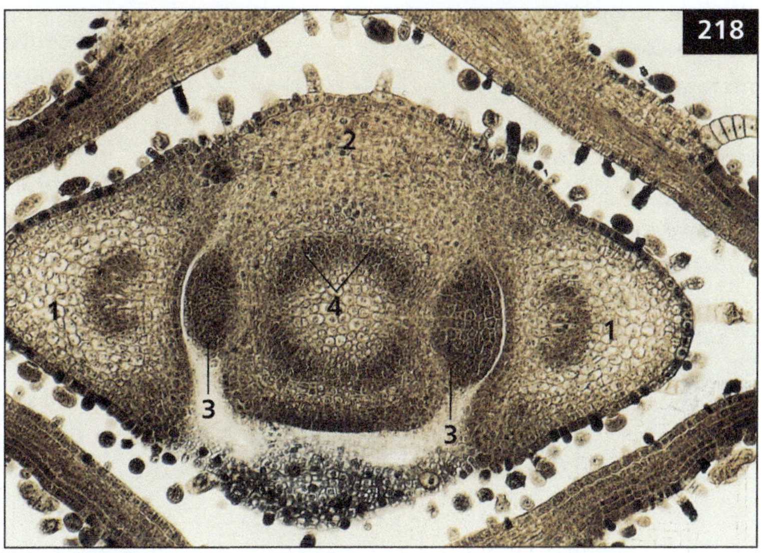

218 TS of the terminal bud of the dicot *Glechoma hederacea* (ground ivy). The petioles (1) of the youngest pair of leaves extend as a parenchymatous collar (2) around the stem. The densely-staining cells of the axillary bud primordia (3) and procambium (4) are also evident in the stem. Note the abundant trichomes which cover the surfaces of the leaves. (LM.)

1 Petioles
2 Parenchymatous collar
3 Axillary bud primordium
4 Procambium

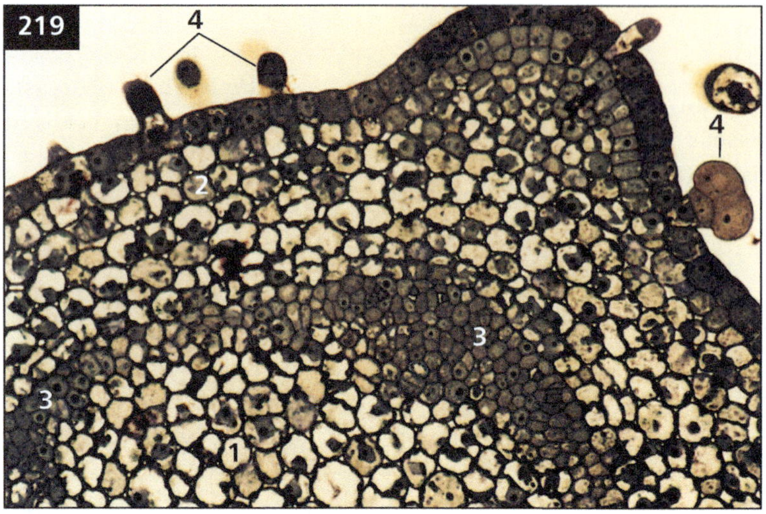

219 TS of the young stem of the dicot *Glechoma hederacea* (ground ivy). Note the large-celled, vacuolated pith (1) and cortex (2); between them, smaller, densely cytoplasmic procambial cells (3) are evident (cf. 49). The epidermis is derived by anticlinal divisions from the single-layered tunica of the apex (cf. 214), except that the glandular trichomes (4) form by periclinal divisions. (G-Os, LM.)

1 Pith
2 Cortex
3 Procambial cells
4 Glandular trichomes

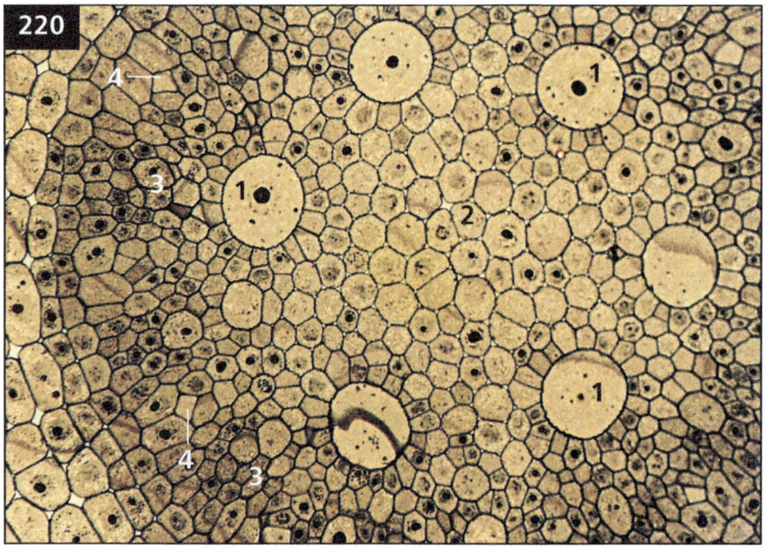

220 TS near the apex of the young root of the monocot of *Zea mays* (maize). A number of wide-diametered, but thin-walled, potential metaxylem elements (1, cf. 217) lie on the periphery of the pith (2). The sites of the potential protoxylem (3) are also evident centrifugally, but the first mature vascular elements to differentiate are the protophloem sieve tubes (4). (G-Os, LM.)

1 Prospective metaxylem
2 Pith
3 Potential protoxylem
4 Protophloem sieve tubes

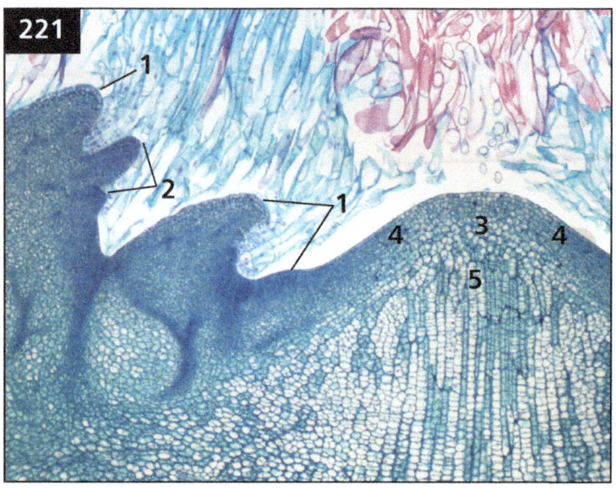

221 *Oroya peruviana*. RLS of a giant cactus shoot apex ca. 1500 μm in diameter; each zone in its corpus is vastly larger than entire shoot apices of most other taxa. Three leaf primordia are visible (1), the two on the left are almost fully developed. The axillary bud of the leftmost leaf has initiated two bud scale primordia (2), both of which will develop as spines. Central mother cells (3), flank meristem (4), rib meristem (5). (LM.)

1 Leaf primordia
2 Bud scale primordia
3 Central mother cells
4 Flank meristem
5 Rib meristem

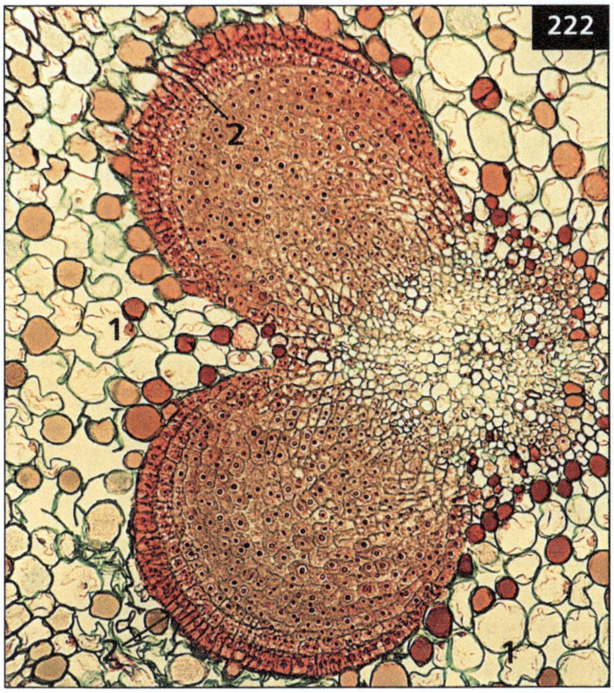

222 TS of the root of the dicot *Salix* (willow). Note the presence in the parent root cortex (1) of a pair of lateral roots covered by root caps (2). These arise from the dedifferentiation and division of pericyclic parenchyma cells at the tips of the pentarch xylem arms in the parent root. (LM.)

1 Parent root cortex
2 Root cap

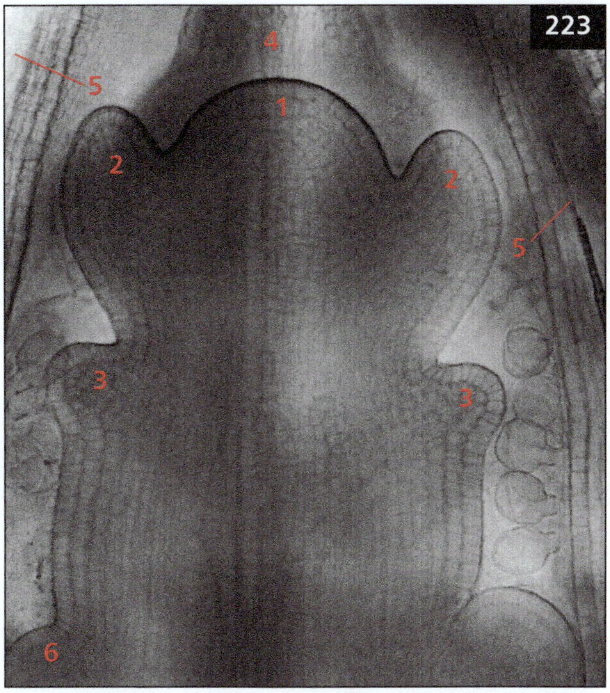

223 Topography of the apical region of a cleared bud of the dicot *Glechoma hederacea* (ground ivy). The hemispherical shoot apex (1) shows its youngest pair of leaf primordia (2) at its margins while the collar (3, cf. 218) of the next leaves is visible at the base of the apex. The tip of one of these pair of leaves is visible in face view (4). An older pair of leaves (5) are seen in side view; these subtend a pair of axillary buds (6). (LM.)

1 Shoot apex
2 Leaf primordia
3 Leaf collar
4 Face view of leaf
5 Older leaf
6 Axillary bud

224 TS of the vegetative bud of the dicot *Solanum tuberosum* (potato). Note the shoot apex (1) and the spiral arrangement of the leaf primordia (2) which invest it. (LM.)

1 Shoot apex
2 Leaf primordia

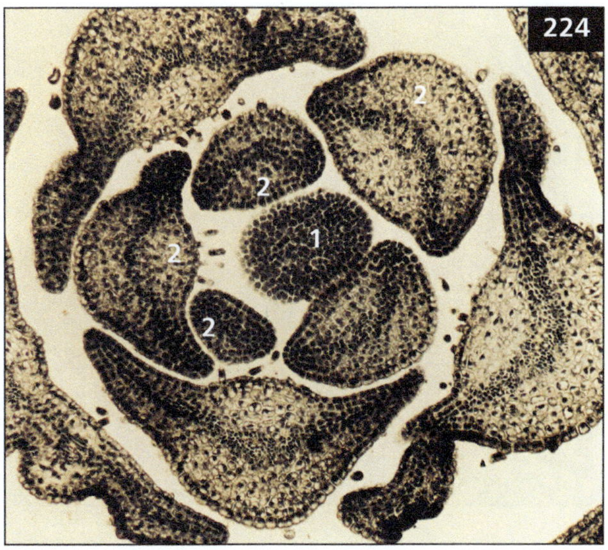

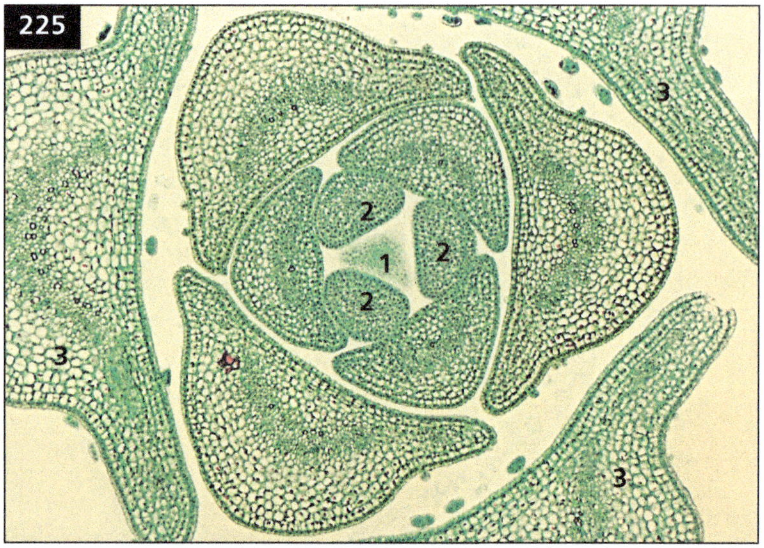

225 TS of the terminal bud of the dicot *Ligustrum vulgare* (privet). Normally privet is decussate but in this bud a whorl of three leaves arises from the shoot apex (1). Note how the primordia of the youngest whorl of leaves (2) lie in the same relative positions as the older ones (3) arising from the axis two nodes below. (LM.)

1 Shoot apex
2 Youngest leaf primordia
3 Older leaves

226 Shoot tip of the succulent dicot *Aeonium*. Note the closely-crowded leaves arranged in a spiral phyllotaxy (cf. 224). (Copyright of T. Norman Tait.)

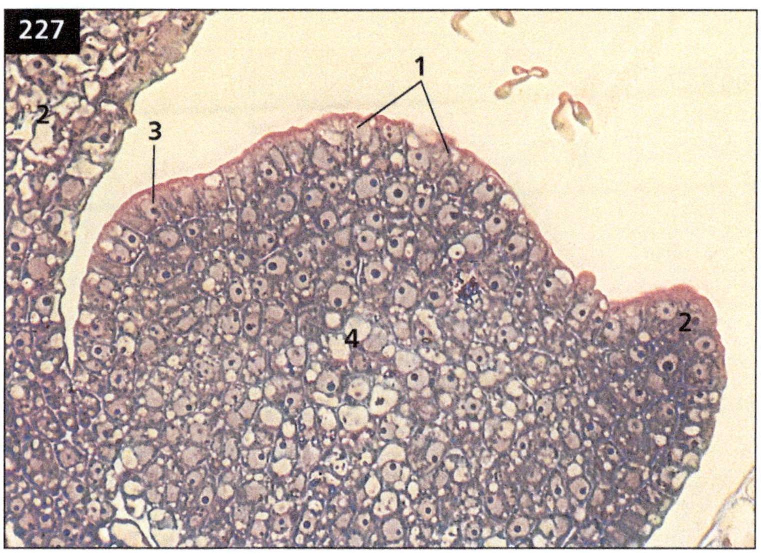

227 RLS of the apex of an axillary bud of the dicot *Phaseolus vulgaris* (bean). The apex has a single tunica layer (1) and two leaf primordia (2) are visible while a leaf buttress (3) is also evident. Rib meristem (4) lies at the base of the apex. (G-Os, Phase contrast LM.)

1 Tunica layer
2 Leaf primordia
3 Leaf buttress
4 Rib meristem

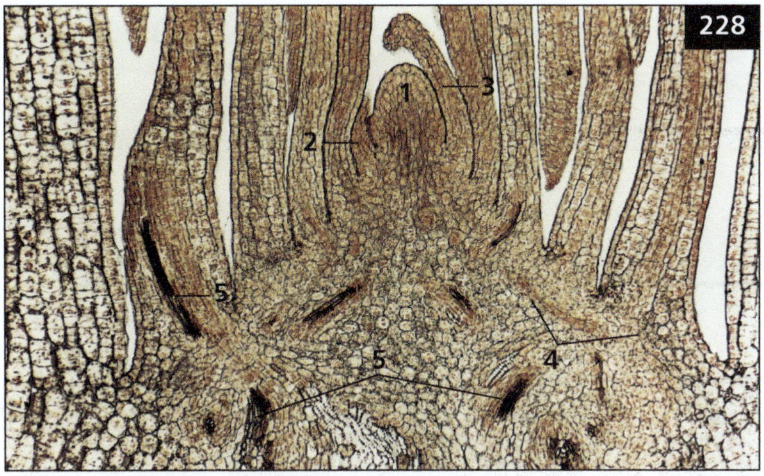

228 LS of the shoot apical region of the monocot *Zea mays* (maize). Although at the shoot apex (1) the leaves form singly, with each new primordium (2) being initiated at 180 degrees from the previous leaf (3), the leaf bases grow laterally and soon encircle the apex (cf. 35). Note the shallow sloping sub-apical margins bearing the leaves; here the cells of the primary thickening meristem (4) divide mainly periclinally with their derivatives differentiating into parenchyma traversed by the procambial (5) strands which supply the leaves. (LM.)

1 Shoot apex
2 New leaf
3 Previous leaf
4 Primary thickening meristem
5 Procambial strand

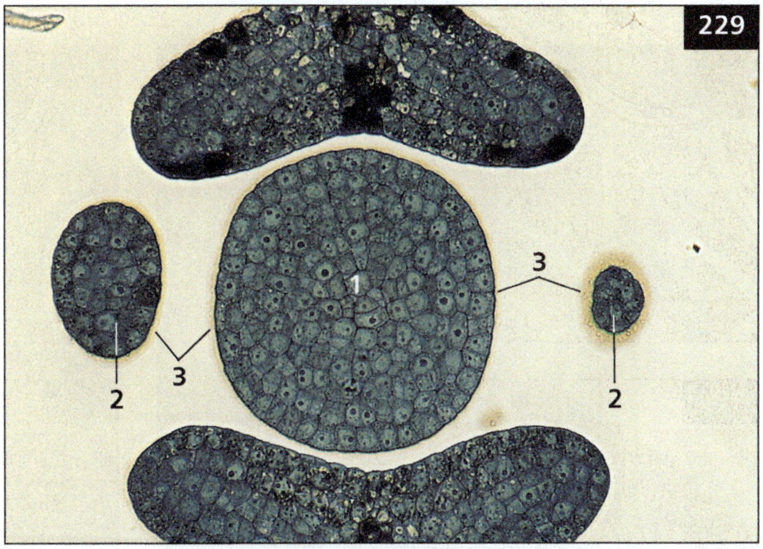

229 TS through the shoot apex of the dicot *Glechoma hederacea* (ground ivy) showing its decussate phyllotaxy. Shoot apex (1), youngest leaf primordia (2), cuticle (3). (G-Os, LM.)

1 Shoot apex
2 Youngest leaf primordia
3 Cuticle

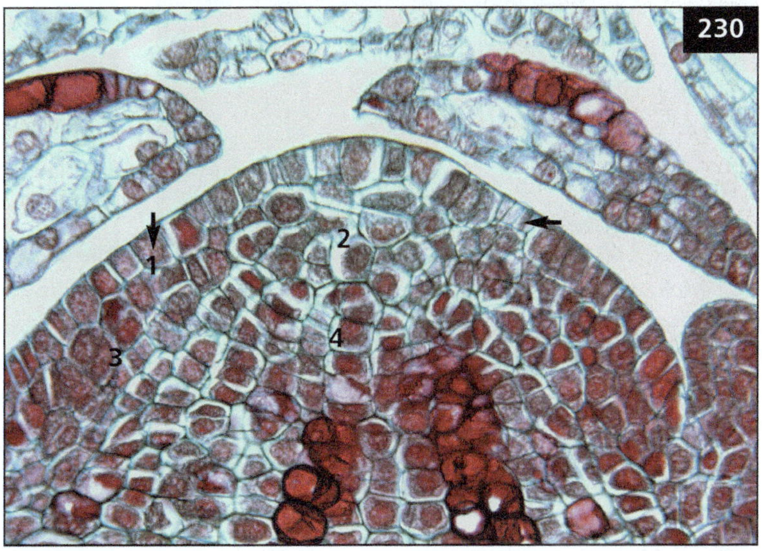

230 RLS of the conifer *Pinus* (pine) shoot apical meristem, showing that cell divisions with periclinal walls (arrows) occasionally occur in the outermost layer of cells. Consequently, the outermost layer contributes cells to the inner tissues of the plant. Periclinal walls (1), central cells (2), flank meristem (3), rib meristem (4). (LM.)

1 Periclinal walls
2 Central cells
3 Flank meristem
4 Rib meristem

231 *Echinocactus platyacanthus*. The giant shoot apex of this cactus is composed of large flank and rib meristems which in turn produce a huge pith and cortex. The width of this cactus is due entirely to the expansion of primary tissues (cf. the narrow primary growth exhibited at the tips of the numerous twigs on the adjacent *Cercidium microphyllum*, palo verde, trees).

232 *Trichocereus pasacana*. The broad shoots produced by giant cactus apical meristems are heavy and initially weak, because they develop little wood until several years old. Consequently, in contrast to the large number of branches typical of most trees, they are infrequently branched and have few apical meristems.

233 *Browningia candelaris*. Each small, dark spot on this cactus stem is an axillary bud, and like their subtending leaves (which are too microscopic to be visible), they occur at the intersections of clockwise and counterclockwise phyllotactic spirals. As in many succulent shoots, the vertically aligned leaf bases have become united into ribs as they develop (these allow the shoots to swell and shrink as they gain or lose water).

234 TLS of the terminal bud of the dicot *Glechoma hederacea* (ground ivy). An axillary bud primordium (1) is linked by densely-staining procambial strands (2) to the vascular system of the internode below. Pith (3), cortex (4). (LM.)

1 Bud primordium
2 Procambial strand
3 Pith
4 Cortex

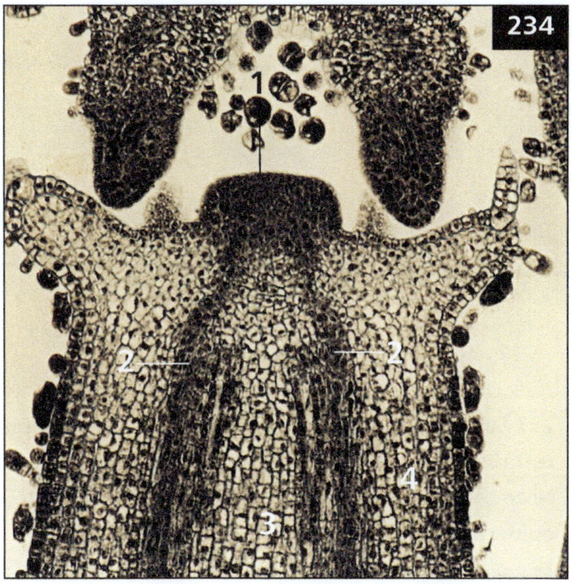

235 Shoot apex of a dicot, the cactus *Pelecyphora aselliformis*, that has just completed a dichotomous division. The white comb-like structures (1) are sets of short, flat spines; each set occurs in the axil of a microscopic foliage leaf, thus the sets indicate the phyllotactic leaf arrangement. The two foci (arrows) indicate the positions of two new shoot apical meristems.

1 Set of spines

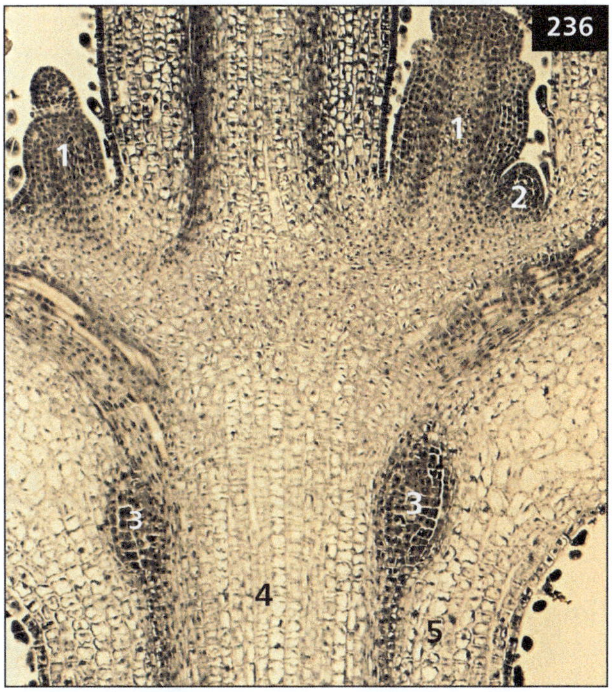

236 LS at the base of the terminal bud of the dicot *Glechoma hederacea* (ground ivy). A pair of axillary buds (1) are present, but their linkage to the vascular system of the main axis is not apparent in this plane of section (cf. 234). The larger bud has an accessory bud (2) at its abaxial margin. Note the pair of adventitious root primordia (3) evident at this node. Pith (4), cortex (5). (LM.)

1 Axillary bud
2 Accessory bud
3 Root primordium
4 Pith
5 Cortex

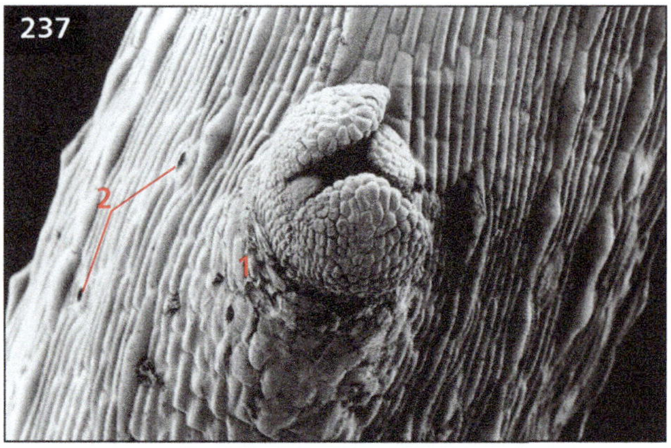

237 Surface view of the hypocotyl of the dicot *Linum usitatissimum* (flax) showing an adventitious bud (1). Note the stomata (2) occurring in the hypocotyl epidermis. (SEM.)

1 Adventitious bud
2 Stoma

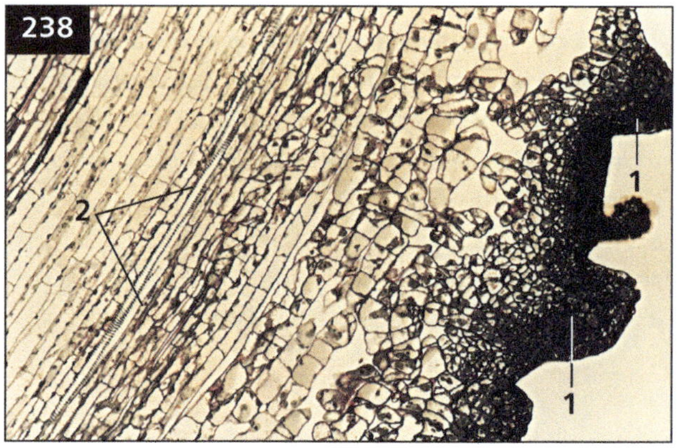

238 LS of the hypocotyl of the dicot *Linum usitatissimum* (flax) bearing an adventitious bud primordium. Note the numerous newly-formed cells in the underlying cortex; further division will give rise to procambial strands to link the leaf primordia (1) to the vascular tissue (2) of the parent hypocotyl. (G-Os, LM.)

1 Leaf primordia
2 Vascular tissue of hypocotyl

239 Base of the trunk of the dicot *Aesculus hippocastanum* (horse chestnut). Note the numerous newly-sprouted leafy twigs; these are of adventitious origin and arise from proliferated parenchyma in the bark.

240 This branch of the dicot *Cercis canadensis* (redbud) is many years old but still produces flowers from very old axillary buds. This axillary bud, and others like it, will flower yearly even after this branch has become massive.

241 Sprouting terminal bud of the deciduous dicot tree *Acer pseudoplatanus* (sycamore). This exerts dominance over the dormant axillary buds (1) at the nodes below. Note the decussate arrangement of the scale leaves (2), which are evanescent and nonphotosynthetic, and the emerging foliage leaves (3). (Copyright of T. Norman Tait.)

1 Axillary bud
2 Scale leaves
3 Foliage leaves

242 Upright flowering shoots of the dicot *Glechoma hederacea* (ground ivy) bearing numerous purple-blue flowers which have developed from axillary floral primordia.

243 RLS of the terminal bud on a vertically growing flowering shoot of *Glechoma hederacea* (ground ivy; a dicot) showing axillary floral primordia (1) but with the terminal shoot apex (2) apparently unchanged from the vegetative apex (cf. 215). (LM.)

1 Floral buds
2 Shoot apex

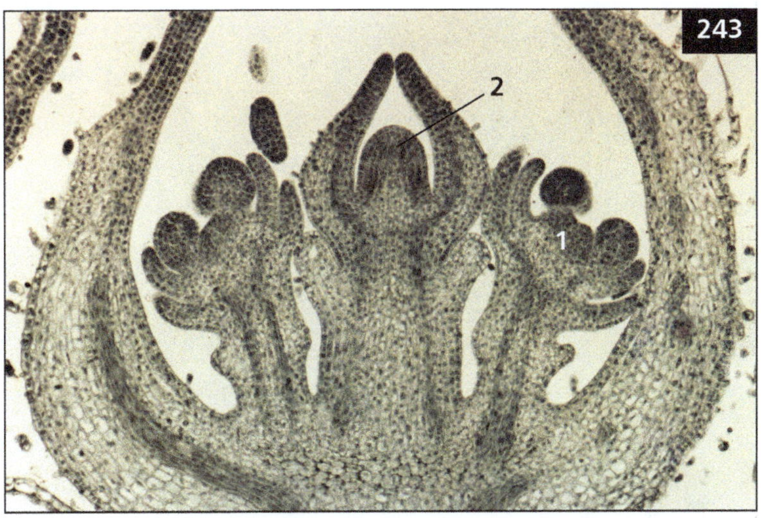

244 The multiple fruit of the monocot *Ananas comosus* (pineapple) borne terminally on the vegetative shoot; note that the fruit bears a tuft of vegetative leaves and this crown can be used to propagate vegetatively a further crop of pineapples.

245 Twig on *Melaleuca* (a dicot tree) showing old, persistent cauliflorous fruits remaining after the shoot has resumed vegetative growth.

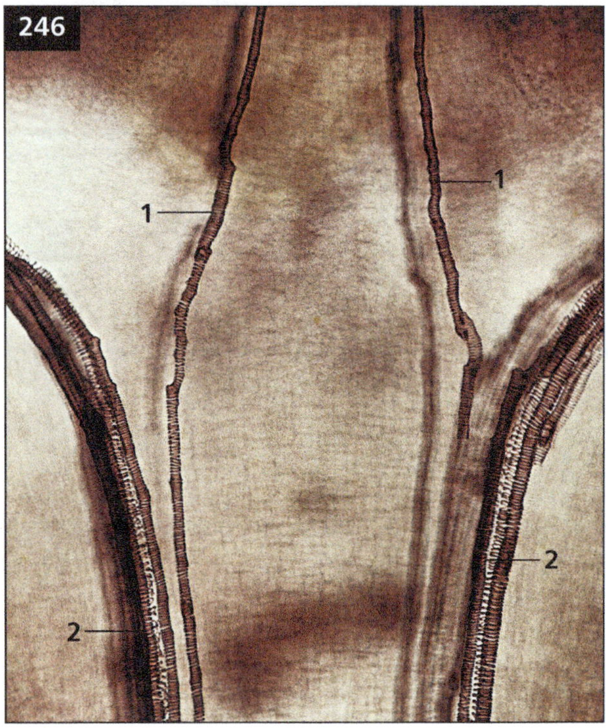

246 Old node from a cleared terminal bud of the dicot *Glechoma hederacea* (ground ivy). Note the linkage of the protoxylem (1) from the younger shoot (composed of single files of short tracheary elements) with the more extensively developed xylem strands (2) of the older stem. (LM.)

1 Protoxylem files

2 Xylem strands

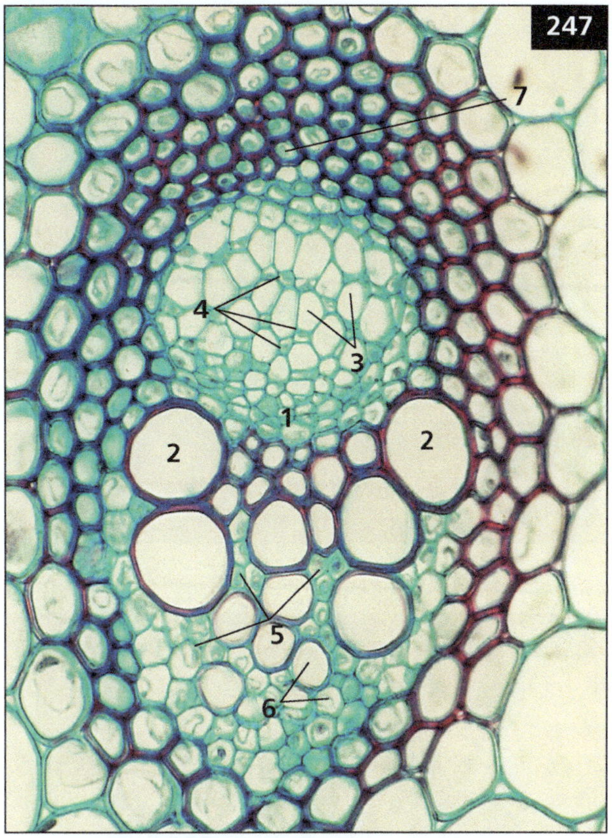

247 TS of a vascular bundle from the stem of *Ranunculus* (buttercup). This herbaceous dicot does not undergo secondary thickening despite the presence of fascicular cambium (1) lying between the metaxylem (2) and metaphloem. The latter shows a distinctive pattern of sieve tubes with wide lumens (3) and smaller companion cells (4) reminiscent of many monocots (cf. 20). Axial parenchyma (5), protoxylem (6), protophloem fibres (7). (LM.)

1 Fascicular cambium
2 Metaxylem
3 Sieve tubes
4 Companion cells
5 Axial parenchyma
6 Protoxylem
7 Protophloem fibres

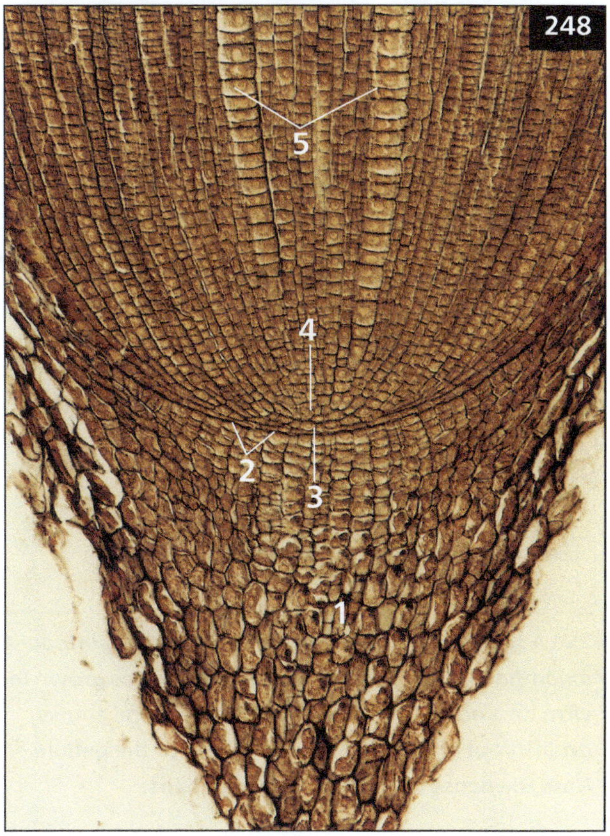

248 RLS of the root tip of the monocot *Zea mays* (maize) showing a prominent cap covering the apex. The cap (1) has its own distinct initials (calyptrogen, 2) while the epidermis and cortex apparently arise from a common tier of initials (3) adjacent to the calyptrogen. The procambial central cylinder has its own initials (4). Note the conspicuous files (5) of enlarged cells within the procambium which represent the future metaxylem elements (cf. 220). (LM.)

1 Root cap
2 Cap initials
3 Initials of epidermis and cortex
4 Procambial initials
5 Potential metaxylem

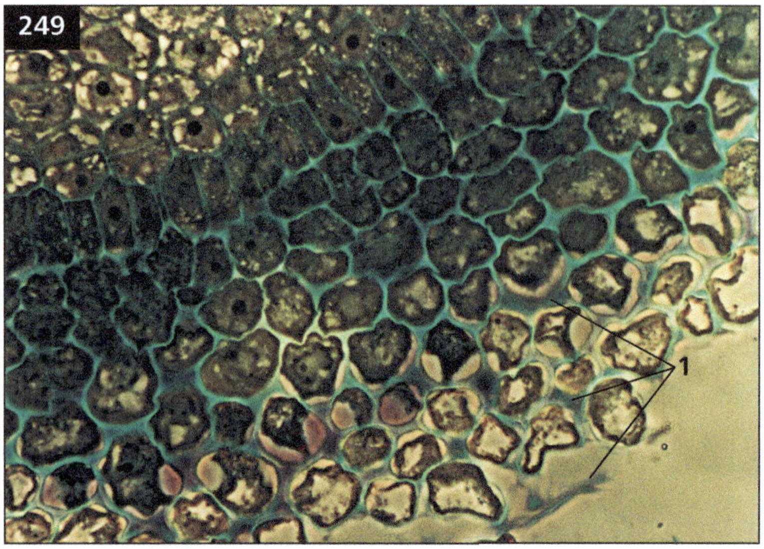

249 TS showing detail of the root cap of the monocot *Zea mays* (maize). Note the progressive increase in thickness of the cell walls towards the margin of the cap. Here the cells are breaking down and sloughing their mucilaginous cell walls and protoplasts to form mucigel (1) which is secreted into the soil. (G-Os, LM.)

1 Mucigel

250 Adventitious roots on a leaf of the dicot *Saintpaulia ionantha* (African violet). The excised leaf was grown *in vitro* on a nonsterile mineral salt medium and shows prolific root production at the cut base of the petiole. Note the dense felty covering of root hairs.

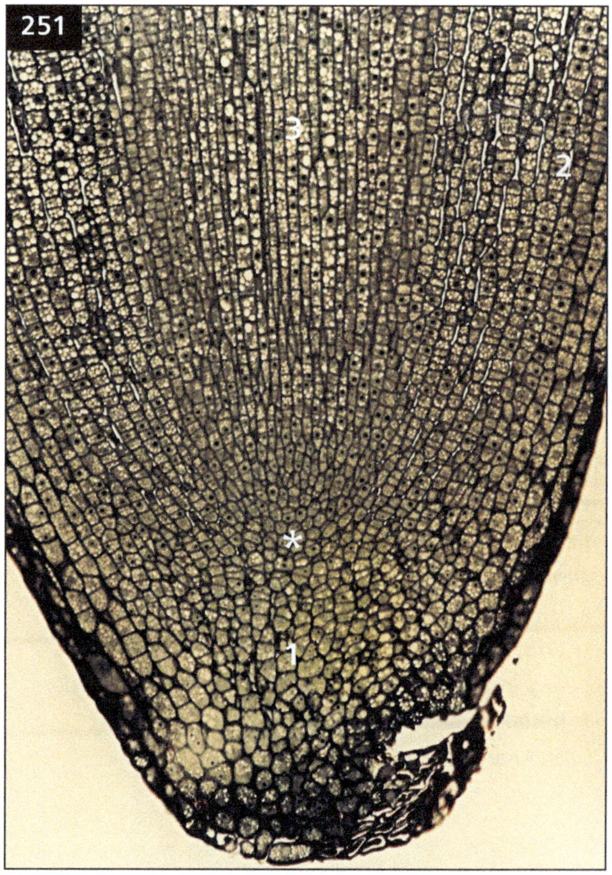

251 RLS of the radicle of the dicot *Phaseolus vulgaris* (French bean). This has an ill-defined group of initials (asterisk) which are common to the root cap (1), cortex and epidermis (2), and central cylinder (3). (G-Os, LM.)

1 Root cap
2 Epidermis and cortex
3 Central cylinder

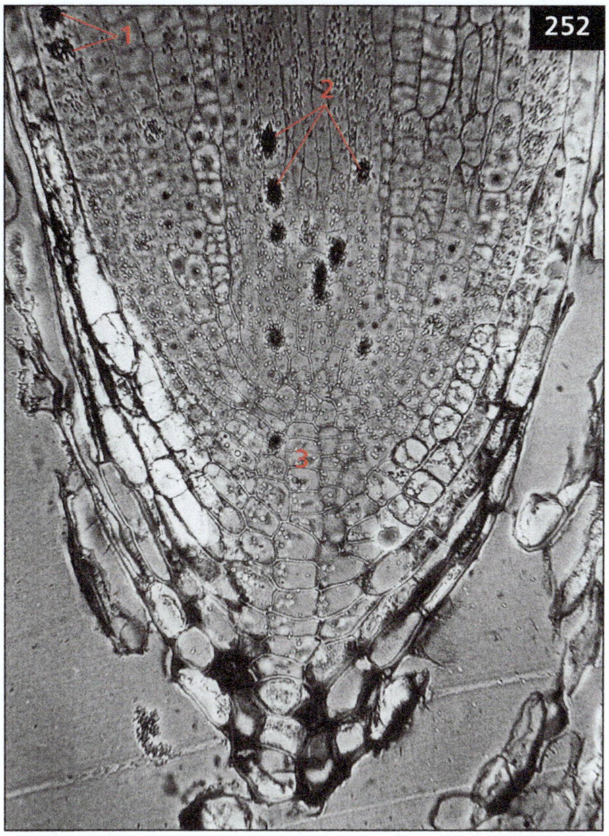

252 Autoradiograph of the root tip of the dicot *Comptonia perigrina*. In this nonleguminous nitrogen-fixing species the root nodules sometimes elongate into normal roots. Such a root was fed with tritiated thymidine, and the subsequent autoradiograph shows heavy labelling of the nuclei in the cortex (1) and the procambium (2). Quiescent zone (3). (G-Os, Phase contrast LM.)

1 Labelled cortical nuclei
2 Labelled procambial nuclei
3 Quiescent zone

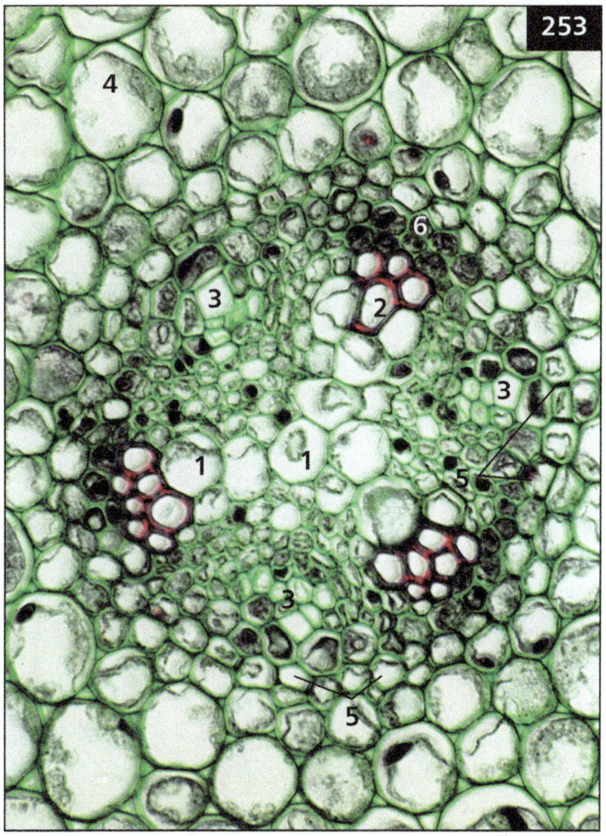

253 TS of the young root of the dicot *Ranunculus* (buttercup). Note the triarch arrangement of its xylem, with the wide, thin-walled cells in the centre representing differentiating metaxylem elements (1); at the poles the protoxylem elements (2) have already undergone secondary wall deposition and lost their protoplasts. Protophloem sieve tubes (3) have differentiated between the protoxylem poles. Cortex (4), immature endodermis (5), pericycle (6). (LM.)

1 Differentiating metaxylem
2 Protoxylem elements
3 Protophloem sieve tubes
4 Cortex
5 Immature endodermis
6 Pericycle

254

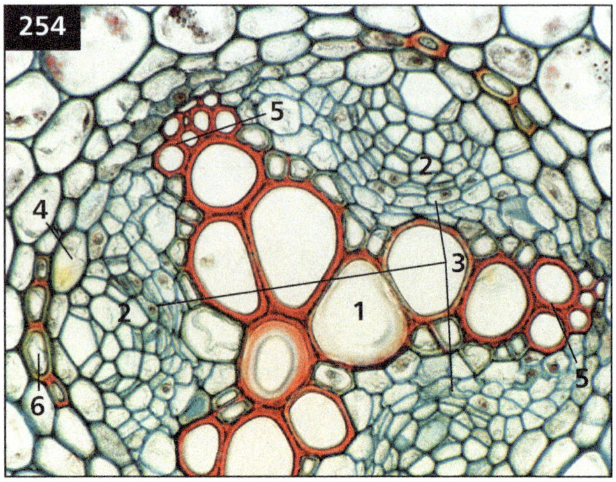

254 TS of the mature vascular tissue in the primary root of the dicot *Ranunculus* (buttercup). Wide, thick-walled metaxylem elements (1) have differentiated from the central procambium (cf. 253) and a number of metaphloem sieve tubes are also evident (2). The vascular cambium (3) occurs as three discontinuous arcs between the xylem and phloem. A single-layered pericycle (4) lies external to the protophloem and protoxylem poles (5). Endodermis (6). (LM.)

1 Metaxylem elements
2 Metaphloem
3 Vascular cambium
4 Pericycle
5 Protoxylem
6 Endodermis

255

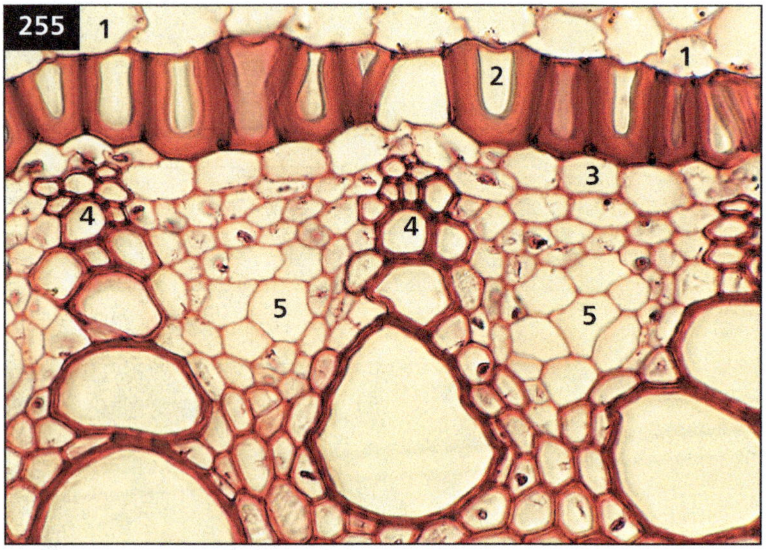

255 TS showing detail of the mature endodermis and vascular tissue of the root of the monocot *Iris*. In the cortex (1) both apoplastic and symplastic transport of water and solutes occur. However, movement across the endodermis (2) is symplastic through plasmodesmata in the outer tangential walls and across the protoplasts to plasmodesmata in the inner tangential walls. Pericycle (3), protoxylem (4), phloem sieve tubes (5). (LM.)

1 Cortex
2 Endodermis
3 Pericycle
4 Protoxylem
5 Sieve tubes

256

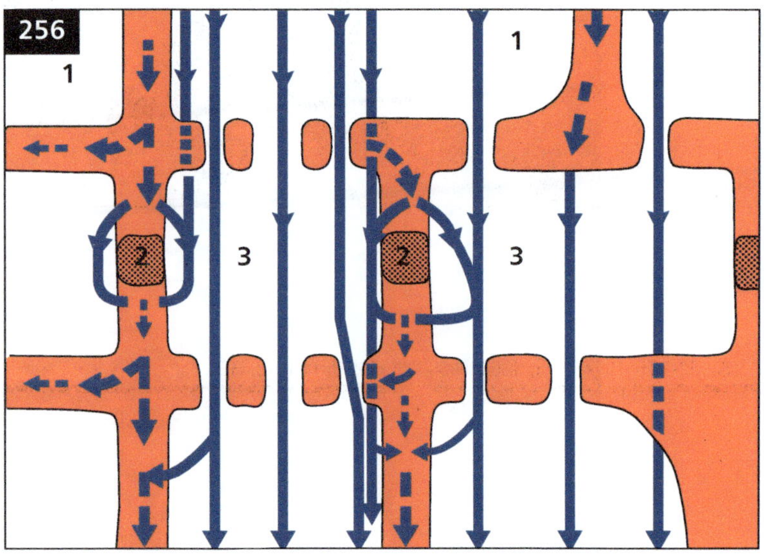

256 Diagrammatic representation of water and solute movement across the young endodermis. Movement from the soil through the cortex (1) is both apo- and symplastic (broken and solid blue arrows respectively). However, the impermeable Casparian bands (2) in the radial walls of the endodermis prevent apoplastic movement; water and solutes must traverse the protoplast (3) but can then move symplastically, via the plasmodesmata, or apoplastically into the vascular cylinder.

1 Cortex
2 Casparian bands
3 Endodermal protoplast

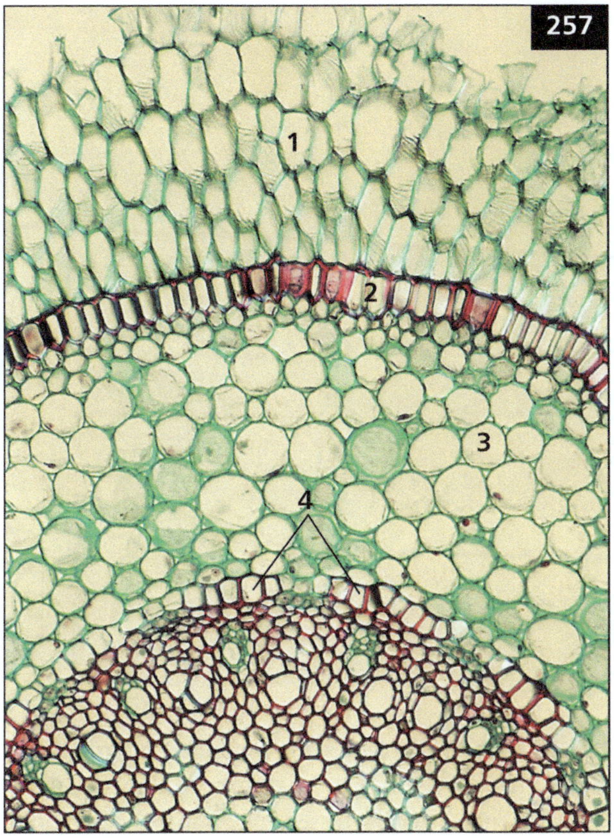

257 TS of the aerial root of the epiphytic orchid *Dendrobium* (a monocot). A multilayered velamen (1) arises from the epidermis; its dead cells are strengthened by bands of cellulosic thickening and this tissue may absorb water from the moist tropical air. Internally a single-layered, thickened exodermis (2) is present and all water entering the cortex (3) must first move symplastically across this barrier. Endodermis (4). (LM.)

1 Velamen
2 Exodermis
3 Cortex
4 Endodermis

258 Root tip of the orchid *Neofinetia falcata* (a monocot). Most epiphytic orchids like this produce aerial roots that dangle freely in moist air or run along the surface of their host tree. The shoot tip is green with chloroplasts, there are no root hairs, and the epidermal cells undergo periclinal divisions and programmed cell death, forming a white multiple epidermis known as a velamen.

CHAPTER 5
The Green Leaf

INTRODUCTION

Although the mature size and form of leaves is very variable (**27, 29, 135, 241, 259, 260**), they all have a common origin from the margin of the shoot apex (**213, 223**). During the life cycle of a flowering plant several types of leaf are generally developed, but only foliage leaves (**261–266**) are considered in detail in this chapter. In plants showing proleptic growth (Chapter 4), scale leaves often invest and protect the immature foliage leaves within the vegetative buds (**241, 268**) while bracts are associated with flowers and inflorescences (**267**). The cotyledons are leaves which become demarcated early in embryogenesis (**46**) and in the mature seed often contain reserve food supplies for the seedling on germination (**89, 104**).

MORPHOLOGY AND VENATION

The foliage leaf usually has a flattened lamina or blade (**259, 261–264**). In many dicots, and some monocots, the blade is borne on a petiole (**262, 270**) whereas in most monocots it is inserted directly on the stem (**266**). The base of the leaf sometimes encircles the stem (**35, 36, 218**) and a distinct leaf sheath occurs in many monocots (**269–271**). In dicots, paired stipules may occur at or near the base of the petiole and these are sometimes large and vascularized. In many grasses a pair of lateral auricles and a median membranous ligule are present at the base of the lamina (**271**). A median hastula is present in a similar position in fan-leaved palms (**272**). A simple leaf has a single blade which is often dissected or lobed. In a compound leaf (**29, 264, 272**) several to numerous leaflets occur and these may be dissected or further divided.

The leaf consists of dermal, ground and vascular tissues with the xylem usually orientated towards the original adaxial surface (**275**). In a transverse section many angiosperm leaves appear bifacial since the ground tissue of the leaf primordium (**224**) differentiates into the adaxial palisade and abaxial spongy mesophyll layers (**273**). Also, the upper and lower epidermis often differ in the thickness of their cuticle and frequency of stomata and trichomes (**273, 274**). By contrast some leaves are isobilateral; here the adaxial and abaxial surfaces are similar in organization but the xylem lies adaxially (**275**). In other species the leaves are unifacial with an investing epidermis of adaxial or abaxial origin and usually have either a cylindrical form (**265**) or are sword-shaped (**34, 266, 276**). However, the bases of such leaves are usually bifacial.

The sword-like blade of *Iris, Sansevieria*, and similar monocots is delimited by an epidermis of abaxial origin since the adaxial surfaces of the folded leaf have become confluent (**34, 276**). Two series of vascular bundles are often distinguishable in thicker regions of the blade and in both series the xylem lies remote from the surface (**276, 277**). Many monocot leaves have basal meristems from which new tissue is formed for a prolonged period (**260**); in thick-stemmed perennials such as *Dracaena, Cordyline,* and *Pandanus*, increasing width of the leaf base is accompanied by the differentiation of new veins within the derivatives of the basal meristem.

In most monocots the main veins run parallel to each other along the lamina (**278**) but eventually merge at the apex, while along the blade smaller veins, usually unbranched, connect the main veins (**37**). By contrast a complex reticulate venation characterizes the great majority of dicots and a few broad-leaf monocots (**262, 279–281**). Many dicot leaves possess a thickened longitudinal mid-rib from which secondary veins branch and often extend to the laminal margins, while tertiary and subsequent order veins form an anastomosing minor network (**281**) in which the finest veins terminate blindly in the mesophyll (**279, 280**).

In gymnosperms the leaf vasculature is variable; the cycads (with the exception of *Cycas*) and their fossil ancestors show a dichotomously branched, parallel venation in the leaflets (**282**); this also occurs in conifers with giant scale leaves (*Araucaria, Agathis*, **283**). In conifers with needle-shaped leaves a single, or sometimes paired, unbranched median vein occurs (**284, 285**). In some gymnosperms (e.g. *Cycas, Taxus*) files of thick-walled and elongate transfusion tracheids branch off from the main vein into the mesophyll (**284, 285**). Most ferns show dichotomous venation (**264**).

ANATOMY OF THE LAMINA

Leaf epidermis

In angiosperms the epidermis is directly derived from the outermost tunica layer of the shoot apex (**215**). In a few genera (e.g. *Ficus, Nerium, Peperomia*) the immature adaxial epidermis undergoes periclinal divisions to form a multiple epidermis, but it is not possible to distinguish this from a normal hypodermis in the mature leaf (**273, 274**). The epidermis is persistent in long-lived, evergreen leaves and it is normally coated externally by the relatively impermeable layers of cuticle and wax (**150, 286, 289**).

The epidermis always contains ordinary parenchyma-type cells together with the guard cells which surround the stomatal pores (**277, 287–289**). However, stomata are often confined to the abaxial surface in bifacial leaves. Additional cell types frequently occur, especially subsidiary cells associated with the guard cells (**290–292**) and various emergences and trichomes (**135, 327**). Chloroplasts are usually only present in the guard cells (**288**), while leucoplasts are normally present elsewhere in the epidermis. However, chloroplasts may differentiate in the ordinary epidermal cells of shade plants and also in aquatics (**293, 294**).

When viewed from the leaf surface the anticlinal (vertical) walls of ordinary epidermal cells are sometimes sinuous. In most monocot leaves the long axes of the epidermal cells are parallel to that of the leaf (**291**). In dicot leaves the cells are more randomly arranged (**292**), but over the mid-rib and main veins the epidermal cells tend to lie parallel to each other. In many xeromorphic grasses (that is, species adapted to dry habitats) large, thin-walled bulliform cells occur in the adaxial epidermis (**295, 296, 299**). These cells preferentially lose water and so contribute to the rolling of the leaf. In xeromorphic plants the stomata often occur in grooves or pits (**146, 297**) and epidermal trichomes are frequently present (**273, 296, 327**). Stomata are usually absent from the epidermis where it overlies the hypodermal sclerenchyma (**297, 300**) which often occurs in extensive tracts at the margins of leaves or associated with large veins. Stomata and subsidiary cells occasionally occur in clusters (**298**).

In bifacial leaves stomata are usually most frequent abaxially. On the other hand, in aquatics with floating leaves (**261**) they are usually confined to the air-exposed adaxial surface, while in submerged leaves stomata are generally absent (**293, 294**). Stomata are often randomly orientated (**292, 298**) but in elongate leaves the guard cells are usually parallel to the leaf's long axis (**291**). In grasses and sedges a pair of dumbbell-shaped guard cells surround the pore (**290, 291**), while crescent-shaped guard cells occur in most other species (**292**).

In crescent-shaped guard cells the anticlinal walls remote from the pore are relatively thin but the walls adjacent to the pore are often thickened (**288, 290**). The differential thickening of these walls, and the radial distribution of cellulose microfibrils in their periclinal walls, is apparently associated with the pore opening when the guard cells are turgid. In the dumbbell-shaped guard cells of grasses the ends are thin-walled while the side walls are thicker (**290**). An increase of turgor in the guard cells causes their ends to swell and the anticlinal side walls are pulled apart at the pore.

A cuticle is always present on the outer periclinal wall of a guard cell (**288, 289**) and wedges of cuticle-covered wall sometimes protrude towards the pore. In most species where subsidiary cells are present they do not differ from ordinary epidermal cells in their cytological appearance, but are normally distinguished by their shape and orientation relative to the guard cells (**290–292, 298**). Subsidiary cells may be derived from the same precursor cell as the guard cells (mesogenous development) or from neighbouring epidermal cells (perigenous origin): the two types are not homologous and their occurrence can be of systematic significance. It is not known whether subsidiary cells have a distinctive role in stomatal movement and they are apparently absent from leaves of a number of species.

Trichomes are common on leaves and stems; they range from unicellular to multicellular structures (**273, 296, 327**) and may be branched. Many are secretory and these often consist of a stalk and glandular head (**135, 218, 219**). Their secretions are diverse and vary from essential oils as in *Lavandula* (lavender) to hydrolytic enzymes in the leaves of carnivorous plants such as *Drosera* (**135**). In some glands the secretion may temporarily accumulate between the outer periclinal wall and the overlying cuticle. Several types of trichomes may be present on the same leaf (**218**). Nonglandular trichomes have several possible functions in the leaf, such as slowing the rate of transpiration by impeding wind movement over its surface, deterring insect attack, and protection against excessive sunlight.

The outer wall of the leaf epidermis is impregnated with cutin, while the cuticle lies externally (**286, 289, 301**). Cutin is a highly hydrophobic lipid polyester of high molecular weight. The cuticle is especially thick in xerophytes (**150, 286, 297, 301**). It can reach 6 mm in thickness and a thin deposit may sometimes be present on the anticlinal and innermost periclinal epidermal cell walls, especially when adjacent to the intercellular spaces of the underlying mesophyll. The cuticle is very thin or absent in submerged shoots of aquatics (**293, 294**). In bifacial leaves the adaxial cuticle is usually thicker than the abaxial (**302**). Wax is usually present on the surface of the cuticle where it occurs as a crust of filaments, granules, or flakes. However, the wax is often dissolved from the surface during the fixation and processing of material for examination under the microscope (see Chapter 3).

Mesophyll and sclerenchyma

In most bifacial leaves a layer of palisade cells occurs adaxially (**274, 302**). These highly vacuolate, cylindrical cells are anticlinally elongated and contain numerous chloroplasts in the peripheral cytoplasm adjacent to their walls (**71, 126**). There is often a well-developed system of intercellular spaces allowing gaseous diffusion through the apoplast in relation to photosynthesis and transpiration. The palisade tissue may be several cells thick and in isobilateral leaves usually occurs both adaxially and abaxially (**275**). Xeromorphic leaves often show a more compact mesophyll in which the intercellular spaces are reduced (**150, 295, 296, 325**).

In bifacial leaves a layer of spongy mesophyll occurs abaxially (**274, 289**). In this nonhomogenous tissue, with its large intercellular spaces, the total volume of apoplast usually exceeds the symplast. However, the surface area of wall in this tissue is often much less than in the palisade mesophyll.

In *Pinus* and several conifers the mesophyll cells are plicate (**146**); invaginations from the main wall protrude into the protoplast and thus increase the surface area of cytoplasm occupied by the chloroplasts. Many tropical grasses and various other taxa photosynthesize by the C4 pathway and these leaves show a 'Kranz' (wreath) anatomy, with the mesophyll cells radiating outwards from the bundle sheaths enclosing the veins (**303**). The chloroplasts within the sheath are usually agranal and commonly larger than the granal mesophyll chloroplasts.

The margins of many leaves are strengthened by tracks of fibres and prominent strands or girders of sclerenchyma may also be interspersed in the mesophyll (**150, 297, 300**). Additionally, sclereids may intrude within the mesophyll (**149, 301**). Sclerenchyma and collenchyma are also frequently present in the ribs developed over the major veins and the mid-rib (**302, 303**).

Vascular tissue

In the lamina of dicots the finest branches of the minor veins delimit the areoles (**279**), but in monocotyledons the minor network is less well defined (**37, 278**). Minor veins are embedded in a sheath of photosynthetic mesophyll (**126**) but the larger veins are often enclosed within ground tissue containing few chloroplasts (**277, 302, 304, 305**). The veins are typically collateral with the xylem normally located adaxially (**276B, 302**) but their orientation may vary (**304**). Limited secondary thickening may occur in the main veins of some dicot and conifer leaves (**284, 302**).

The minor veins (**279–281**) are concerned with the loading of sugars formed by photosynthesis into the phloem (**171**) and the unloading of water from the xylem into the mesophyll. Vascular parenchyma and transfer cells (**113**) are especially well developed in relation to the phloem and some represent companion cells. The minor veins are enclosed within compact bundle sheaths (**126, 280**) so that the vascular tissue is not normally directly exposed to the intercellular space system of the mesophyll. The sheath is often composed of parenchyma but sometimes a several-layered sheath is formed from parenchyma and sclerenchyma. The minor veins usually contain protoxylem tracheids (**279, 280**) but not vessels, while sieve elements are often lacking in the finest veins.

ANATOMY OF THE PETIOLE

In many leaves the vascular supply from the stem passes to the leaf blade via the petiole (**262, 263, 270**). The simplest vascular arrangement is a single, large crescent-shaped vein with adaxial xylem but several veins are often present (**306**). The peripheral ground tissue commonly contains collenchyma and sclerenchyma. The leaves of many plants lie extended in daylight (**307**) but undergo a slow drooping at night (**308**) in response to various external or internal stimuli. This movement results from loss of turgor in the extensive zone of peripheral parenchyma located in a joint-like thickening (pulvinus) present at the base of the petiole, or at the bases of petiolules in compound leaves. The more rapid response of the leaf of *Mimosa pudica* (sensitive *Mimosa*) to touch is also due to the activity of the numerous pulvini.

In both deciduous and evergreen species an abscission zone occurs at the base of the petiole (**309**). Here the xylem is often restricted to tracheids or narrow, often discontinuous, vessels (**310**). Such tracheary elements reduce the risk of embolism when the blade is wind damaged or eaten and also when the senescent leaf abscises. At the basal face of the abscission zone a suberized protective layer of cork develops which covers the scar (**311**) formed when the petiole becomes detached. In some perennial monocots several abscission zones occur, and in palms the leaf base may persist after the petiole and lamina are detached (**312**).

MODIFICATIONS OF THE LEAF

Leaves as insect traps

In members of several angiosperm families that grow in environments where nitrates and ammonia are in short supply (for example bogs and moors, **313**), various modifications of their leaves allow them to catch insects (**314–318**). Nevertheless, normal photosynthesis also occurs with stomata and mesophyll present in the leaf. Glands on the leaf surfaces of insectivorous plants (**135**) secrete proteolytic enzymes onto the insect prey which is then digested and the resultant soluble amino acids and nucleotides are absorbed by the leaf. In some genera, e.g. *Drosera*, *Dionaea*, and *Utricularia*, the leaf traps are active (the leaves move as they trap insects; **315**) but in others such as *Pinguicula*, *Nepenthes*, and *Sarracenia* they are passive (**314, 316–318**).

In *Drosera* the adaxial leaf surface is covered by long, complex multicellular hairs (**135**) whose glandular heads secrete a viscous, fragrant coating in which attracted insects become stuck. As the insect struggles, it touches other hairs and becomes more firmly held while the leaf blade bends inwards and the insect is securely trapped. The sticky fluid originates in a peripheral layer of densely cytoplasmic cells of the glandular head and migrates to the surface via numerous pores in the gland's cuticle. The gland contains a central file of tracheids which is separated from the secretory cells by an endodermis with Casparian bands of the type present in the young root (see Chapter 4).

In *Dionaea* (Venus fly-trap) the flat adaxial leaf surface bears several trigger-like hairs while the leaf margins have a number of long tooth-like projections (**313**). When an insect walks across the leaf and touches the trigger hairs, the two halves of the leaf spring together and become locked, sometimes for several weeks, by their interdigitating marginal teeth; meanwhile, the trapped insect is subjected to hydrolytic enzymes secreted by numerous short epidermal glands on the leaf surface. The mechanism for the 'springing' of the leaf trap is due to the almost instant loss in turgidity of motor cells situated on the upper side of the mid-rib which, in the flat 'unsprung' leaf, were distended with water. Later the motor cells again become turgid and the leaf resumes a flattened form.

Each pitcher-like leaf of *Sarracenia* and *Nepenthes* (**316–318**) and other such genera contains a digestive fluid secreted by glands on the leaf's inner surface. In *Sarracenia* insects entering the pitcher are trapped by the numerous downward sloping hairs at its entrance and by the smooth inner leaf surface, while in *Nepenthes* the rim and inside of the pitcher are very slippery. In *Pinguicula* small insects are trapped by the sticky secretions from glands on the adaxial leaf surface, while other glands secrete proteolytic enzymes. However, unlike *Drosera*, the leaf margins of *Pinguicula* (**314**) show little incurving subsequent to catching an insect prey.

Chimeral leaves

Differential patterns of colouring commonly occur in leaves, particularly in decorative foliage plants and may arise for a variety of reasons. However, variegated chimaeral leaves (**266**, **319**) usually originate from a nuclear mutation in the shoot apex which prevents derivatives of the mutated cell from developing chloroplasts. If the mutation occurs in the inner tunica or corpus, it may result in the formation of achlorophyllous tissue (**320**, **321**) in place of normal green mesophyll.

Chimeras may also arise from grafting and result in a composite plant consisting of tissues derived from two different species or even genera (**322**). The small tree +*Laburnocytisus adamii* originated in 1825 as an abnormal shoot arising from the grafting of a bud of *Cytisus purpureus* onto a stock of *Laburnum anagyroides*. The distribution of stomata and epidermal hairs on the leaves of this 'sport' (now known to be a periclinal chimera) closely resemble those of *C. purpureus*, but its tree habit is more like that of *L. anagyroides* than of the bushy *C. purpureus*. The chimera is conventionally designated as CLL; at the shoot apex T1 (the outer tunica layer) is derived from *Cytisus purpureus* (C) while T2 (the inner tunica) and corpus layers consist of *Laburnum anagyroides* tissues (L). The epidermis of the leaves and stem is derived from T1, while T2 and the corpus give rise to the leaf mesophyll, stem cortex and inner tissues. However, the chimera is not very stable and sometimes yellow *Laburnum* (LLL), purple *Cytisus* (CCC) and dingy red +*Laburnocytisus* (CLL) blossoms may all occur on the same tree (**322**, **323**). It seems that occasional periclinal divisions in T1 may lead to a T2 layer which is also (C), while periclinal divisions in this transformed T2 layer may eventually result in a nonchimeral shoot apex of CCC constitution (i.e. pure *Cytisus purpureus*); correspondingly reversions to a pure LLL shoot apex may arise from a periclinal division in the original T2 (L) layer displacing the T1 (C) component.

Succulent leaves

Many desert plants store water in succulent leaves, stems, or roots, using either primary or secondary tissues. The role of stored water and the site of storage vary depending on severity of the habitat: during brief droughts, water stored in leaves keeps the chlorenchyma hydrated and maintains photosynthesis. With severe drought, the leaves abscise and water stored in stems or roots sustains the buds and narrow vessels.

Succulent leaves differ from ordinary photosynthetic leaves in simple ways. They are thicker (**324**) and have more numerous mesophyll layers with larger cells, reduced intercellular spaces, and their vascular bundles may occur in two or more layers (**325**). In moderately thick leaves, the palisade mesophyll may be indistinct; instead all cells are chlorenchymatous and water is stored directly in photosynthetic cells, providing storage volume for carbon dioxide. Very thick, partially subterranean, leaves have an upper window-like epidermis and a chlorophyllous outer layer surrounding a transparent centre that transmits light with photosynthesis occurring underground (**326**). Perennial leaves (many monocots) tend to have fibres underlying the epidermis and ensheathing the vascular bundles but short-lived leaves (many dicots) have almost no sclerenchyma.

Monocot bulbs store carbohydrates and water in subterranean nonphotosynthetic succulent leaves with a uniform mesophyll of parenchyma cells of moderate size and ordinary intercellular spaces; but stomata are sparse.

Alpine plants and leaves

Alpine habitats are dominated by a small number of growth forms (the same is true for lowland habitats such as marshes, prairies, and seashores) but the following are common characters: the leaves are often small, thick, leathery, and perennial (**327**). Woody plants have a low or prostrate stature, keeping leaves and buds in the microhabitat near soil level which tends to be less windy or variable and more often protected by snow (**328**). Herbaceous perennials also frequently grow as rosettes keeping their leaves close to the soil level while their flower buds are protected by being nestled among the leaf bases at or below soil level (**329**). Tussock-forming grasses and sedges have a slightly higher stature, but they too keep buds, leaf basal meristems and living portions of leaves close to the soil; much of their tallest, exposed portions are dead leaf tips (**330**). Cushion plants are shrubs less than 1.0 m tall which are so highly branched that the small, thick leaves of any branch are crowded tightly against those of neighbouring branches (**331**). The cushion's canopy is typically such a tight, solid mass of leaves that wind is kept away from the trunk and branches; this is beneficial on cold winter days but actually prevents solar warming of the soil and roots on summer days when leaves are transpiring. Alpine plants often store starch underground in enlarged roots or thickened stems; this nutrient reserve allows them to begin growing quickly at the beginning of the brief growing season. Annuals are rare.

Alpine plants grow above the natural tree-line but below altitudes permanently covered by snow or ice. On mountains in far northern or southern latitudes they occur as low as a few hundred metres elevation but may grow above 4000 m on tropical mountains (the record height: *Saussurea gnaphalodes* [Asteraceae] occurring at 6400 m on Mt. Everest). Alpine habitats make up 3% of Earth's terrestrial surface area and have several universal features: short growing seasons, low air pressure, low amounts of carbon dioxide, and high evaporation rates. They may be cold, windy, cloudy, and dim but others have clear skies, intense sunlight, and UV irradiation. Some are extremely dry, others are wet, even sodden. Microhabitats add variability, some being warm on clear summer days, mild in winter if protected by a translucent layer of snow.

Because alpine habitats are extensive and diverse, plants from many families live above the tree-line. Between 8000 and 10 000 species in about 100 families occur in alpine habitats, some having adaptations that permit them to thrive there better than anywhere else, others merely being the few surviving individuals of species adapted to lower altitudes. Three very different families have the greatest number of alpine species—the Asteraceae (daisies), Poaceae (grasses), and Brassicaceae (mustards)—but the Caryophyllaceae (pinks or carnations), Cyperaceae (sedges), Rosaceae (roses), Ranunculaceae (buttercups), and Cactaceae (cacti) are also important floral constituents.

Anatomy of cactus spines

Cactus spines are examples of leaf structure modified to a function other than photosynthesis (**332**). Each spine develops from a leaf primordium initiated by an axillary bud (**333**), and cell divisions occur throughout it at first, causing it to become taller and wider. Cells at the primordium tip elongate into fibres, deposit secondary walls, lignify, and then die. Meanwhile, the basal meristem cells continue to divide, pushing new cells upward where they too elongate into fibres and die. A leaf trace runs just to the basal meristem but does not extend into it (**334, 335**). Once a spine reaches its full length, cell divisions cease and the basal cells also elongate and die. A mature spine consists only of dead, sclerified epidermal cells, without stomata, and mesophyll fibres; there is no palisade or spongy mesophyll, nor vascular tissue. Cells below the axillary bud differentiate into a corky base that prevents spines from being pushed into the shoot parenchyma by passing animals.

259 *Gunnera manicata*. The huge leaf blades of these plants (a native of South America but now cultivated besides ponds and streams as a garden plant) grow up to 2 m wide and are supported on stout spiny petioles that bear nodules containing the nitrogen-fixing cyanobacterium *Nostoc*.

260 *Xanthorrhoea preisii* (grass tree). The very long xerophytic leaves of this indigenous Australian shrubby monocot have newly emerged from a persistent apical bud; the bud is protected by surrounding leaf bases from frequent forest fires that ravage the dry regions of this continent.

261 Large floating leaves of the water lilies *Victoria amazonica*, which are capable of bearing the weight of a moderate-sized human child, are connected by long petioles to the rootstock growing in the mud on the bottom of the pond. The leaf stomata are confined to the adaxial surface which is exposed to the atmosphere, and the spongy mesophyll on the abaxial leaf surface has very extensive air spaces which assist with aeration. Large branched sclereids strengthen this tissue. Spines occur on the ribs of the abaxial epidermis, possibly deterring animal browsing.

262 Petiolate, simple leaf of the dicot *Populus deltoides* (cottonwood) has a prominent mid-rib, several lateral leaf veins, and numerous smaller veins not visible here. The leaf margin is notched rather than smooth.

263 Large fan-shaped leaf of the monocot *Licuala grandis*. This palm is indigenous to southeast Asia and Australia and bears simple leaves on long petioles.

264 Part of the compound leaf of the fern *Angiopteris*. Note the longitudinal mid-ribs in each pinna from which lateral, dichotomously-branched veins supply the lamina.

265 Hanging mass of the tropical epiphytic flowering plant *Tillandsia usneoides* (Spanish moss). This is rootless but the numerous fine, cylindrical leaves absorb water from the moist atmosphere.

266 Variegated plant of *Chlorophytum comosum* (spider plant). The elongated, sword-shaped leaves are typical of a monocot. Note the leafy plantlet (1) which has developed adventitiously on the inflorescence stalk.

1 Leafy plantlet

267 Inflorescence of the dicot *Digitalis purpurea* (foxglove). Each flower is subtended by a small green bract while five green sepals lie at the flower's base. The petals are united into purple bells, the spots of which probably act as honey-guides for visiting insects.

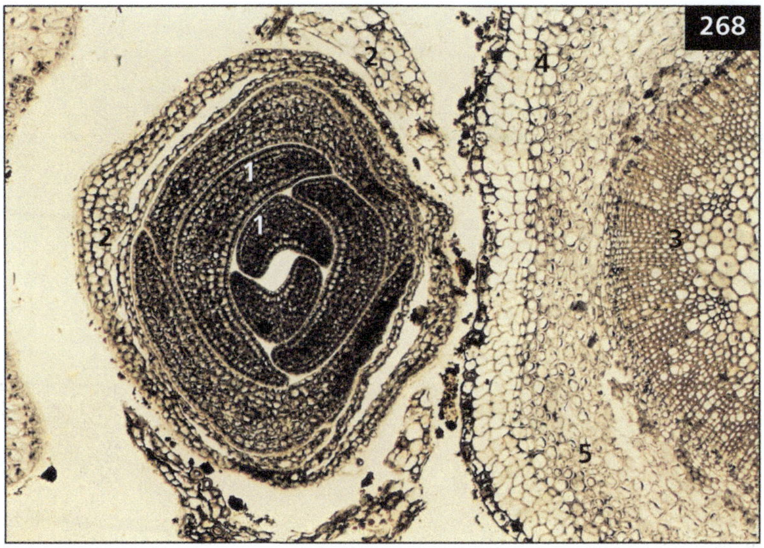

268 TS of the shoot of the dicot *Ligustrum vulgare* (privet) showing a dormant axillary bud. The decussately arranged foliage leaf primordia (1) are invested by scale leaves (2). The stem has some secondary xylem (3), and cork (4) has already developed. Cortex (5). (LM.)

1 Foliage leaf primordia
2 Scale leaves
3 Secondary xylem
4 Cork
5 Cortex

269 Stump of an old shoot of the monocot *Musa* (banana). The stem (1) is ensheathed by the bases of the large foliage leaves; the numerous parallel and longitudinally orientated fibre strands in the sheaths greatly strengthen the inflorescence axis which at maturity reaches several metres in length and bears at its tip a very heavy crop of bananas.

1 Stem

270 Trunk of the monocot *Archontophoenix alexandrae* (Alexander palm). Note the compound nature of the large leaf blade, the petiole (1) of which expands into a leaf sheath (2) encircling the trunk.

1 Petiole
2 Leaf sheath

271 At the junction of the blade of a grass leaf with its sheathing base (1), there is often a pair of lateral auricles (2) present. A median membranous ligule (3) projects from the upper surface of the junction and presses against the stem (4), preventing dirt and spores from being washed off the blade into the space between the sheath and the stem.

1 Sheathing base of leaf
2 Auricles
3 Ligule
4 Stem

272 A hastula (1) is a projection of tissue located on the upper surface of a palm petiole near its junction with the leaf blade.

1 Hastula

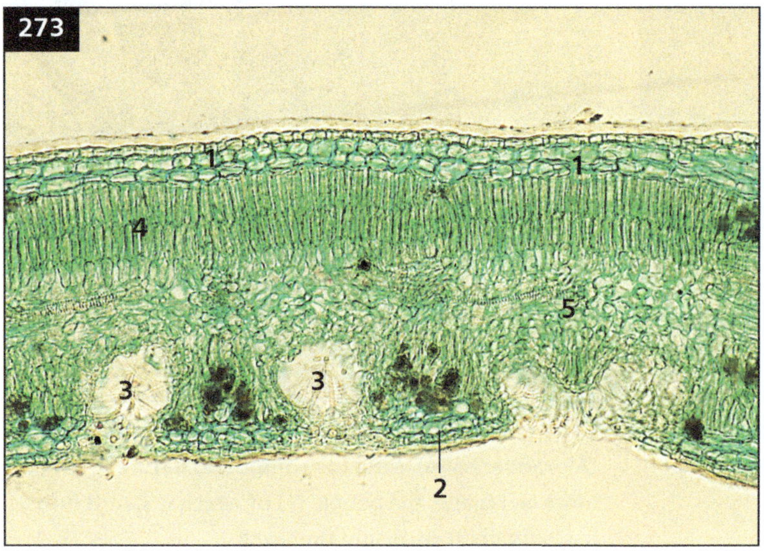

273 TS of the lamina of the bifacial leaf of the dicot *Nerium oleander*. In this xeromorphic species stomata are absent from the adaxial surface and the multiple epidermis (1) is covered by a very thick cuticle. The single abaxial epidermal layer (2) has a thinner cuticle and its stomata are confined to hair-lined crypts (3). The mesophyll is differentiated into adaxial palisade (4) covering a layer of spongy tissue (5). (LM.)

1 Adaxial multiple epidermis
2 Abaxial epidermis
3 Hair-lined crypts
4 Palisade mesophyll
5 Spongy tissue

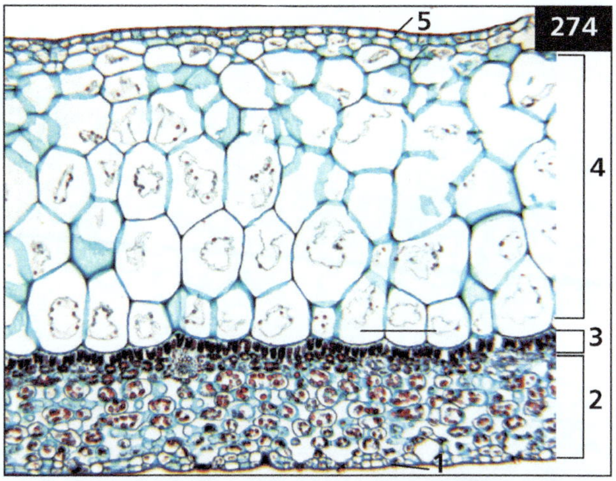

274 TS of leaf of the magnoliid *Peperomia*; this has an ordinary lower epidermis (1), spongy mesophyll (2), and palisade mesophyll (3). All the clear tissue constituting the upper two-thirds of the leaf is multiple epidermis (4): the large water-storage cells and what appear to be ordinary upper epidermal cells (5) were all derived from periclinal divisions in the protoderm. (LM.)

1 Lower epidermis
2 Spongy mesophyll
3 Palisade mesophyll
4 Multiple epidermis
5 Upper epidermis (part of multiple epidermis)

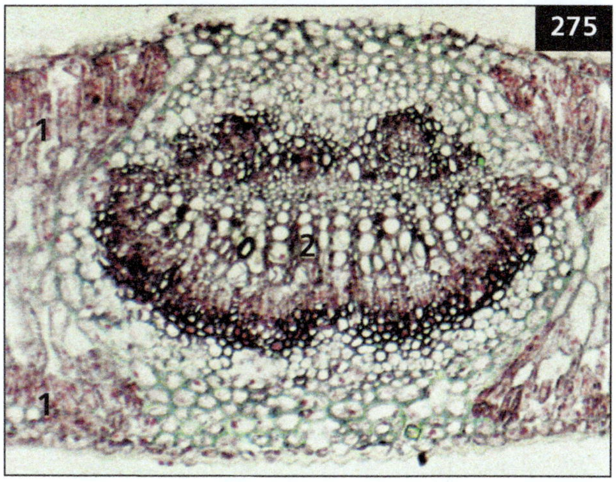

275 TS of the isobilateral leaf of the dicot *Eucalyptus*. Palisade mesophyll (1) occurs at both the adaxial and abaxial surfaces. The xylem (2) is adaxial. (LM.)

1 Palisade mesophyll
2 Xylem

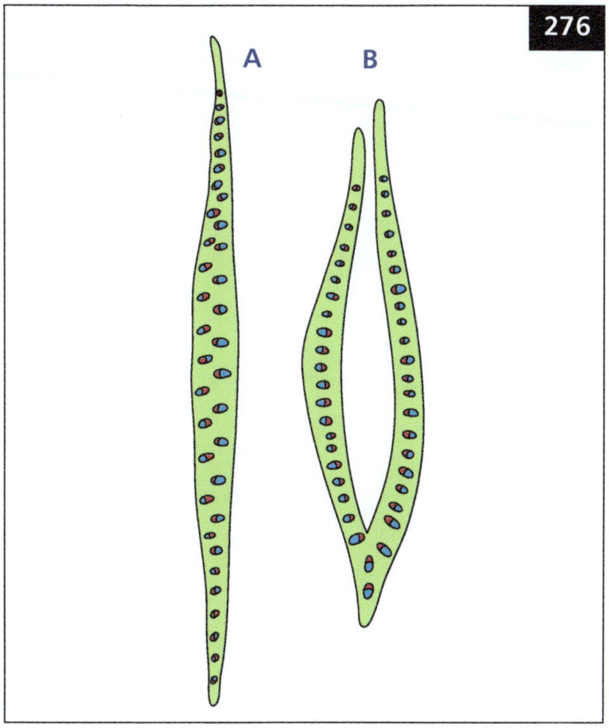

276 TS of the leaf of the monocot *Iris* showing its unifacial blade (A) and bifacial sheath (B). Note that in the numerous veins the phloem (blue) lies nearest the abaxial surface, while the xylem (red) faces the adaxial surface (B) or lies towards the confluent adaxial surfaces (A).

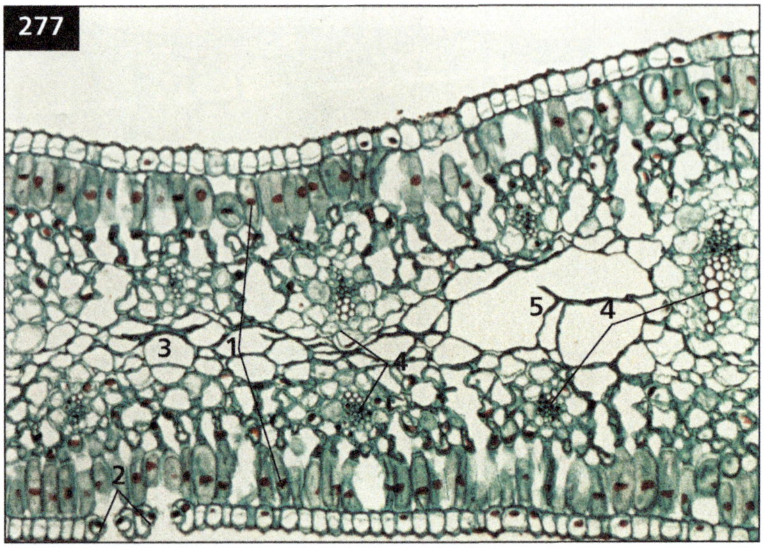

277 TS of the unifacial leaf blade of the monocot *Narcissus* (daffodil). Note the single layer of palisade mesophyll (1) underlying both surfaces, and the abaxial epidermis with numerous stomata (2). Within the central layer of nonphotosynthetic parenchyma (3) two series of veins occur, with the xylem (4) of opposed series facing towards each other (cf. 280). Mucilage cavity (5). (LM.)

1 Palisade mesophyll
2 Stomata
3 Nonphotosynthetic parenchyma
4 Xylem
5 Mucilage cavity

278 Leaf of the monocot *Hosta* showing its venation. The main parallel longitudinal veins are connected by abundant, obliquely transverse, branched commisures.

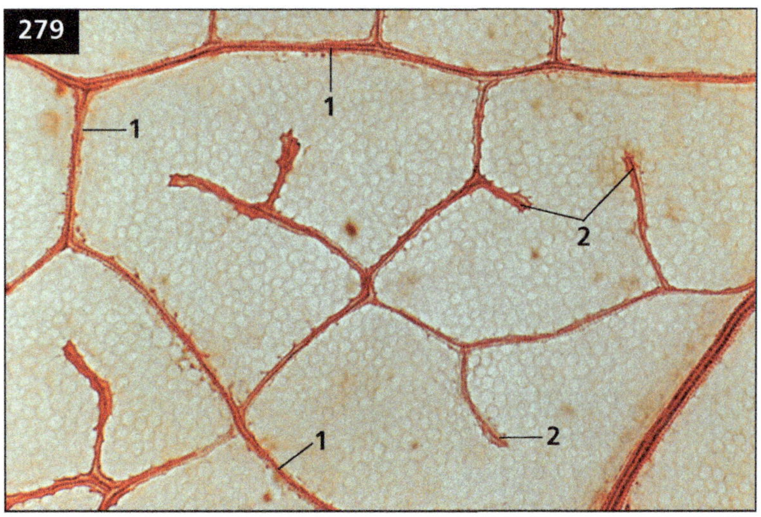

279 Cleared dicot leaf showing the xylem and its reticulate venation. Note the demarcation of areoles in the mesophyll which are enclosed by veinlets (1), the branches of which end blindly (2). (LM.)

1 Veinlets bounding an areole
2 Blindly-ending veinlets

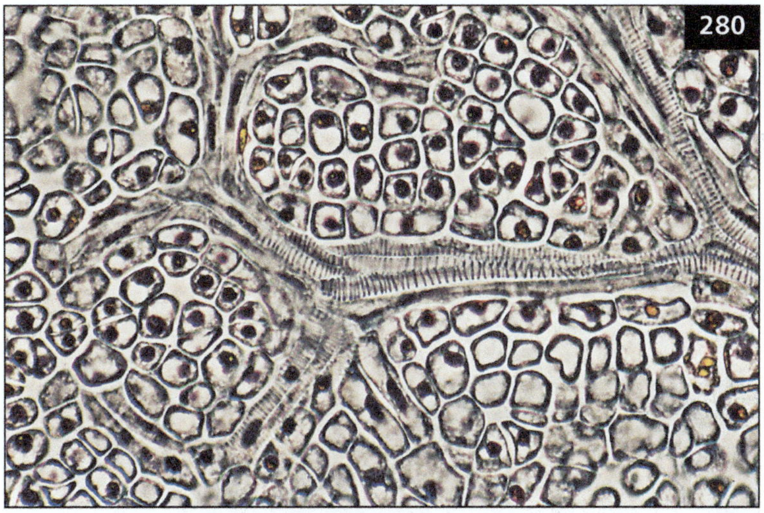

280 Section parallel to the surface of the leaf of the dicot *Ligustrum vulgare* (privet). Note the prominent protoxylem elements in the veinlets and the numerous air spaces in the mesophyll of the areoles. (Phase contrast LM.)

281 Leaf of the magnoliid *Magnolia* showing its venation. The pinnately arranged laterals are interconnected by smaller tertiary veins which further branch to form a reticulum.

282 Fossilized remains of a compound leaf of the cycad *Nilsonia*. Note the parallel venation of the leaflets; these veins show some dichotomous branching.

283

283 Giant scale leaves in the conifer *Agathis silbai*. Although most conifers have needle leaves or small scale leaves, a few have very large leathery scale leaves. (Photographed at the Montgomery Botanical Center.)

284

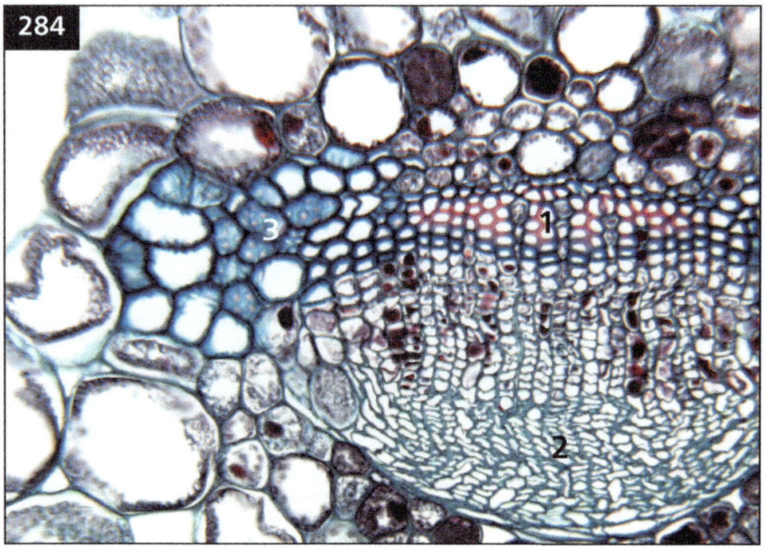

284 TS of the vascular bundle running down the centre of the leaf of the conifer *Taxus baccata* (yew), showing adaxial xylem (1), abaxial phloem (2), and peripheral transfusion tissue (3). The phloem is abundant because this leaf has a unifacial vascular cambium producing secondary phloem but not secondary xylem. (LM.)

1 Xylem
2 Phloem (mostly secondary)
3 Transfusion tissue

285

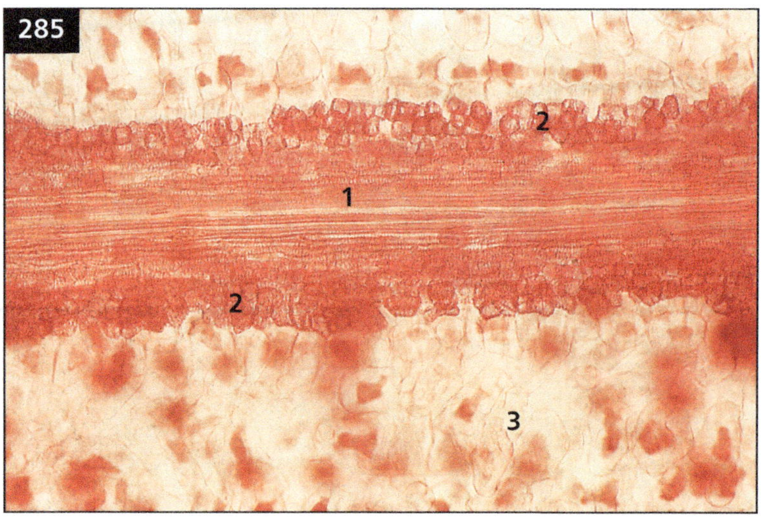

285 Cleared leaf of the conifer *Taxus baccata* (yew) showing detail of its midrib. Note the median vein (1) and the transfusion tracheids (2) on its lateral margins (cf. 284) which extend into the spongy mesophyll (3). (LM.)

1 Median vein
2 Transfusion tracheids
3 Spongy mesophyll

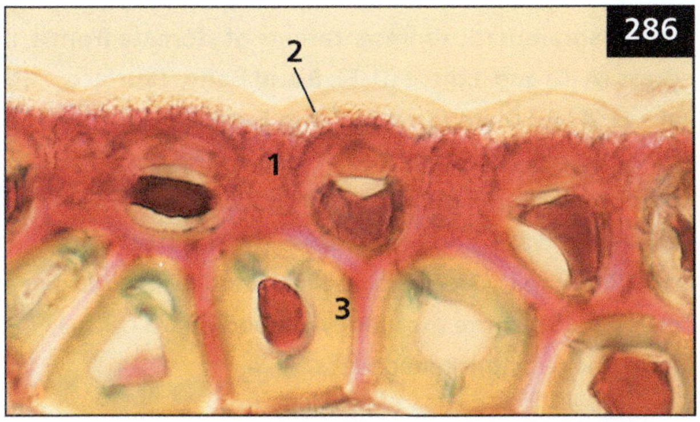

286 TS showing surface detail of the xeromorphic leaf of the conifer *Pinus monophylla* (pine). Note the thick-walled, lignified epidermal cells (1) which are coated externally by a thick cuticle (2) and the hypodermal sclerenchyma (3). (LM.)

1 Epidermal cells
2 Cuticle
3 Sclerenchyma

287 The ordinary epidermis cells of this desert-inhabiting dicot *Mesembryanthemum crystallinum* swell enormously and become bubble-like. Guard cells and stomatal pores are located in depressions between the giant cells.

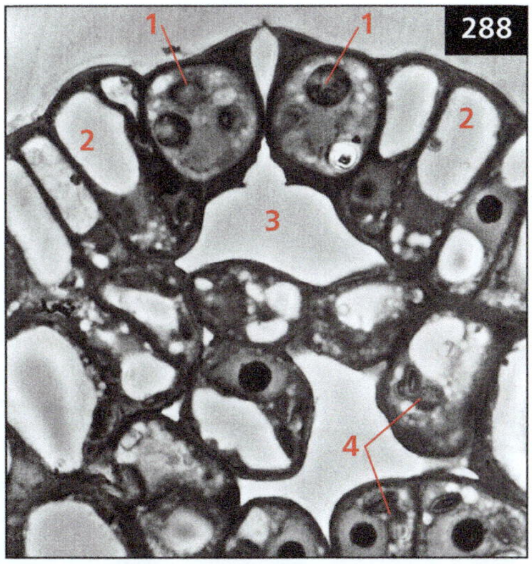

288 TS of the young stem of the dicot *Phaseolus vulgaris* (bean) showing detail of a stoma. The guard cells have dense cytoplasm and prominent amylochloroplasts (1) whereas the ordinary epidermal cells have large vacuoles (2) and are without chloroplasts. The anticlinal guard cell walls adjacent to the stomatal pore and the periclinal walls are thickened, and the outer periclinal walls extend into prominent ledges over the pore. Note the sub-stomatal space (3) and the chloroplasts (4) in the mesophyll. (G-Os, LM.)

1 Amylochloroplasts
2 Vacuoles
3 Sub-stomatal space
4 Mesophyll chloroplasts

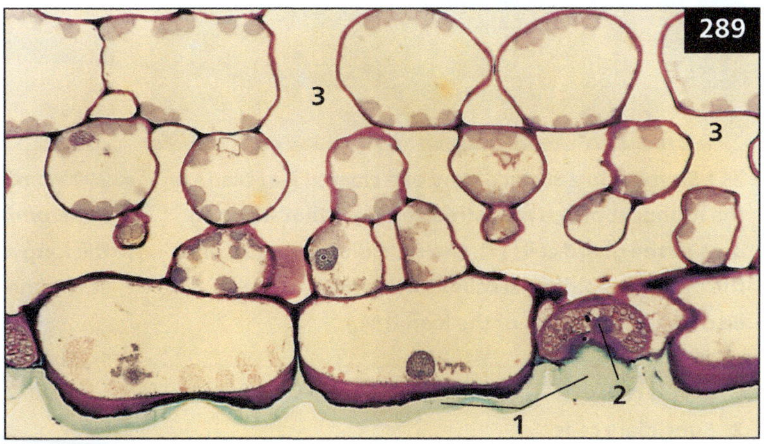

289 LS of the abaxial surface of the leaf of the monocot *Clivia miniata*. The epidermis is covered by a thick cuticle (1) which is also present over the guard cell seen in LS (2). The spongy mesophyll shows large intercellular spaces (3). (G-Os, LM.)

1 Cuticle
2 Guard cell
3 Intercellular spaces

290

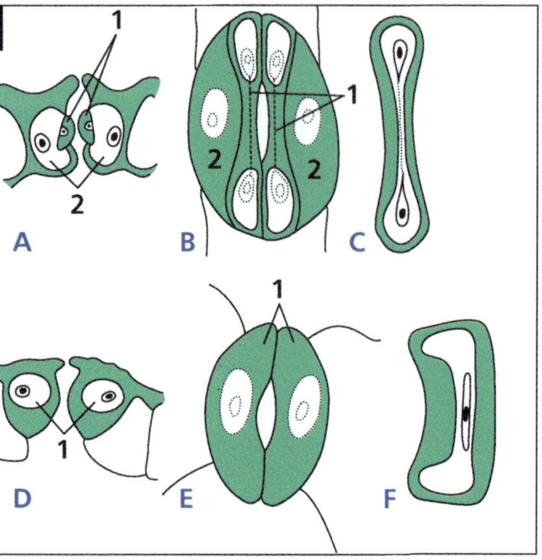

290 Diagrammatic representations of stomata from a grass (A–C) and a dicot (D–F). B and E show surface views, A and D are TS while C and F are cut longitudinally through a guard cell. In the grass the dumbbell-shaped guard cells (1) have unevenly thickened walls, and are dwarfed by the larger subsidiary cells (2). The dicot illustrated lacks subsidiary cells (cf. 292); note the kidney-shaped guard cells (1) with unevenly thickened walls.

1 Guard cells
2 Subsidiary cells

291

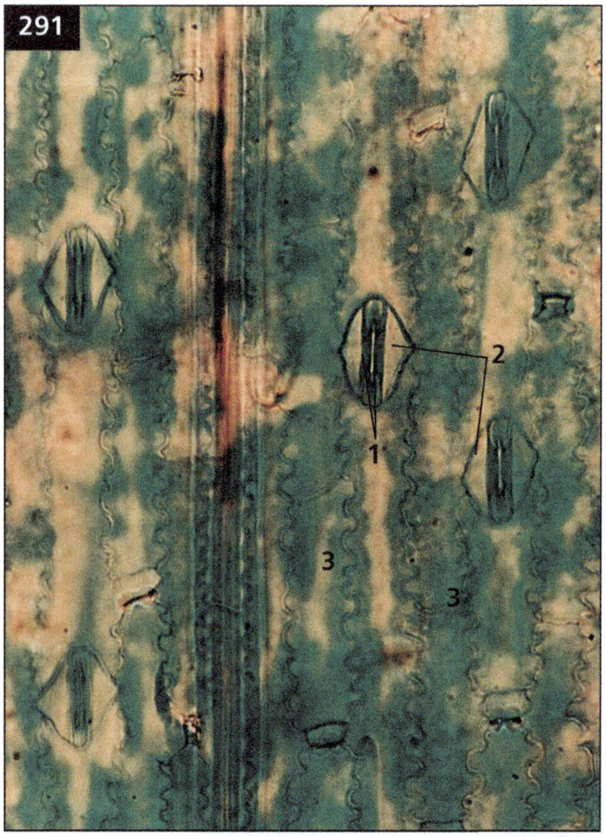

291 Stomata in a surface view of a cleared leaf blade of the monocot of *Zea mays* (maize). Note that the long axes of the guard cells (1), the subsidiary cells (2), and the ordinary epidermal cells (3, with sinuous walls) all lie parallel to the axis of the leaf. (LM.)

1 Guard cells
2 Subsidiary cells
3 Epidermal cells

292

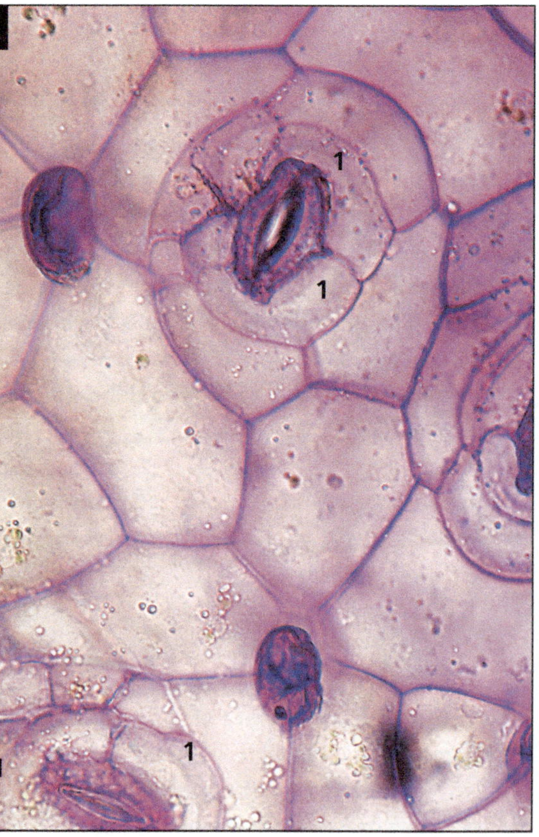

292 Abaxial surface view of the leaf of the magnoliid *Peperomia argyreia*. Note the subsidiary cells (1) and that the long axes of the stomata lie in varying orientations to each other (cf. 291). (LM.)

1 Subsidiary cells

293 Submerged leaf of the monocot *Potamogeton illinoensis*. The leaf has parallel venation with unusually prominent interconnecting veins, and the lamina between veins is both exceptionally thin and lacks stomata (cf. 294).

294 Close up view of submerged leaf of the monocot *Potamogeton illinoensis*, showing an interconnecting vein (1) and the lamina (2) which has no stomata. Although not detectable from this view, this lamina is only three cells thick: an upper epidermis, one layer of mesophyll, and a lower epidermis. All cells in all layers have well-developed chloroplasts.

1 Interconnecting vein
2 Lamina

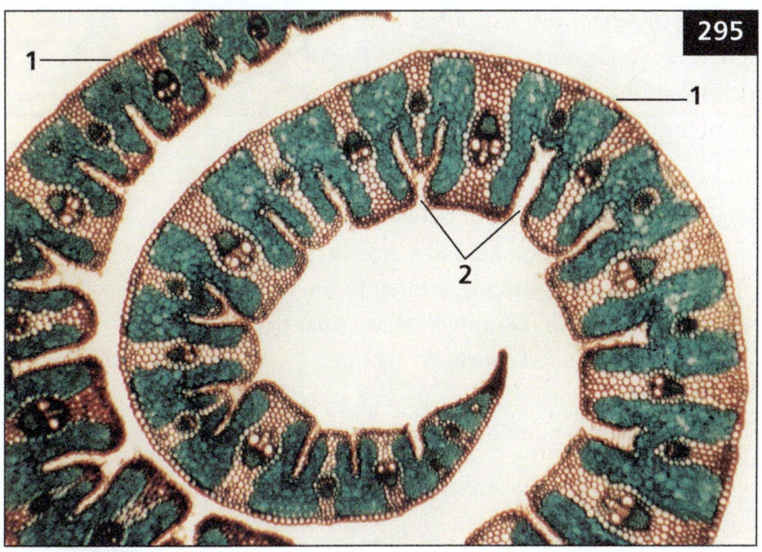

295 TS of the xeromorphic, bifacial leaf blade of the monocot *Ammophila arenaria* (marram grass). In life the leaf blade is variably rolled, according to the humidity of the atmosphere and availability of water, with the smooth abaxial surface (1) outermost, while the inrolled adaxial surface has longitudinal grooves (2, cf. 296). (LM.)

1 Abaxial surface 2 Adaxial grooves

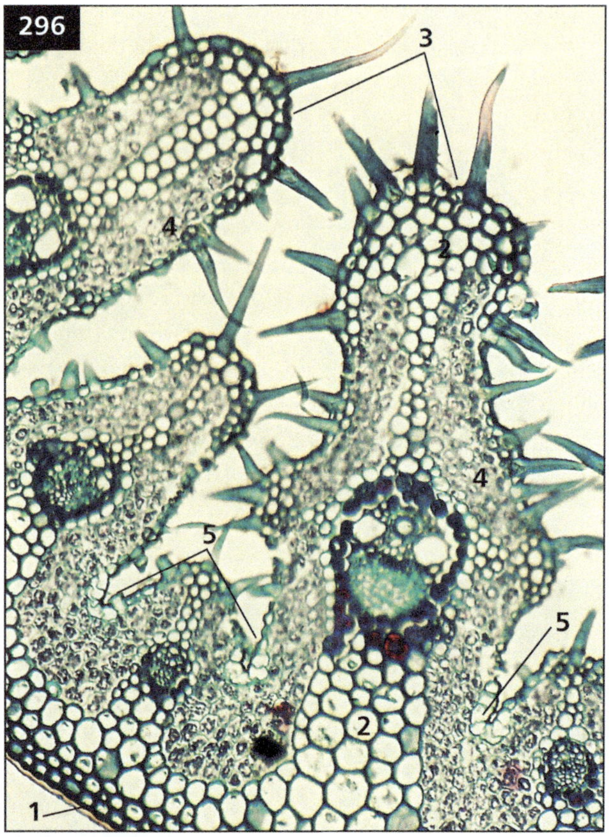

296 TS of *Ammophila arenaria* (marram grass) showing detailed structure of the lamina. The abaxial epidermis (1) of this monocot is covered by a thick cuticle and lacks stomata. A hypodermal layer of sclerenchyma (2) extends into the ridges (3) on the adaxial surface. The mesophyll (4) is confined to the adaxial surface, while numerous long hairs extend from the epidermis. The adaxial cuticle is thin. The stomata occur on the margins of the adaxial ridges and in the grooves conspicuous bulliform cells (5) are present. (LM.)

1 Abaxial epidermis
2 Sclerenchyma
3 Adaxial ridges
4 Mesophyll
5 Bulliform cells

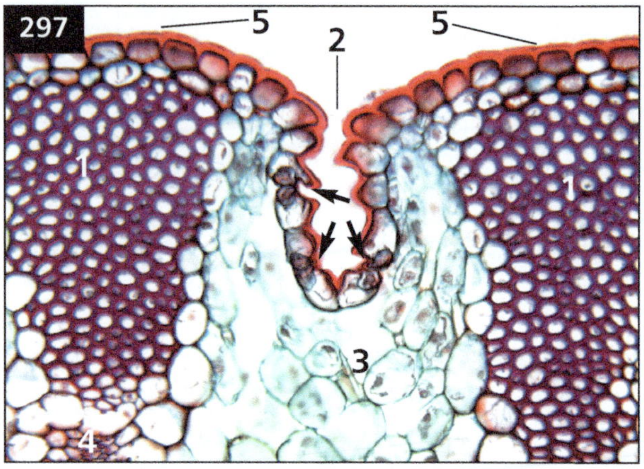

297 TS of leaf of the xerophytic monocot *Yucca* sp. There are no stomata where the epidermis is underlain by bundles of fibres (1), but between fibre bundles, the epidermis is invaginated as a groove (2) that has abundant stomata (arrows). Interior to the stomata is an aerenchymatous mesophyll (3). Vascular tissue (4), and thick, red-stained cuticle (5). (LM.)

1 Sub-epidermal bundles of fibres
2 Groove lined by epidermis with stomata
3 Aerenchymatous mesophyll
4 Vascular tissue
5 Cuticle

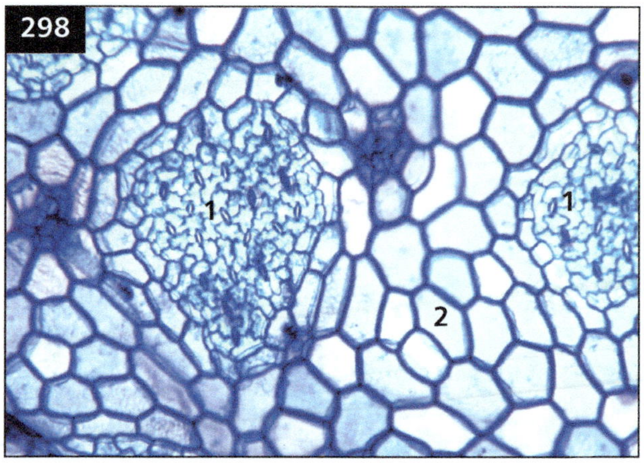

298 Epidermal peel of the abaxial epidermis of the dicot *Saxifraga sarmentosa*. In this species, guard cells and subsidiary cells occur in compact clusters (1), separated from other clusters by patches of ordinary epidermal cells (2) lacking stomata. (LM.)

1 Clustered stomata and subsidiary cells
2 Epidermis lacking stomata

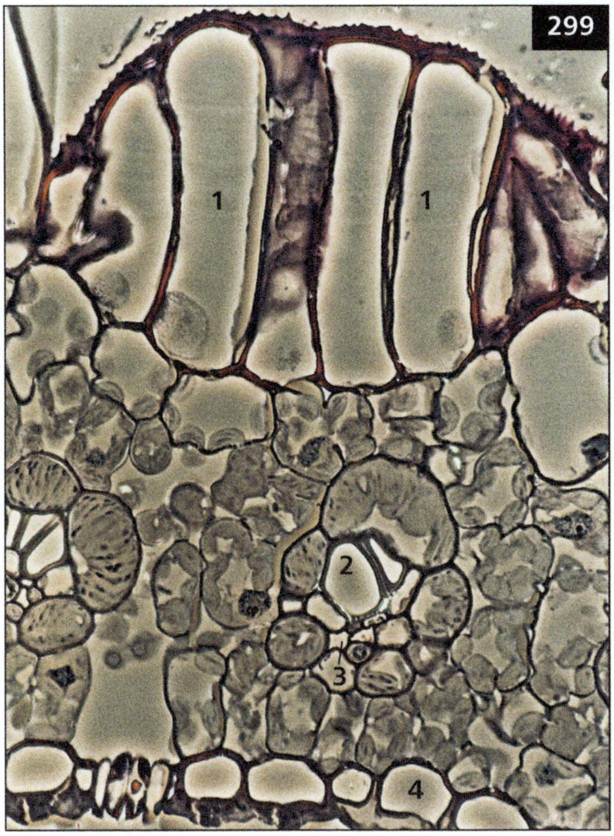

299 LS of bulliform cells (in a transverse section of the lamina) of the grass *Zea mays* (maize). The large, thin-walled and highly vacuolate bulliform cells (1) of this monocot are confined to the adaxial epidermis. When they lose water, they contribute to the rolling of the leaf. Tracheary elements (2), sieve tubes (3). (Phase contrast LM.)

1 Bulliform cells
2 Tracheary elements
3 Sieve tube
4 Abaxial epidermis

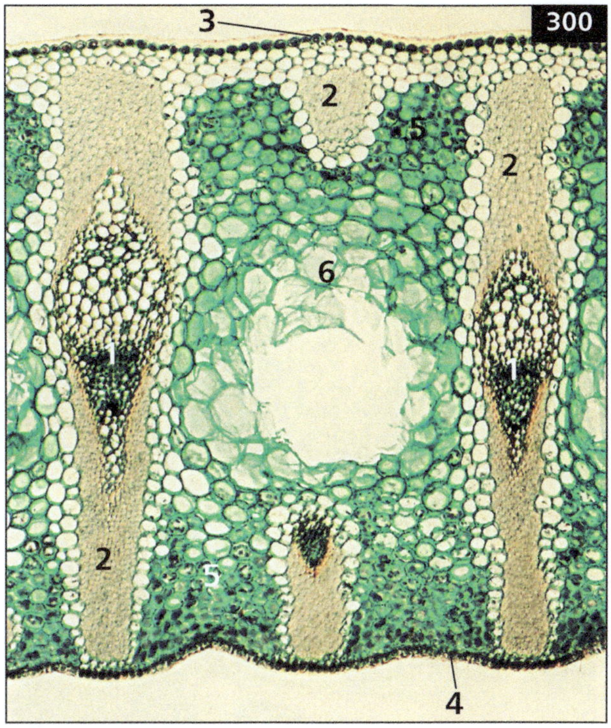

300 TS of the isobilateral leaf of the monocot *Phormium tenax* (New Zealand flax). The veins (1) are embedded in thick-walled, heavily lignified fibres (2) which form a series of 'girders' linking adaxial (3) and abaxial (4) surfaces. The fibres are used commercially for cordage and individual fibres may reach up to 15 mm in length. Mesophyll (5), nonphotosynthetic parenchyma (6). (LM.)

1 Veins
2 Lignified fibres
3 Adaxial surface
4 Abaxial surface
5 Mesophyll
6 Nonphotosynthetic parenchyma

301 TS of the xeromorphic leaf of the dicot *Olea europaea* (olive). Note the numerous thick-walled sclereids (1) ramifying in the mesophyll and the very thick cuticle (2) coating the adaxial epidermis. (LM.)

1 Sclereids
2 Cuticle

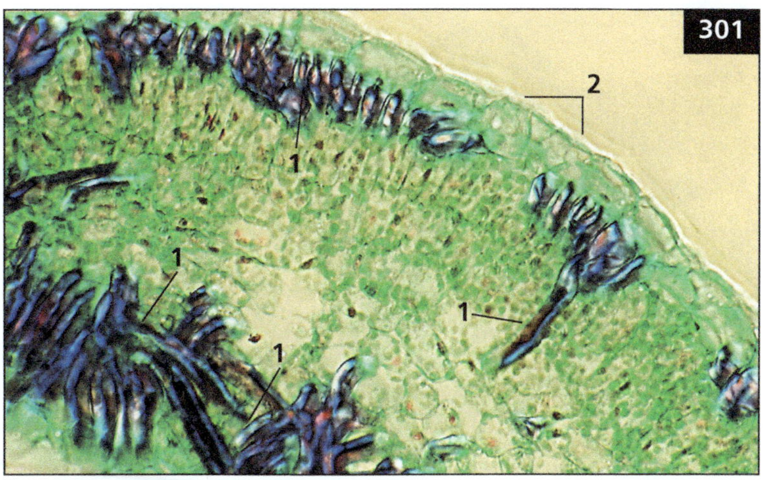

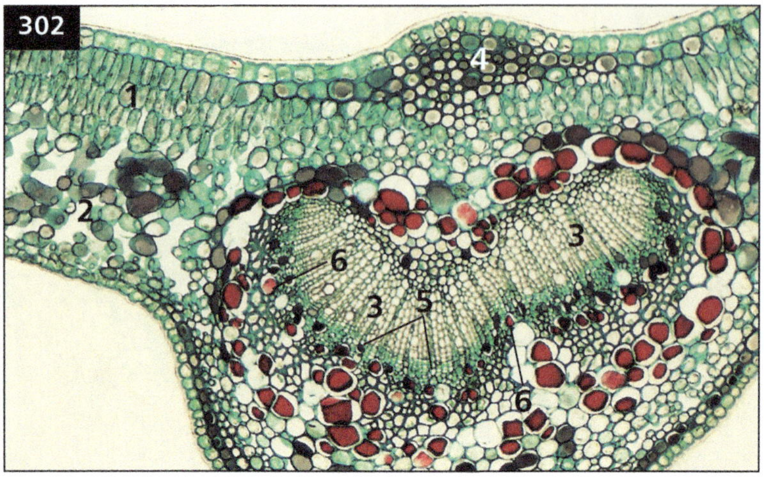

302 TS of the mid-rib of the bifacial leaf of the dicot *Prunus laurocerasus* (cherry laurel). The lamina consists of a compact adaxial palisade mesophyll (1) and an irregular spongy layer abaxially (2). Both surfaces are covered by a conspicuous cuticle. The mid-rib shows a single large vein with adaxial xylem (3), while a strand of collenchyma (4) causes a slight ridge on the adaxial surface. Cambial-like layer of parenchyma (5), phloem (6). (LM.)

1 Adaxial palisade mesophyll
2 Spongy layer
3 Xylem
4 Collenchyma strand
5 Cambial-like layer
6 Phloem

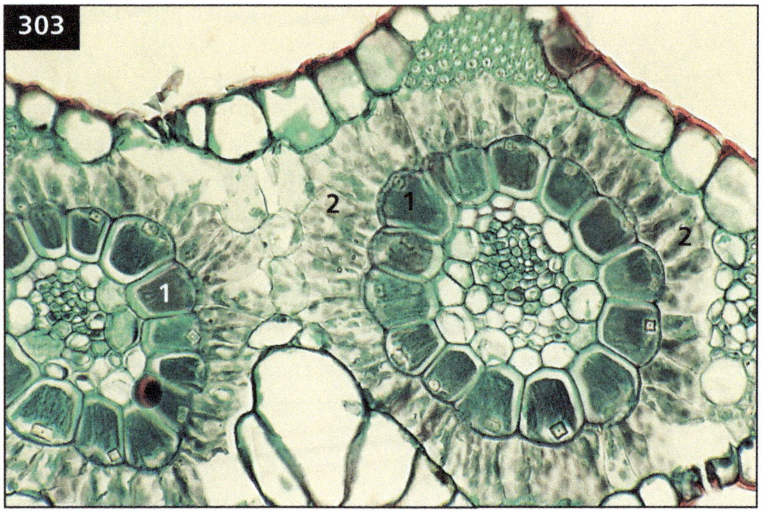

303 TS of the leaf of the grass *Panicum turgidum*. This desert species shows typical 'Kranz' anatomy, with the parenchyma cells of the prominent bundle sheath (1) containing large aggregated chloroplasts. The mesophyll cells (2) radiate out from the sheath and contain smaller, discrete chloroplasts. (LM.)

1 Bundle sheath
2 Mesophyll cells

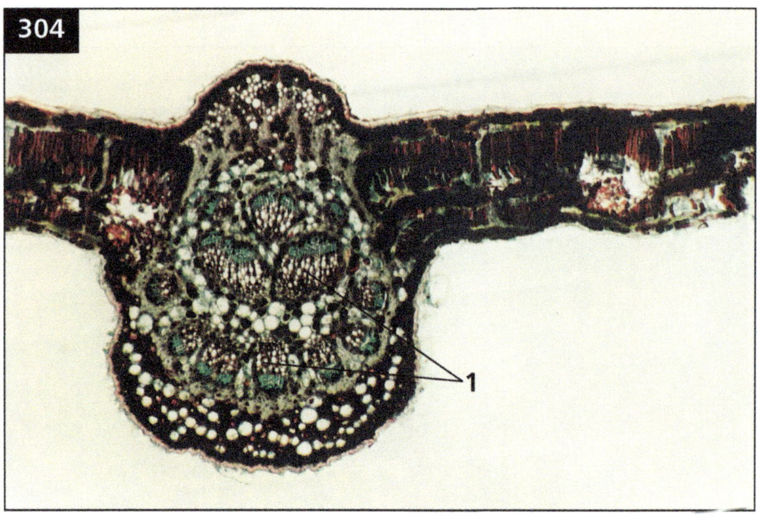

304 TS of the bifacial leaf of the dicot *Banksia*. In this indigenous Australasian genus the xeromorphic leaves are covered on both surfaces by a thick cuticle and much tannin is present in the mesophyll. In the mid-rib a number of discrete veins are present and the orientation of their xylem strands (1) is variable. (LM.)

1 Xylem strands

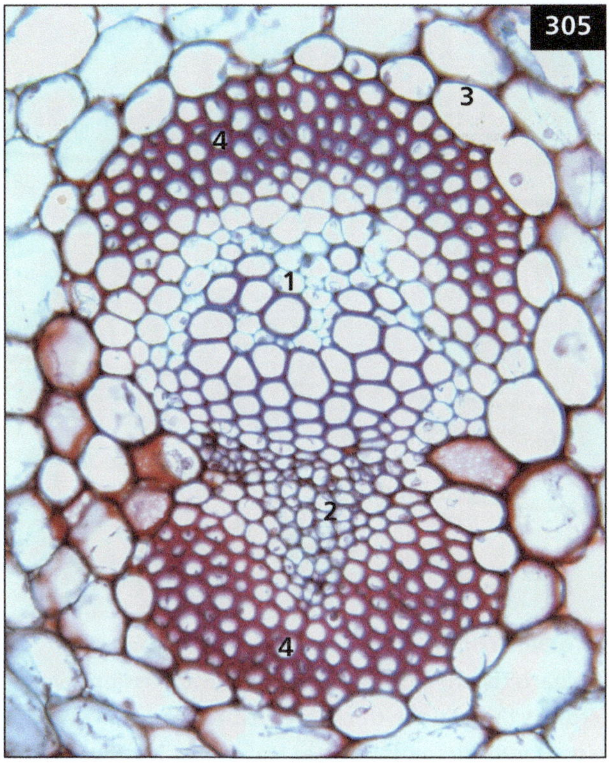

305 TS of a large vein from the leaf of the xerophytic monocot *Yucca* sp. Vascular cambium is lacking between the adaxially situated xylem (1) and abaxial phloem (2). The vein is partially separated from the mesophyll (3) by a sheath of fibres (4). (LM.)

1 Adaxial xylem
2 Abaxial phloem
3 Mesophyll
4 Fibres

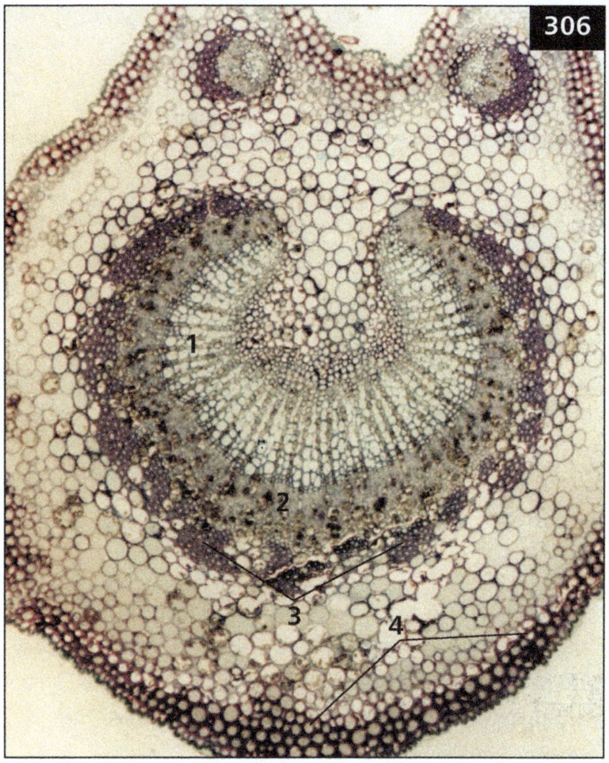

306 TS of the petiole of the dicot *Phaseolus vulgaris* (bean). A large crescent-shaped vein and a pair of small lateral veins are present. The extensive xylem (1) of the large vein lies adaxially while the phloem (2) is delimited by strands of fibres (3). An extensive hypodermal band of sclerenchyma (4) is evident and the remaining ground tissue is parenchymatous. (G-Os, LM.)

1 Xylem
2 Phloem
3 Fibre strands
4 Sclerenchyma

307, 308. Potted specimen of the dicot *Oxalis angularis* showing sleep movements. In the daytime (307) the three leaflets of each leaf are extended but they droop at night (308). This movement is caused by loss of turgor in specialized tracts of parenchyma tissue located in a joint-like thickening (pulvinus) situated at the top of the petiole just beneath the leaflets.

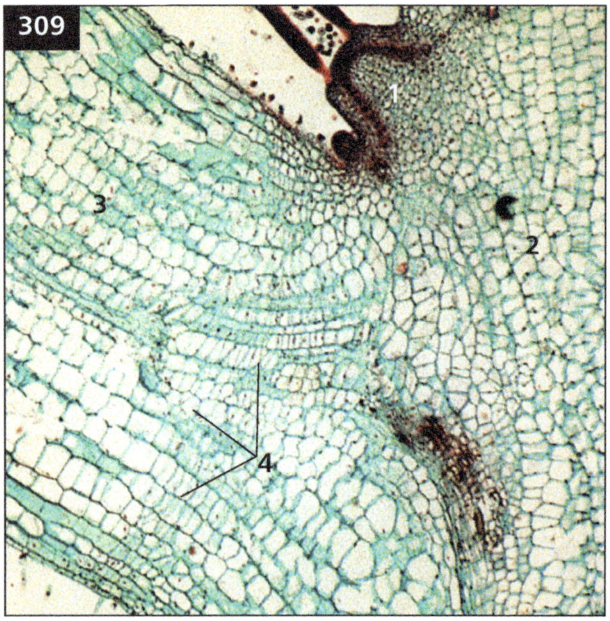

309 LS of the node of the dicot *Coleus*. Note the axillary bud (1) between the stem (2) and the leaf petiole (3); a well-developed abscission zone (4) runs across the petiole. (LM.)

1 Axillary bud
2 Stem
3 Petiole
4 Abscission zone

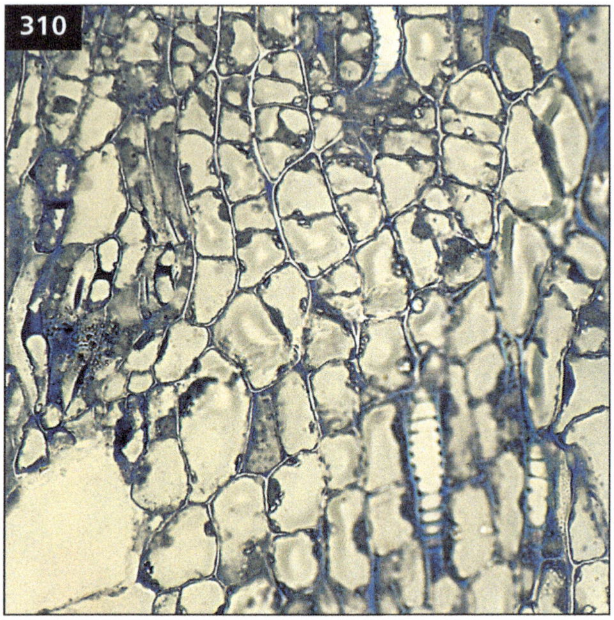

310 Detail of the abscission zone in the *Phaseolus vulgaris* (bean) cotyledon. Where the abscission layer crosses the vein the tracheary elements are narrow and apparently discontinuous; at leaf-fall such features probably reduce the danger of xylem embolism. The abscission zone cells are derived from parenchyma precursor cells by a series of transverse cambial-like divisions. (G-Os, LM.)

311 Detail of a corky leaf scar on the stem of the monocot *Monstera*. Note also the smaller vein scar (1).

1 Vein scar

312 Base of the mature trunk of the palm *Corypha elata*. Note the spiny leaf bases which persist for many years after the withered leaves (fronds) of this monocot have been removed.

313 Specimens of *Pinguicula vulgaris* (butterwort) and *Drosera rotundifolia* (sundew, arrow) growing together on boggy acid moorland in Cumbria, England. Both plants supplement their intake of nitrogenous and mineral substances by trapping and digesting insects.

314 Specimen of *Pinguicula vulgaris* in flower. Note the several small, trapped insects lying on the adaxial surface of the right-hand leaf.

315 Specimen of *Dionaea muscipula* (Venus fly-trap). This insectivorous species is endemic to the United States, in acidic, moist but freely draining soils of coastal pine savannas of North and South Carolina. If an insect contacts the trigger hairs on the adaxial leaf surface, turgid cells of the mid-rib rapidly lose water, the two halves of the leaf trap close, and the prominent marginal tooth-like hairs interlock, trapping the insect.

316 Specimen of *Sarracenia*. Several species of this insectivorous pitcher plant occur in North America on acidic soils which are often boggy or waterlogged.

317 *Sarracenia*. Close up of a pitcher showing the downward sloping hairs at its mouth, which encourage insects to walk downward, deeper into the trap. The inner surface is glossy and slippery, causing insects to lose their grip, fall into the trap's pool of water and drown.

318 Specimen of *Nepenthes*. A number of species of this insectivorous pitcher plant occur in various humid regions of India and southeast Asia. They grow either as tree epiphytes (as here) in mountain forests or on the ground in peat swamp forests.

319 Periclinal chimaeral leaves of the dicot *Pelargonium* (geranium). Their variegated appearance is due to a mutation in a single corpus cell at the extreme tip of the shoot apex; derivatives of such a cell contain proplastids which are incapable of differentiating into chloroplasts. The mesophyll of the leaf primordium develops from the inner tunica layer(s) and corpus but only the former are capable of giving rise to green cells: proliferation of this tissue at the leaf margins gives a green border to the leaf.

320 TS showing detail of a variegated bifacial leaf of the dicot *Glechoma hederacea* (ground ivy). In this chimaeral leaf the photosynthetic palisade mesophyll is reduced to a single layer (1) of squat cells while the hypodermal palisade layer (2) is devoid of chloroplasts. The abaxial spongy mesophyll (3) contains chloroplasts but is more compact than in the nonvariegated leaf (cf. 126). Note the absence of chloroplasts from both epidermises (G-Os, LM).

1 Photosynthetic palisade mesophyll
2 Hypodermal nonphotosynthetic palisade
3 Spongy mesophyll

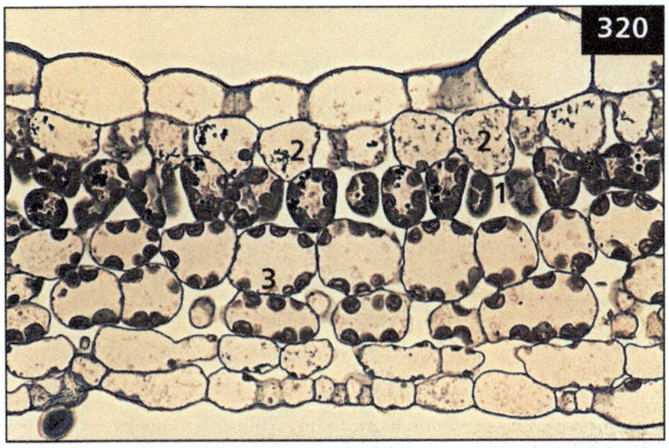

321 TS of the lamina of *Glechoma hederacea* (ground ivy, a dicot) through a pure white region of a variegated leaf; no chloroplasts are evident in the mesophyll and this tissue shows no differentiation into spongy and palisade layers (cf. 126, 320). (LM.)

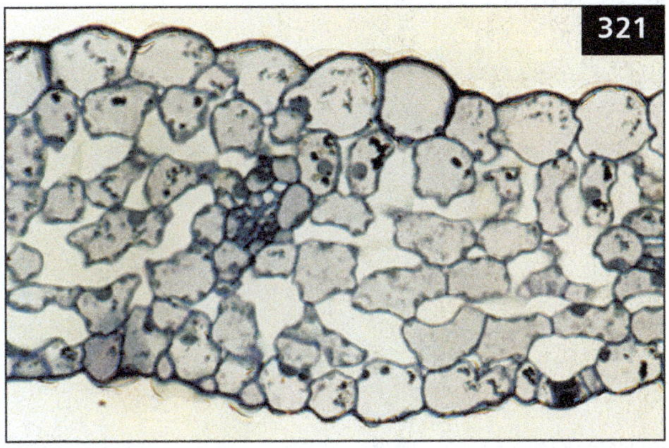

322 Specimen of +*Laburnocytisus adamii* growing in Glasgow Botanic Gardens showing tresses of yellow flowers of *Laburnum anagyroides* and purple-red flowers of *Cytisus purpureus* both borne on the same tree.

323 Details of the flowers of *Laburnum anagyroides* (left), *Cytisus purpureus* (centre) and *+Laburnocytisus adamii* (right) which were all borne on the same tree of *+Laburnocytisus adamii*.

324 *Pleiospilos nelii*. At any one time its shoots bear only two very thick, mature succulent leaves; the old leaves die as new ones begin to expand. The shoot and its apical meristem are located below ground but the mature leaves lie mostly above ground.

325 TS *Aloe vera* leaf section. This leaf, about 15 mm thick, has thick chlorenchyma (1) along both upper and lower surface, and its vascular bundles (2) occur throughout the mesophyll (3). Succulent leaves have few intercellular spaces, which reduces water evaporation and increases light transmission. (Light refracts as it passes repeatedly from cells to intercellular spaces in nonsucculent leaves, causing pith, cortex, and mesophyll [and similarly whipped egg whites] to appear white.)

1 Chlorenchyma
2 Vascular bundles
3 Mesophyll

326 *Haworthia cooperi* and several other genera have subterranean window leaves. All of the vegetative plant is subterranean except for the leaf tips. Their tips are transparent, the central mesophyll is glassy, and photosynthesis occurs in the chlorenchyma located below soil level.

327 *Espeletia*. The young leaves of this high altitude, fog-zone plant are extremely hairy and form a loose cluster around the shoot apical meristem; their hairiness keeps fog droplets away from the epidermis, preventing a film of water from occluding stomatal pores and from drowning the shoot apex. (Coastal Range, Venezuela.)

328 *Salix arctica*. This mature tree grows completely prostrate, appressed to the surface of soil or rock, covering an area about 1 m^2. Although its small, thick leaves project slightly into the air, all buds and meristems lie in the warmer, less windy zone right at soil level. A cover of mosses provides a bit more protection for the buds. (Rocky Mountains, Colorado.)

329 *Ranunculus glacialis*. This shows a common alpine growth form with all leaves clustered in a rosette on a shoot that does not extend above soil level. Its flowers are slightly elevated but form in relatively mild seasons. (Alps, Switzerland.)

330 Some alpine habitats are broad, relatively level plains with few obstructions to mitigate high winds. Tussock grasses such as *Stipa* shown here (growing in tight clumps rather than spreading by rhizomes) are common. Most leaf material here consists of their dead tips while photosynthesis occurs in the more protected basal leaf portions. Cushions of the cactus *Maihueniopsis* are also abundant here (arrows). (Andes, Argentina.)

331 *Azorella*. This cushion plant (growing with an alpine grass, arrow) forms a small upright tree, rather than being prostrate. However, it is so highly branched that the leaves of any twig are pressed firmly against those of its neighbours, forming an almost solid surface that cold winds do not penetrate. The tiny yellow dots are flowers. (Andes, Chile.)

332 *Epithelantha bokei*. These spines, modified leaves composed entirely of dead cells when mature, affect plant biology in more ways than just deterring large herbivores. They protect living tissues from sunburn and UV damage, retain transpired water near the shoot surface, and protect the shoot apex. Cactus spines perform little or no photosynthesis.

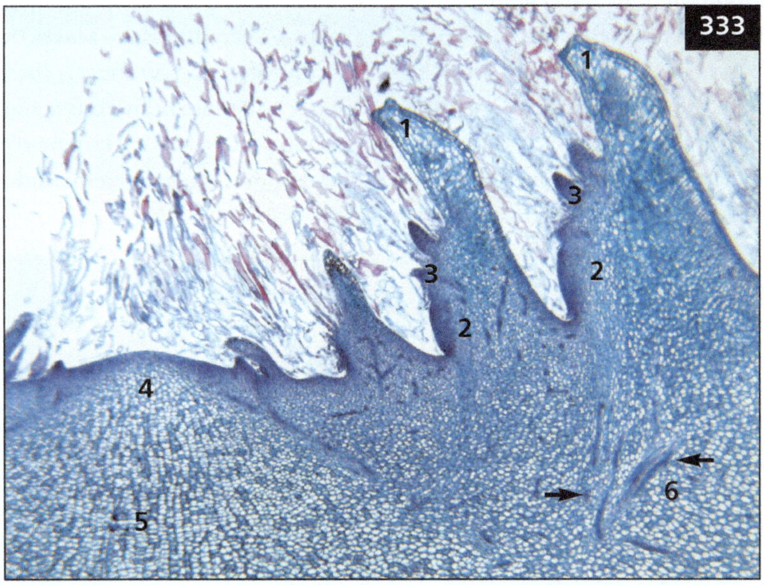

333 RLS of *Matucana* shoot apex, a cactus with microscopically small foliage leaves. Each leaf primordium (1) has an axillary bud meristem (2) that produces bud scale primordia (3) which develop into spines. Note the large shoot apex (4), the broad pith (5) developing from a wide rib meristem, and the thick cortex (6) forming from a broad flank meristem. Although microscopic, foliage leaves have chlorenchyma and vascular tissue. In the cortex, leaf/bud traces and cortical bundles (arrows) occur. (LM.)

1 Foliage leaves
2 Axillary bud meristem
3 Bud scale primordia (spine primordia)
4 Shoot apical meristem
5 Pith
6 Cortex

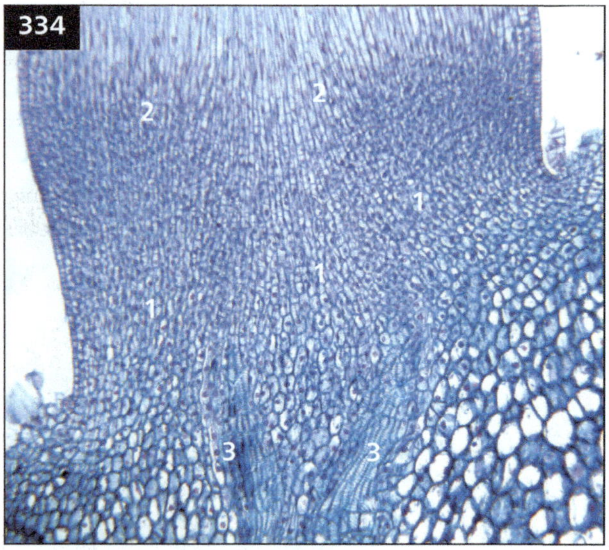

334 *Morawetzia doelziana*. RLS of basal meristem (1) and lowermost part of zone of elongation/fibre formation (2) of a growing spine. The basal meristem has a uniseriate protoderm, all other cells are formed from ground meristem. Almost all cell divisions are transverse, contributing to the spine's elongation, but a few longitudinal divisions cause the spine to be slightly tapered. A leaf trace (3) divides just at the base of the basal meristem but does not enter it. (LM.)

1 Spine basal meristem
2 Region where cells begin to elongate and differentiate into fibres
3 Leaf trace

335 *Gymnocalycium chiquitanum* shoot apex with young, developing axillary buds. Spine basal meristems are visible as the reddish basal region while fully mature, dead, fibres are black. The youngest axillary buds have just a few short spines, but older buds have longer and more numerous spines. After making spine primordia, one axillary bud meristem has produced floral primordia and an emerging flower bud (1) is now visible.

1 Flower bud

CHAPTER 6
The Stem

PRIMARY GROWTH

The aerial stem typically bears the green photosynthetic leaves and the reproductive organs (**26**, **336**, **337**), whereas underground stems are frequently perennating and food storage organs (**338**). Most unthickened stems are cylindrical, but ridged and rectangular forms (**339**, **340**) are common, and in a few species stems are flattened, leaf-like structures called phylloclades (**341–343**). In many plants of very dry habitats, leaves are reduced to scales and the stem's persistent, chlorophyllous cortex is the main site of photosynthesis (**344**, **345**, **347**, **362**). Succulent stems store water as well as carrying out photosynthesis, and are often protected by spines that are modified leaves in cacti (**362–364**), and modified axillary branches in *Euphorbia* (**347**, **368**). Starch is commonly stored in the parenchymatous ground tissue of the stem and is particularly abundant in the swollen stems of succulents and the underground stems of corms, tubers, and rhizomes (**338**). On the condensed shoots of rosette species (**226**) the leaves are crowded and the internodes short, but at flowering the internodes commonly become much more widely spaced as is dramatically shown in *Agave* (**337**).

ANATOMY OF THE MATURE PRIMARY STEM

The vascular system in the young internode usually consists of separate vascular bundles (**339**, **340**, **348**, **349**) that typically form a peripheral cylinder in dicots and gymnosperms (**41**) but are scattered in monocots (**40**). The cortex lies external to the vascular tissue and is bounded by an epidermis which often bears stomata and trichomes (**340**, **344**). The ground tissue in which the vascular tissue is embedded is basically parenchymatous and the cortex is often photosynthetic (**343**, **344**, **349**, **362–366**). In dicots a parenchymatous pith is usually present, but vascular bundles occasionally may be present centrally (**348**, **350**). In the majority of monocots the bundles occur throughout the ground tissue (**40**), but sometimes a pith is present (**339**). Sclerenchyma fibres are often present in the ground tissue (**339**, **344**, **348**) and the parenchyma may become lignified. Collenchyma frequently occurs just beneath the epidermis, especially at the angles of the stem (**340**). In some stems a prominent starch sheath occurs in the innermost cortical layer and in underground stems this may develop thickening to form an endodermis.

Vascular bundles in the stem are commonly collateral, with the phloem lying nearest to the epidermis and the xylem situated internally and on the same axis (**343**, **348**). Bicollateral bundles, in which the phloem lies both external and internal to the xylem, may also occur (**349**). In many monocots the bundles are amphivasal with a central strand of phloem surrounded by xylem (**351**). Amphicribral bundles, in which xylem is surrounded by phloem (**352**), occur in ferns and a few angiosperms, while in others the bundles may lack xylem. In the great majority of dicot stems, a cambial layer is located between the xylem and phloem (**343**, **349**) but in monocots this is absent (**351**).

The vascular anatomy at the node is more complex than in the internode due to the vascular traces that pass outwards from the axial vascular bundles to the leaves and axillary branches (**234**, **236**, **246**, **353**). Apart from branching at the nodes, the axial bundles normally interconnect with adjacent vascular bundles at various levels along the internodes. In monocots, axial bundles often run obliquely for some distance in the internode and have frequent interconnections (**354**) with numerous veins (leaf traces) passing outwards to each leaf. In dicots there are usually fewer leaf traces. In species with few interconnections between axial bundles, damage to one part of the axial system may severely disrupt the supply of water and nutrients to parts of the plant lying above or below the injury site.

In most stems the protoxylem and protophloem elements (**49**) are damaged during elongation and expansion growth (**20**), so that in the older primary stem (**128**, **184**, **348**, **349**) only the metaxylem and metaphloem are normally functional. Pericyclic fibres often develop in the outer procambium (**10**, **125**) and replace the crushed, isolated files of protophloem, while the protoxylem is sometimes represented by lacunae after the primary walls (**117**) of the tracheary elements become over-extended. In dicots the metaxylem vessels are frequently arranged in radial files separated by parenchyma or sclerenchyma (**128**). In monocots the relatively few vessels are usually wider and parenchyma or sclerenchyma often occurs between them.

MODIFICATIONS OF THE PRIMARY STEM

Aquatic (hydrophytes) and salt marsh/shoreline (halophytes) plants

Hydrophytes are those plants for which the habitats have standing or flowing water much or most of the time. Whereas many nonaquatic plants will survive only brief flooding, the bodies of most hydrophytes must either be partially or completely immersed for most of their lives; alternatively, the plants float freely on the water's surface, as does *Lemna* (**3**) and *Eichhornia* (**355**). Floating hydrophytes allow their roots to hang in the water and most have aerenchymatous floatation devices such as the swollen petioles of *Eichhornia,* which consist almost entirely of enormous intercellular spaces.

Anchored hydrophytes often have rhizomes buried below the water's surface (**356**). This affects several leaf features such as petiole length, which varies depending on species and water depth. Some species have short petioles and are always immersed (*Potamogeton*, **293**); petioles or leaf bases of others are held stiffly erect above the water's surface (*Typha*, *Sagittaria*); and water lilies allow their leaves to float on the surface of slowly flowing or still waters (**261**). Lamina position affects stomatal distribution: stomata have normal distribution on erect leaves, occur only on the exposed adaxial surface of floating leaves of water lilies, but are lacking in submerged leaves (surprisingly, immersed leaves typically have chloroplasts in their epidermis). Aquatic plants growing in rapidly flowing or disturbed waters often have elongate (**356**) or dissected leaves while those of calm waters may be huge and entire, such as the floating leaves of *Victoria amazonica* (**261**). Leaf support is typically at least partially supplied by buoyancy due to large air chambers, but sclerenchyma may also be present; for example the giant leaves of *V. amazonica*, which grow up to 2 m in diameter, are mostly aerenchymatous and buoyant, but they are also greatly strengthened by numerous prominently ribbed abaxial veins which radiate outwards from the insertion of the petiole.

Immersed organs exist in hypoxic environments and typically need supplemental oxygen. This diffuses down from leaves by broad air chambers present in leaves, petioles, and the rhizome itself (**357**). Rhizome vascular tissue is concentrated in a central cylinder with the xylem greatly reduced and restricted to annular or spiral tracheids while protoxylem lacunae are common (**358**).

Aquatic plants known as sea grasses total some 60 species of saltwater monocots. These mostly thrive in clear shallow coastal waters of warm-temperate to tropical areas. For example, in certain areas of the Mediterranean coast *Posidonia oceanica* grows profusely (**356**). The plants are anchored to the silty sea bottom by much branched rhizomes but the force of the waves often tears the plants free (**359**). Its flowers are inconspicuous and their thread-like grains increase the likelihood of successful pollination occurring in the water. Similarly, shallow brackish bays off the Parramatta river near Sydney, Australia, support a rich flora of the sea grasses *Zostera capricorni* and *Halophylla ovalis*; each hectare of this flora produces several tonnes of vegetation per year while each square metre of it evolves some 10 litres of oxygen a day.

Halophytes grow in saline terrestrial marsh or swamp environments. The soils are low in oxygen or even anaerobic when flooded, salty at all times. Tropical–sub-tropical mangrove communities are coastal species subject to regular inundation by high tides and have well developed aerenchyma that provides an oxygenation pathway (see Chapter 7). Many species of the Chenopodiaceae (goosefoots) are salt marsh herbs or shrubs. *Salicornia* is a typical example (**360**); its shoots are succulent and semi-translucent and its small fused, scale-like, leaves closely invest the stem (**361**). Palisade mesophyll is located peripherally while excess salts are stored in the vacuoles of its fleshy parenchymatous core—the plant is edible and tastes distinctly salty.

Shoot dimorphism in cacti

The succulent, spiny, 'leafless' bodies of many cacti appear exotic (**346**) but they have the same organization as any other seed plant. The key is that each cactus body is dimorphic, consisting of long-shoots (green and succulent) and short-shoots (the clusters of spines). The long-shoot apical meristem produces leaf primordia, nodes, and internodes, but its leaf primordia typically stop developing while still microscopic (**362, 363**). The axillary buds of the microscopic foliage leaves initiate their own leaf primordia, which become spines in cacti whereas in most other taxa these would differentiate as overlapping bud scales (**156, 333**). The bud meristem then either produces flower parts (**364**) or prepares to develop as a vegetative branch, then it becomes dormant (**232, 365**). The bud and its spines constitute a short-shoot (often called an areole in cacti). Most vascular plants have just long-shoots without short-shoots, so are simply called 'shoots'. The terms 'long-shoot' and 'short-shoot' are used only for plants in which some shoots differ from ordinary long-shoots, as in cacti, *Larix* (larches), or *Malus* (apples).

Cactus long-shoot anatomy is like that of most other plants, differing mostly in that the nodes and internodes become exceptionally broad, mostly due to an enlarged cortex although the pith may also be unusually wide (**366**). The surface is an epidermis with cuticle, stomata, and guard cells, but only rarely trichomes. The next several layers are often a tough hypodermis made of cells with very thick walls but this is absent interior to stomata. The cortex of many cactus long-shoots (but not short-shoots) has palisade chlorenchyma, cortical bundles, and collapsible cortex (**333, 369**). In the centre are collateral vascular bundles, with phloem and xylem, surrounding the pith. Bark formation is delayed indefinitely, allowing the epidermis and photosynthetic cortex to be retained and functional for years, even decades.

Cactus short-shoot anatomy is simple. The bud scales are spines, consisting of a central mass of fibres (but no vascular tissue or chlorenchyma) covered by sclerified epidermal cells without guard cells (**334**). The epidermis between spine bases mostly grows out as trichomes (**156, 362**) and there are no stomata. Short-shoot nodes and internodes are microscopically short and narrow while dormant. If the bud grows out as either a vegetative (**365**) or a floral (**364**) shoot, then cortex, stele, and pith develop. In one genus (*Neoraimondia*), short-shoots flower perennially, gradually elongating to have a more recognizable shoot structure.

Succulent desert plants

Succulent shoots allow desert plants to store water (**337**, **341**, **346**, **347**, **366**, **367**), but their structure varies with site of water storage (cortex, pith, wood) and whether the stems are also photosynthetic. If plants have persistent photosynthetic leaves, then succulent stems mostly just store water (**337**, **367**), but if leaves are ephemeral or vestigial then stems must both photosynthesize and store water (**333**, **347**, **368**). Nonphotosynthetic succulent stems differ little from ordinary stems, merely having a thicker cortex or pith, but photosynthetic succulent stems are more complex. They may have a long-lived epidermis providing stomata and a translucent surface for many years (**232**, **364**, **366**); outermost cortex cells may occur as rows of photosynthetic chlorenchyma (palisade cortex), and innermost cortex cells often have thin flexible walls that allow them to shrink easily and release water during drought before chlorenchyma, vessels, or axillary buds become stressed (**369**). Succulent stems are protected by spines or poisons: laticifers (**161**, **212**), tannin cells (**370**), raphide cells (**97**), or druses (**98**) are abundant in many species.

Cacti have a fundamental modification – cortical bundles – that hydrate the cortex and epidermis by mass flow through xylem, so the cortex is free to evolve to almost any thickness (**231**). Noncactus stem-succulent plants lack cortical bundles so their outermost tissues are hydrated only by diffusion: if the cortex becomes too thick, diffusion is too slow and outer tissues die.

A succulent stem changes volume as it absorbs or loses water but its surface is constant. Most have a pleated surface folded into a set of vertical ribs (**347**, **363–366**) or into helical sets of tubercles (cone-like projections): as a stem absorbs water, ribs swell, widen and push outward, then shrink as a stem loses water. Epidermis and hypodermis must be flexible at the apexes and bases of the ribs but can be tough and protective along the rib sides. Species with soft epidermis and hypodermis forego ribs and merely wrinkle as they lose water.

Because pith is confined by wood, it has little ability to swell when water is available or to shrink and release its water during drought. *Pachypodium* and some cacti have pith several centimetres in diameter, and some species have vascular bundles (medullary bundles; **350**).

Whereas cortex volume is limited by lack of cortical bundles in most species, wood is automatically vascularized and is often the basis of stem-succulence. Rays may be tall and wide with large, thin-walled parenchyma cells. Vessels may be surrounded by abundant axial parenchyma (**370**). Water stored within wood reduces the risk that vessels will cavitate and it can refill vessels once water stress is ended. Water stored in wood is also readily available to axillary buds by means of bud trace xylem. However, water stored within wood is not easily accessible to the outermost cortex except in cacti with cortical bundles, so wood-based stem-succulents often are trees with large, persistent photosynthetic leaves.

SECONDARY GROWTH

Most dicots and all gymnosperms undergo some degree of secondary thickening (**43**, **125**, **151**, **172**). The amount of thickening produced depends upon whether the mature plant is herbaceous (**371**, **372**) or arborescent (**4**, **172**). The fascicular vascular cambium develops from a narrow strip of procambium between the xylem and phloem which remain meristematic after the primary vascular tissues have matured. At the onset of cambial activity the divisions are normally localized within the individual vascular bundles (**41**, **128**, **371**). They then spread laterally through the adjacent interfascicular parenchyma cells so that a continuous cylinder of vascular cambium eventually results (**43**). The vascular cambium normally commences activity by the end of the first season's growth.

Anatomy of the ordinary woody stem

The vascular cambium consists of fusiform and ray initials (**373**, **374**). The cambium is often storied, with the fusiform initials arranged in approximately horizontal layers when viewed in tangential longitudinal section. However, in a nonstoried cambium the fusiform initials tend to be longer and their end walls taper more acutely than in the storied cambium. Fusiform initials give rise to the axial components of the woody stem: vessels, tracheids, fibres, sieve tubes, companion cells, and axial parenchyma cells. These may be storied or otherwise (**192**) according to the pattern of the cambium from which they are formed. The tangential (periclinal) walls of fusiform initials are wider than the radial walls (**373**). The initials divide tangentially, cutting off xylem centripetally and phloem centrifugally (**375, 376**). During the growing season the cambial initials are actively dividing. They are highly vacuolate cells and the expanding cell plate formed after mitosis is invested at its periphery by a prominent phragmoplast (**111, 373, 375**).

Each cambial initial produces radial rows of derivatives and in an active cambium a fairly wide cambial zone is apparent (**371**). In this zone tangential divisions also occur within the potential xylem and phloem elements (**375**). In gymnosperm and some angiosperm xylem the radial seriation is retained as the tracheary elements mature (**142, 185, 376, 377**). In most angiosperms this radial pattern is more or less severely disturbed by the maturation of vessels with very wide lumens (**201, 202, 371**). The ray initials are approximately isodiametric and they divide tangentially (**373**) to form the rays which run radially through the secondary vascular system (**376, 377**). As secondary thickening progresses, the increasing circumference of the stem is accommodated by the fusiform initials sometimes dividing radially (anticlinally) to form additional fusiform initials (**376**). They also give rise to new ray initials which form new rays to further supply water and nutrients to the expanding stem.

The expansion of the stem brought about by secondary growth is accompanied by various changes. The primary phloem and xylem cease to function in translocation and transpiration. The pith often remains more or less intact over a number of years, but the cortex is usually replaced by bark rather quickly (**336**); in succulent plants, cortex may persist for decades (**346, 347, 366**). The primary phloem and the early secondary phloem (**151, 377**) frequently become crushed by expanding xylem cylinder, and even in older stems the secondary phloem remains a relatively narrow layer (**172**). However, if axial fibres develop abundantly in the secondary phloem the older tissue may remain discrete (**165**). The intervening ray parenchyma cells divide periclinally so that the rays flare outwards due to dilatation growth, which accommodates the increasing circumference of the stem (**165**).

Anomalous secondary growth

Secondary growth occurs in either the ordinary pattern described above or as one of several anomalous types. An ordinary vascular cambium is single, bifacial, continuous, and composed of both fusiform and ray initials. It produces secondary xylem from its inner face, secondary phloem from its outer face, and both tissues consist of two systems, one of ray cells, the other of axial cells. Some secondary bodies appear anomalous at first glance but in fact the cambium is functioning in the ordinary manner. For example, wood may have alternating bands of parenchyma and fibres (**196**), or ray initials may outnumber fusiform initials greatly (**378**), but these are merely extreme forms of ordinary secondary growth.

However, evolutionary diversification has resulted in various types of anomalous vascular cambia and secondary growth:

Raylessness

The vascular cambia of some desert shrubs lack ray initials and produce rayless wood and phloem.

Unequal activity

All regions of an ordinary vascular cambium produce the same amount of wood, so trunks, branches, and roots are radially symmetrical (**172, 173**). There are ordinary exceptions: branches may have thicker wood on their lower or upper side (**205, 206**), and in roots the newly-formed vascular cambium follows the contours of the primary xylem and is initially star-shaped but becomes cylindrical by temporarily producing more wood in some areas (**427**). Radially symmetrical wood provides both increased conducting capacity and increased rigidity, but a plant may need flexibility instead. Thus unequal wood production in some tropical trees (*Ficus*, *Xylocarpus*) provides maximum buttressing strength with minimum use of wood (**409**). *Bauhinia divaricata*, *B. rubiginosa*, and *B. sericella* have ribbon-like, flexible stems (**379, 380**) because the vascular cambium cells on the two protruding sides undergo many periclinal divisions resulting in many broad vessels. At right angles to these two regions, the cambial cells undergo almost no divisions and the stem remains narrow. Between these four regions the cambial cells have intermediate rates of cell division; these also undergo enough anticlinal divisions to prevent the cambium from being torn apart as the wood accumulates differentially.

The rigidity of ordinary wood is counterproductive for woody vines that must withstand torsion (twisting). The vascular cambium of species of *Ambrosia*, *Passiflora*, and *Tinospora* has a type of unequal activity in which some regions produce ordinary fibrous wood, other regions produce axial parenchyma that fractures when twisted (**381, 382**). Wound bark forms next to the fractures and the stem soon has the appearance of a rope: irregularly interwoven strands incapable of columnar support but able to resist years of twisting.

Death of parts of the cambium

Faced with severe water shortage, some plants such as *Artemesia* and *Fumana* channel their small amount of available water to just a few strips of vascular cambium and vessels and several axillary buds. Most leaves die, followed by large segments of wood and vascular cambium but, when rains return, the surviving buds and cambia revive and grow. Wood is not produced as complete cylinders but rather as long, slender strips reflecting the pattern of vascular cambium survival (**383, 384**). The alternative—distributing water uniformly to all living cells—would result in the death of the entire plant.

Unifacial vascular cambia

In the long-lived leaves typical of conifers, unifacial, phloem-producing vascular cambia occur. Bifacial cambia are unnecessary because tracheary elements either function indefinitely, or are refilled if they cavitate, whereas sieve elements function only briefly and must be constantly replaced (**284**). Perennial leaves are rarer in dicots but some of these also have unifacial cambia. The long-lived, long-functioning cortical bundles of cacti have vascular cambia that produce large amounts of phloem and an occasional cell or two of xylem. These alternate between long periods of unifacial activity and brief episodes of being bifacial.

Very exotic unifacial vascular cambia occur in cross-vines (*Clytostoma callistegioides* and *Macfadyena unguis-cati* in the Bignoniaceae and *Machaerium purpurascens* in the Fabaceae). An ordinary cambium arises in young stems, but soon four vertical cambial strips become unifacial as they stop producing secondary xylem but continue producing large amounts of secondary phloem with prominent sieve tube members (**385, 386**). Because the four strips no longer produce secondary xylem, they remain stationary as intervening regions of ordinary bifacial cambia are forced outward by their own production of secondary xylem. The four unifacial cambia become buried within the wood, as are the new sieve tube members they produce: the actively transporting sieve tube members are well protected. Continued production of secondary phloem pushes old phloem outward, sliding past the stationary xylem. After a year or two, more strips of bifacial vascular cambium become unifacial.

Multiple vascular cambia

Most plants have a single vascular cambium that functions continuously except for dormant seasons. Several forms of anomalous secondary growth involve additional cambia, which may form in primary tissues (exterior to the primary phloem, around cortical bundles or leaf traces) or in secondary tissues produced by a preceding vascular cambium. They may co-occur simultaneously or instead may act successively. Because only one cambium can form in the ordinary position (as fascicular and interfascicular cambia), all others have an anomalous placement.

Simultaneous cambia. The storage root of *Beta* (and storage roots of many other taxa) grows rapidly because as its ordinary vascular cambium produces groups of vessels surrounded by a parenchymatous wood, new cambia arise around these vessels and they too produce parenchymatous wood within the pre-existing wood (**387, 432**). Such a storage root has one ordinary vascular cambium and multiple cambia with anomalous placement, all producing wood simultaneously. Many cycads produce vascular cambia in the phloem parenchyma produced by preceding cambia and all continue to function even after newer cambia have arisen.

Despite having concentric multiple cambia, cycads produce only small amounts of parenchymatous wood (manoxylic) with just a few tracheids.

Successive cambia are multiple cambia that are not present and active at the same time. In *Iresine*, an ordinary vascular cambium initially produces secondary xylem on its inner side and secondary phloem on its outer side but then it disorganizes. A new bifacial vascular cambium arises as a complete cylinder in the outermost secondary phloem parenchyma. The secondary xylem it produces is located exterior to the bulk of the secondary phloem of the first cambium. The phloem located between two masses of xylem is termed included phloem (not internal phloem). The second cambium too stops and disorganizes, then a third cambium arises in the former's secondary phloem and so on. These cambia are not active simultaneously (**388**). In *Bougainvillea* stem the vascular cambium is similar but only narrow strips of it become disorganized, not the entire cambium. New bifacial cambia arise in the outermost phloem of the disorganized regions, then join with the continuing parts of the ordinary cambium as it is pushed outward. Later, other strips disorganize. The included phloem here exists as narrow strips, not as complete cylinders (**389**).

THICKENED MONOCOT STEM

Primary thickening meristem in monocots

The majority of monocots are herbaceous and lack a vascular cambium, but in bamboos and some other species the stem is relatively wide (**390**) due to the activity of the primary thickening meristem. Here the cells are aligned in a transverse or oblique sheet and undergo periclinal divisions (**354**), with the internal derivatives differentiating into the axial vascular bundles and ground parenchyma (**228**). During early growth of most palms the internodes remain short, while diffuse growth and division within the ground parenchyma leads to the stem becoming progressively thicker. When the stem reaches its adult diameter (**312**), internodal elongation occurs and some species may attain great heights and life span. In palms and *Pandanus* (**39, 270**) diffuse growth continues throughout the stem due to persistent divisions in the ground parenchyma.

Cambial zone in monocots

In a small number of monocots (*Xanthorrhoea*, *Cordyline*, *Dracaena*, some *Yucca* species, and others; **33, 391**), cortex cells just exterior to the primary vascular bundles divide and form a cambial zone. Divisions are predominantly periclinal, resulting in a thick layer of new cells, most of which mature into secondary ground tissues. Narrow, vertical strands of cells undergo divisions in various planes and develop into secondary vascular bundles, each having xylem and phloem and surrounded by a sheath of fibres (**392, 393**). This cambium differs greatly from all others; the monocot cambial zone has little in common with the ordinary dicot type of vascular cambium and perhaps did not evolve from it.

PERIDERM

In gymnosperms and woody dicots the epidermis of the stem is normally replaced by the protective cork and associated tissues (**137, 142, 167, 172, 394, 395**). The outermost region of this periderm comprises the phellem (cork, **43**) derived from the phellogen (cork cambium). This is a meristematic layer of tangentially flattened cells which commonly arises hypodermally. It may also be of epidermal origin (**396**) or may form deeper in the cortex. In some species the cork cambium cuts off a little parenchymatous tissue (phelloderm; **397**) internally, and it may produce sclereids externally. Cork cells often retain their shape but they can collapse at maturity (**398**). The cork consists of radially aligned, tightly packed cells; they are dead and have thickened walls which are suberized and impermeable. Periderm also forms over wounded surfaces (**310, 311, 399**) and occurs in thickened stems of some monocots (**400**). Cork cambium is usually equally active in a stem or root, producing rather uniform bark. However, cork cambium may be formed precociously in some areas of a shoot but be delayed indefinitely in adjacent areas, producing tall, thin 'wings' of cork or elevating spines on mounds of cork (**401, 402**).

Commercial cork is harvested from *Quercus suber* (**136, 137**) and is stripped off the trees in cycles of about 10 years (**395**). In this species each phellogen produces several millimetres of cork and is then replaced by new phellogens which arise in successively deeper regions of the cortex and eventually in the outer secondary phloem (**167**). The removal of the cork crop does not harm the vascular cambium of the oak tree. In most woody species the outer dead bark (rhytidome) is periodically sloughed off the trunk and main branches. The rhytidome consists of successively deeper-formed, discontinuous but overlapping periderms and intervening patches of nonfunctional phloem (**167**).

The cork is impermeable to gaseous diffusion but the numerous lenticels (**403–406**) facilitate the movement of oxygen into the living tissues within this barrier and also allow the exit of carbon dioxide. Lenticels arise from less tightly-packed regions of the phellogen, and the cork (complementary tissue) produced consists of rounded cells with large intercellular spaces between them. In many woody species layers of more compact cork are produced periodically and these retain the loose complementary cells within the lenticel. Viewed macroscopically, lenticels may be dot-like or elongated either horizontally or vertically; in thick bark with prominent furrows, lenticels are located at the bottom of the cracks (**405**, **406**). Most stem-photosynthetic succulents such as cacti and euphorbias retain their epidermis and cortex for years, producing a periderm only when many years old or not at all (**232**, **346**, **364**, **366**, **368**); the same is true for long-lived succulent leaves of monocots (**337**).

336 *Carya illinoinensis* (pecan). The nonsucculent, ordinary dicot shoot has long narrow internodes (1) and nodes (2) where large, persistent leaves are attached by petioles (3). Its axillary (4) and terminal (5) buds are covered with bud scales, and its epidermis is quickly replaced by bark.

1 Internode
2 Node
3 Petiole of leaf
4 Axillary bud
5 Terminal bud

337 Flowering shoots of the monocot *Agave americana* (century plant). This perennial monocot grows vegetatively as a rosette bearing numerous sword-shaped, xeromorphic perennial leaves. However, after many years of vegetative growth, reproduction occurs and an inflorescence axis (several metres tall) arises with numerous small bracts, in the axils of which a number of flowers are borne on short lateral branches.

338 Swollen underground stem tuber of the dicot *Solanum tuberosum* (potato). This perennating organ is almost entirely composed of compact, large parenchyma cells containing considerable quantities of starch. In nature the thin stolons that connect the tubers to the mother plant die off and the isolated tubers give rise to daughter plants. Note the numerous sprouting axillary buds (several of which arise from each 'eye' of the tuber) bearing leaf primordia (1) at their tips. Adventitious root primordia (2).

1 Leaf primordia   2 Root primordia

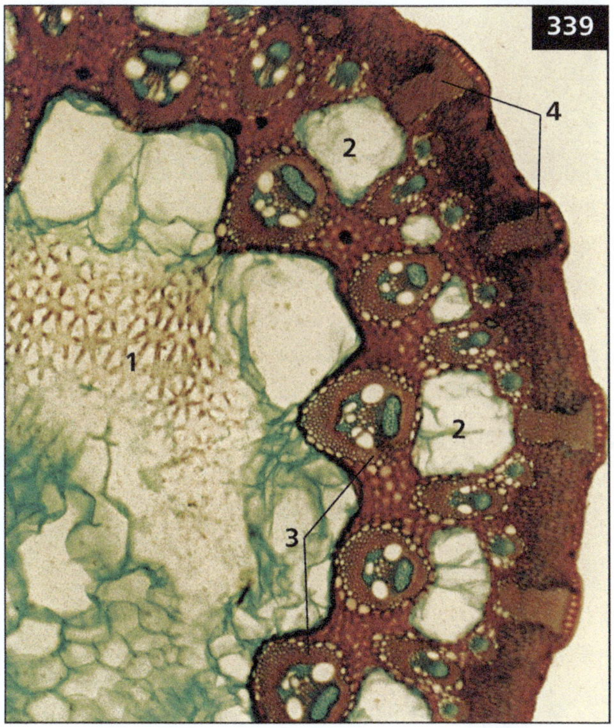

339 TS of the hydrophytic stem of *Juncus communis* (rush). This shows a wide pith (1) and the large air cavities that occur in the outer pith and cortex (2). Unlike the majority of monocots the vascular bundles (3) are peripherally distributed; however, the absence of a vascular cambium between the xylem and phloem distinguishes this stem from that of a dicot. Fibre bundles (4). (LM.)

1 Pith   3 Vascular bundles
2 Cortex   4 Fibre bundles

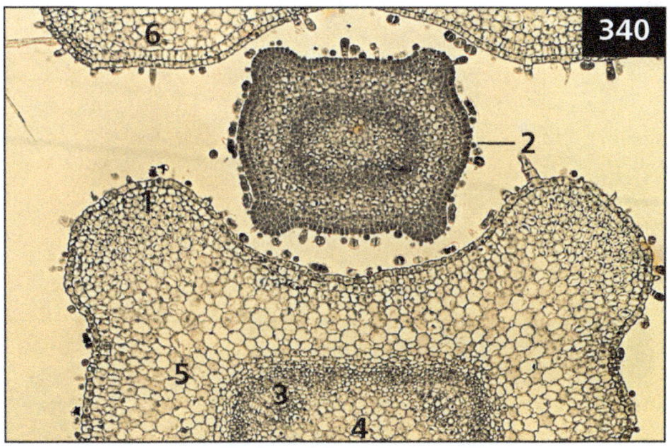

340 TS of a young node of the dicot *Glechoma hederacea* (ground ivy). The corners of its rectangular stem are swollen by peripheral collenchyma (1) and the axillary bud (2) shows a similar form. Vascular tissue (3), pith (4), cortex (5), subtending leaf (6). (LM.)

1 Collenchyma   4 Pith
2 Axillary bud   5 Cortex
3 Vascular tissue   6 Subtending leaf

341 Cladodes of *Opuntia engelmannii* (a dicot) are shoots flattened in one direction and exceptionally broad in the perpendicular direction. The cladode at (1) is viewed edge on, that at (2) shows the full width. These shoots had small, ephemeral leaves but they have all abscised already, so the cladodes are the main photosynthetic organs.

1 Cladode viewed edge on
2 Cladode in face view

342 Shoot of the monocot *Semele* showing its green leaf-like phylloclades. These represent flattened shoots of limited growth which are developed from buds borne in the axils of scale leaves situated on the stems (1) which are cylindrical.

1 Main stem

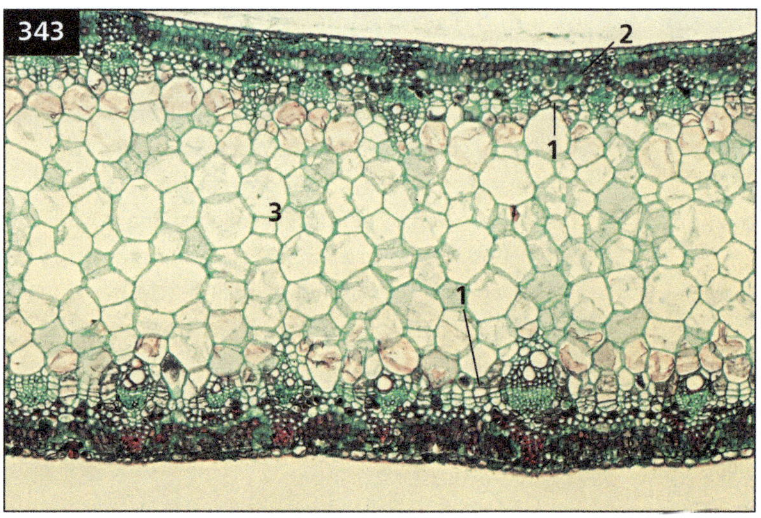

343 TS of the phylloclade of *Muehlenbeckia platyclados*. In this plant the flattened stem acts as a photosynthetic organ but the peripheral distribution of vascular cambium (1), and discrete vascular bundles, demonstrate that this is a dicot stem. Photosynthetic cortex (2), pith (3). (LM.)

1 Vascular cambium
2 Photosynthetic cortex
3 Pith

344 TS of the stem of the dicot *Casuarina*. The stem of this nitrogen-fixing plant bears only scale leaves and the photosynthetic function of the plant is assumed by the xeromorphic stem. This shows hair-lined grooves (1) in which stomata occur, and chlorenchyma (2) at the margins of the grooves. The epidermis has a thick cuticle, and tracts of sclerenchyma fibres (3) occur both hypodermally and internally. Note the ring of vascular bundles (4). (LM.)

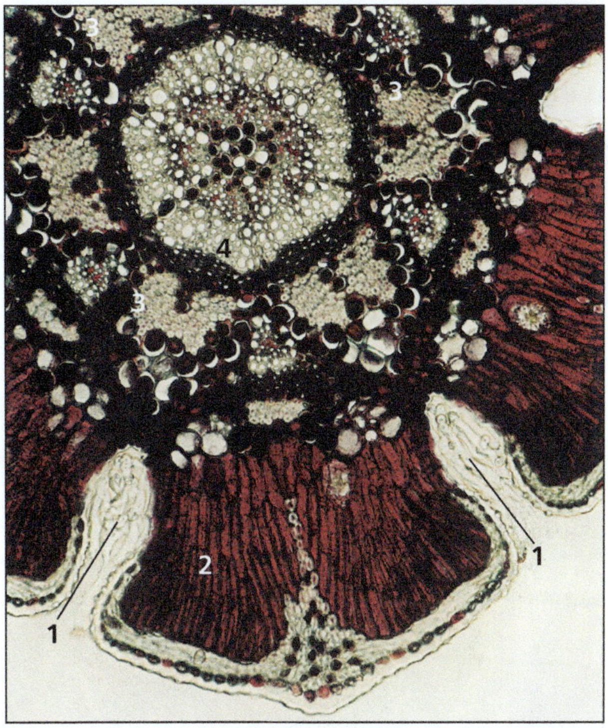

1 Hair-lined grooves	3 Sclerenchyma fibres
2 Chlorenchyma	4 Vascular bundles

345 Specimen of the small dicot tree *Casuarina glauca* (c.f. 344).

346 Several large specimens of *Echinocactus*. In this xeromorphic dicot the highly modified leaves occur as spines while the green succulent stem is the photosynthetic organ. Each vertical ridge on the stem shows a row of areoles (spine clusters) which represent axillary buds bearing several spines.

347 Large tree-like specimen of a *Euphorbia*, a dicot. There are nearly 650 large succulent species of this genus, most of which occur in sub-tropical regions of Africa and the Canary Islands. The spines on the water-storing photosynthetic stems represent highly modified leaf stipules. Note the terminal clusters of fruits, each formed from three fused carpels.

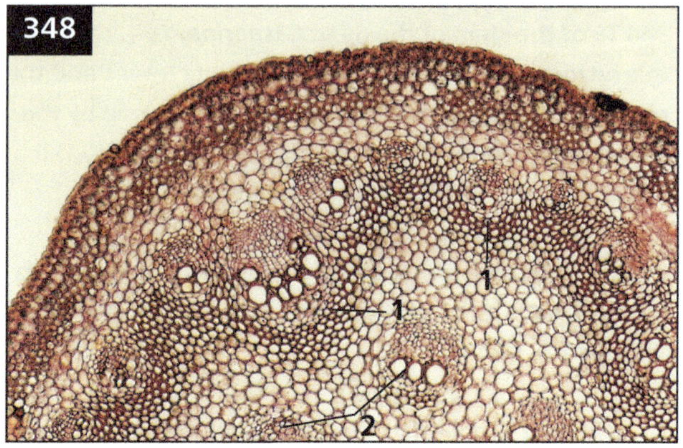

348 TS of the young stem of the dicot *Piper*. This shows an outer ring of variably-sized vascular bundles, demarcated internally by a sheath of sclerenchyma (1), while in the pith a number of medullary vascular bundles occurs (2). Such diffuse distribution of bundles is uncommon in anything other than monocots. (LM.)

1 Sclerenchyma sheath
2 Medullary vascular bundles

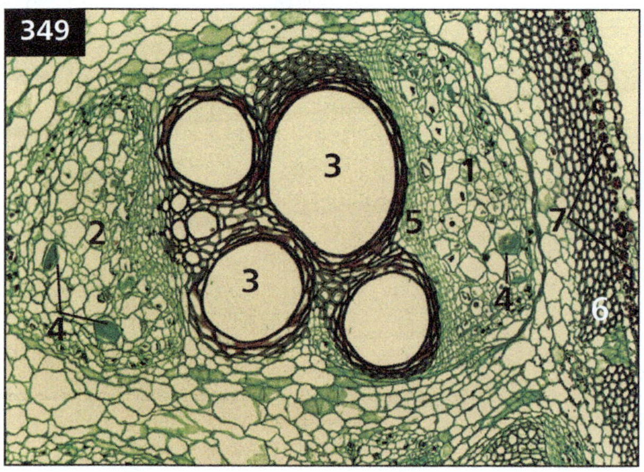

349 TS of a vascular bundle in the stem of the gourd *Trichosanthes*. This dicot shows bicollateral bundles with strands of external (1) and internal phloem (2) on either side of the xylem. Note the very wide xylem vessels (3) and the conspicuous sieve tubes with sieve plates (4). Vascular cambium (5), sclerenchyma (6), photosynthetic cortex (7). (LM.)

1 External phloem
2 Internal phloem
3 Xylem vessels
4 Sieve plates
5 Vascular cambium
6 Sclerenchyma
7 Photosynthetic cortex

350 TS of innermost wood (1), pith (2) and medullary bundles (3) in *Buiningia aurea* (a cactus, a dicot); each medullary bundle has a vascular cambium, wood, and secondary phloem. (LM.)

1 Wood
2 Pith
3 Medullary bundles

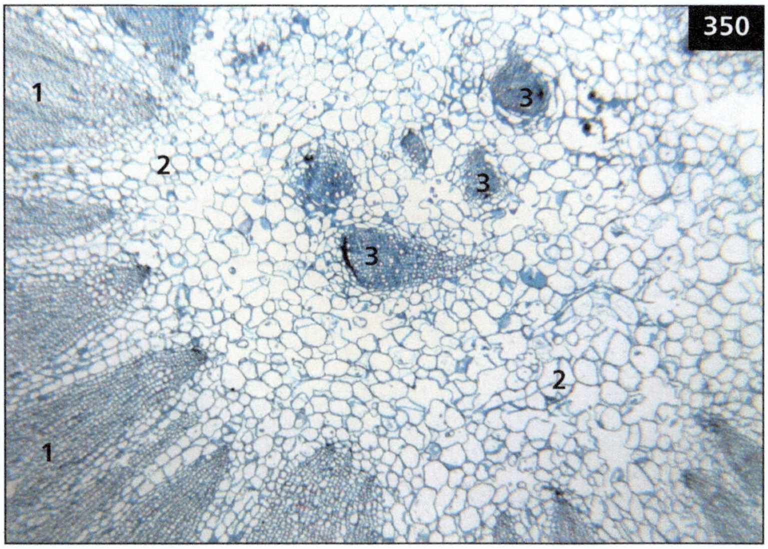

351 TS of the rhizome of the monocot *Convallaria majalis*. A one- to several-layered and uniformly thickened endodermis (1) delimits the wide cortex (2) from the pith in which scattered vascular bundles occur. The inner bundles are amphivasal with xylem (3) surrounding the phloem (4). (LM.)

1 Endodermis
2 Cortex
3 Xylem
4 Phloem

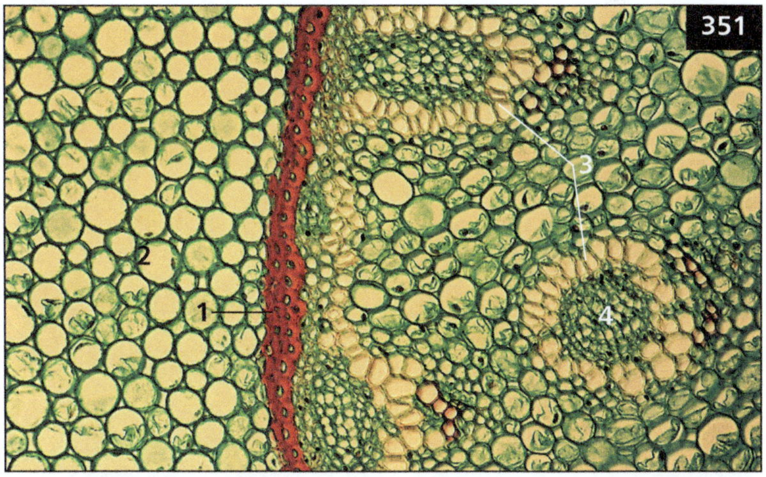

352 TS of the rhizome of the fern *Pteridium aquilinum* (bracken). An extensive vascular strand is separated by an endodermis (1) from the ground parenchyma (2). The large-diametered xylem elements (3) are surrounded by smaller phloem parenchyma and sieve cells (4). (LM.)

1 Endodermis
2 Ground parenchyma
3 Xylem elements
4 Sieve cells

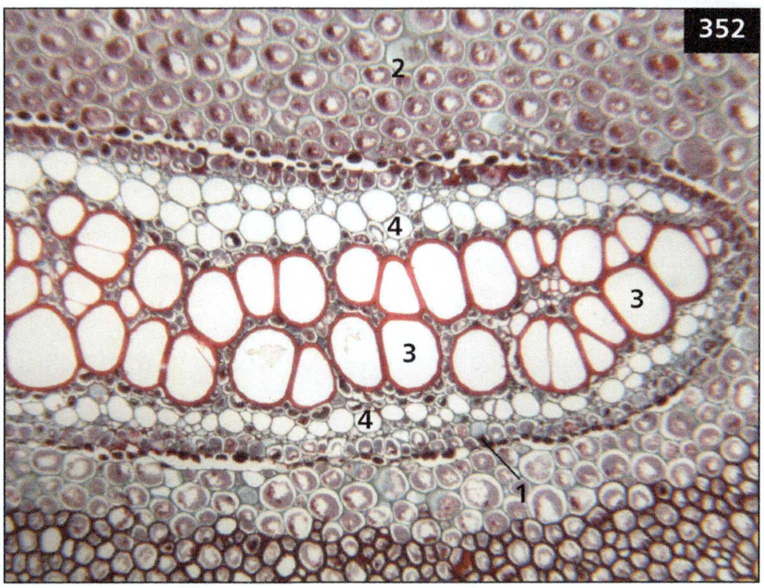

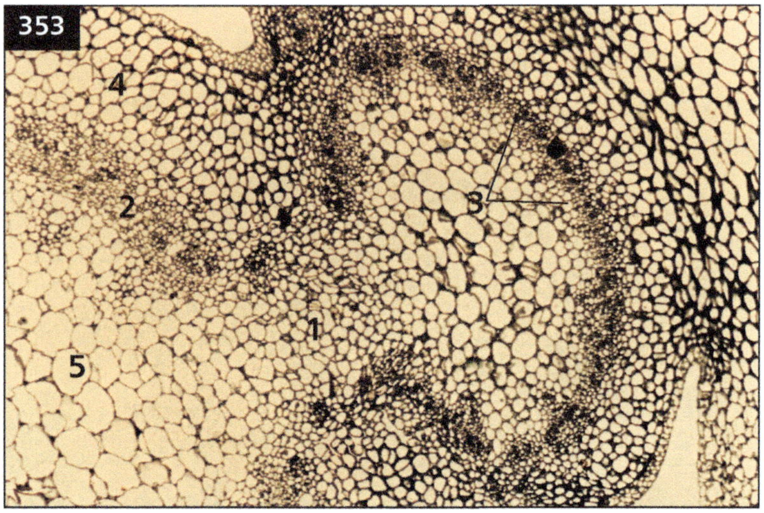

353 TS of the young node of the dicot *Phaseolus vulgaris* (bean). Note the gap (1) in the vascular cylinder of the stem (2) where it branches to supply the vascular system (3) of the axillary bud. Main stem cortex (4), pith (5). (G-Os, LM.)

1 Vascular gap
2 Stem vascular cylinder
3 Vascular system of bud
4 Stem cortex
5 Pith

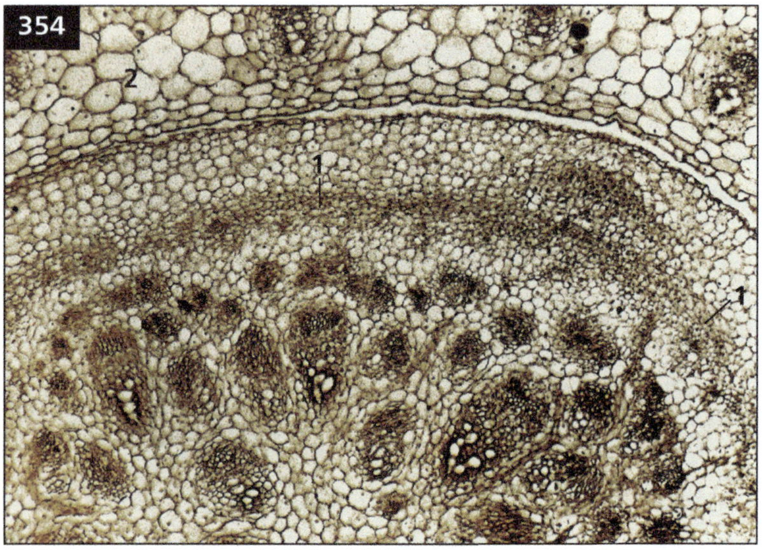

354 TS of the sub-apical region of *Zea mays* (maize) stem. Note the primary thickening meristem (1) in this monocot (cf. 228) from which numerous, scattered vascular bundles are derived internally, while peripheral parenchymatous derivatives lead to increase in stem thickness. Bundles frequently anastomose and their branches run obliquely outwards and upwards to supply the numerous veins of each leaf. Sheath of leaf primordium (2). (LM.)

1 Primary thickening meristem
2 Leaf sheath

355 Specimens of *Eichhornia crassipes* (water hyacinth), a dicot, free-floating due to their air-filled petioles, growing in clusters on a busy Bangkok canal. These plants can grow densly and may impede or block waterways, but in Thailand they are often cultivated to provide a green manure.

356 Washed-up specimen of *Posidonia oceanica*, a marine flowering plant occuring in shallow coastal regions of the eastern coast of Spain. Note its long strap-like leaves (which minimize resistance to the buffeting sea) and the brown remnants of its stem.

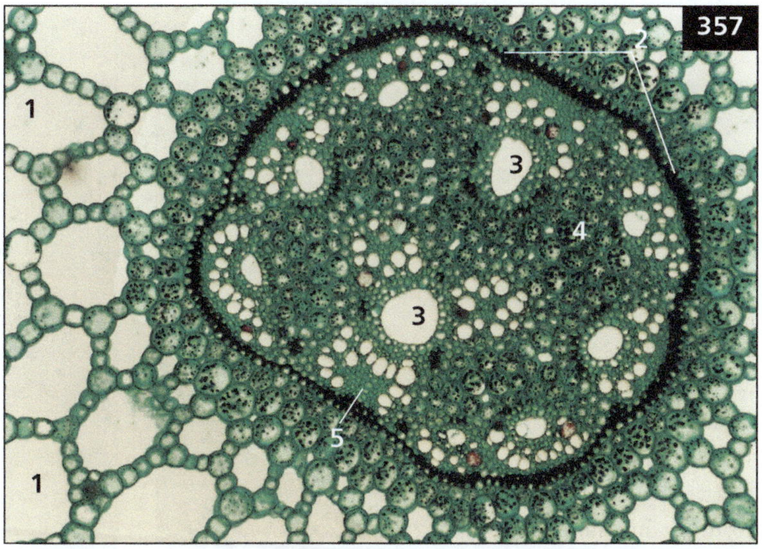

357 TS of the aquatic stem of the monocot *Potamogeton* (pondweed). Extensive air chambers (1) occur in the cortex while the vascular tissue is confined to a narrow central cylinder demarcated externally by an endodermis (2). Eight variably-sized vascular bundles are present, with the xylem of each represented by a protoxylem lacuna (3). Pith (4), phloem (5). (LM.)

1 Air chambers
2 Endodermis
3 Protoxylem lacuna
4 Pith
5 Phloem

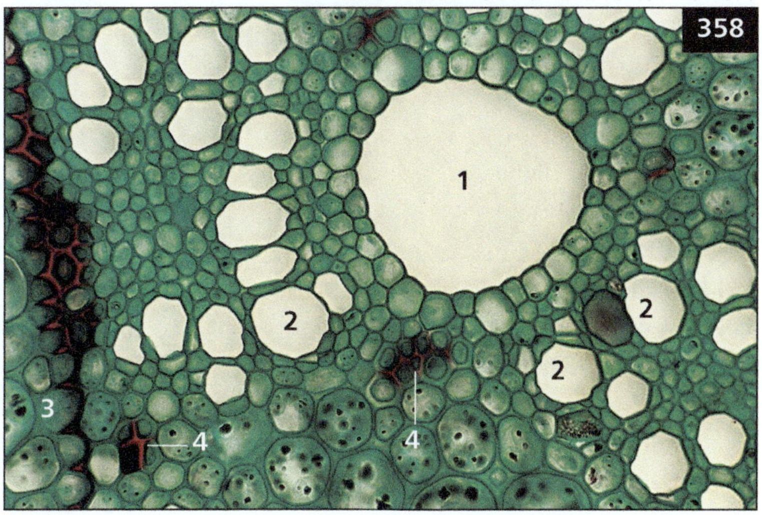

358 TS illustrating detail of the vascular tissue of the monocot *Potamogeton* (pondweed) stem. This shows a prominent protoxylem lacuna (1) with wide sieve tubes (2) on either side (cf. 357). Endodermis (3), fibres (4). (LM.)

1 Protoxylem lacuna
2 Sieve tubes
3 Endodermis
4 Fibres

359 Ball-like debris created by the pounding waves on the plants of *Posidonia oceanica* which have broken free from their bases. They occur on the beaches of Santa Pola and elsewhere in eastern Spain are regularly bull-dozed to clear the way for sunbathing holiday-makers.

360 Pioneering plant of the dicot *Salicornia* growing in a dried-up salt pan on the coast of eastern Spain.

361 Specimens of the herbaceous dicot *Salicornia* in flower, growing on a salt marsh. Note the succulent stems invested by swollen, adhering leaves.

362 *Austrocylindropuntia subulata*. The green bodies of cacti are long-shoots with foliage leaves (1), which are large here but usually microscopic; the axillary buds (2) immediately produce bud scales that develop as spines (3; the yellow-green bases of larger spines are basal meristems). Masses of dead trichomes surround spine bases and protect axillary bud apical meristems.

1 Foliage leaf
2 Axillary bud (short-shoot, areole)
3 Spine (modified bud scale)

363 *Cereus huilunchu*. Most cactus long-shoots have monopodial growth, as here. (Thick wax on last year's shoot is cracking due to tissue expansion.) Long-shoot apical meristems periodically become dormant but do not form terminal buds. The foliage leaves are microscopic, but their axillary buds are visible as the spine clusters. (In the opuntias [prickly pears and chollas; c.f. 341], each long-shoot apical meristem is determinate and growth is sympodial.)

364 *Echinocereus enneacanthus*. After producing spine primordia, short-shoot apical meristems may become dormant or act as floral shoot apices. The flower bud (arrow) is emerging from a 1-year-old short-shoot.

365 *Notocactus coccineus*. Many short-shoot apical meristems produce only spines and flower organs, then become suppressed by apical dominance. Some, however, later become active, convert themselves into long-shoot apices and produce branches (arrows).

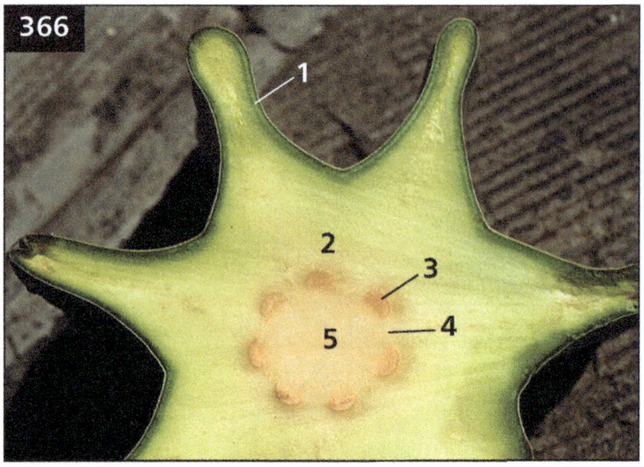

366 TS of a long-shoot of *Neoraimondia herzogiana*. The outermost cortex is chlorenchyma (1), the rest of the cortex provides a voluminous water-storage capacity (2). Seven fascicular vascular cambia have produced regions of fibrous wood (3) and the intervening interfascicular cambia have produced wide rays (4). Pith (5) is unusually broad.

1 Chlorenchymatous cortex
2 Water-storing cortex
3 Fibrous wood
4 Wide, parenchymatous rays
5 Pith

367 This *Pelargonium carnosum* has large, persistent leaves, so its succulent stems perform little photosynthesis. Stem succulence here is due to parenchymatous wood, not a thick cortex or pith.

368 This *Euphorbia mammillaris* has small green foliage leaves (arrows) that die quickly (are ephemeral), so the succulent stem must both store water and be photosynthetic. Unlike cactus spines (modified bud scales; 362), *Euphorbia* spines (1) are modified axillary branches that bear tiny vestigial leaves.

1 Spines (modified branches)

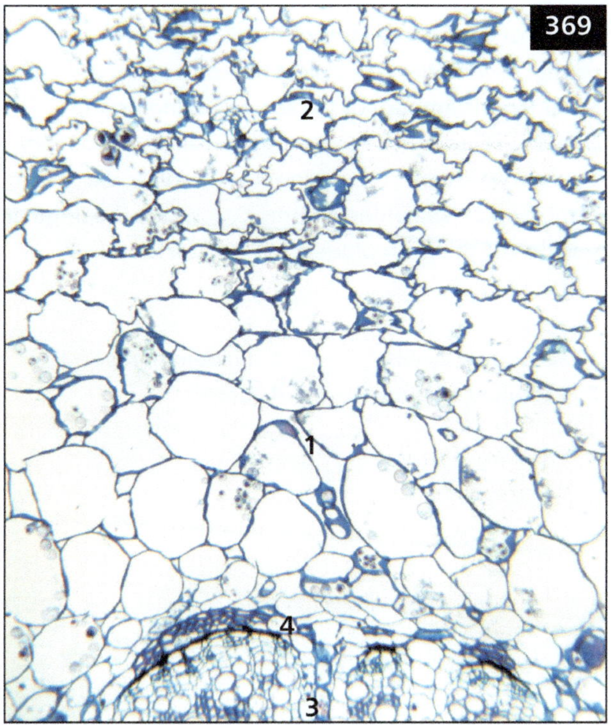

369 TS of cortex of *Haageocereus australis*, showing innermost, turgid cortex cells (1) near outer cortex cells with plicate walls (2). Cells with plicate walls can shrink and give up their water easily without plasmolyzing, allowing water to be transferred to other tissues, keeping them turgid. Secondary xylem (3), secondary phloem (4). (LM.)

1 Turgid inner cortex
2 Collapsible cortex with plicate cell walls
3 Secondary xylem (wood)
4 Secondary phloem

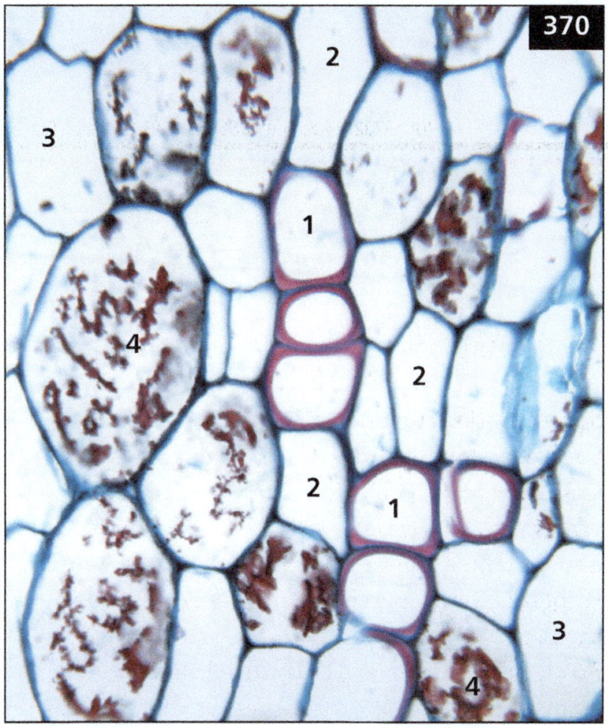

370 TS of succulent wood of *Crassula argentea*. Vessels (1) are surrounded by abundant axial (2) and ray (3) parenchyma; during drought, water can be easily transferred into vessels, minimizing the likelihood of cavitation. Brownish deposits (4) in some cells are tannins. (LM.)

1 Vessels
2 Axial parenchyma
3 Ray parenchyma
4 Tannins

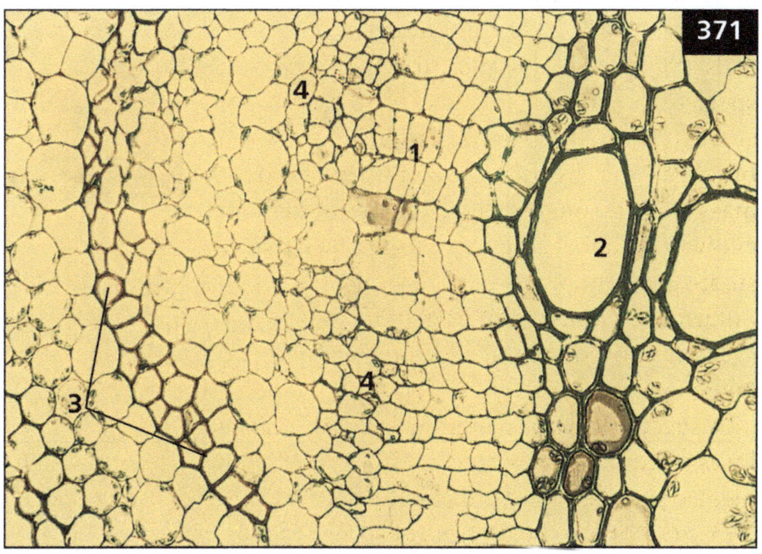

371 TS of a vascular bundle from *Phaseolus vulgaris* (bean) stem. Note the wide cambial zone (1) in which the cells are generally tangentially flattened; their radial alignment is abruptly distorted by the wider-diametered secondary xylem elements (2). The phloem is demarcated externally by the fibre cap (3) and, adjacent to the vascular cambium, secondary elements (4) are differentiating. (LM.)

1 Cambial zone
2 Secondary xylem
3 Fibre cap
4 Secondary phloem

372

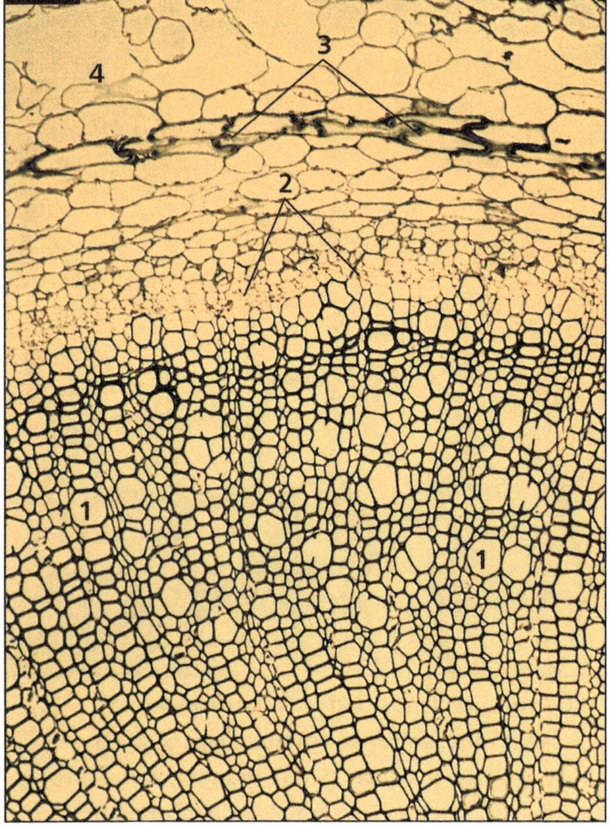

372 TS of the mature stem of the dicot *Linum usitatissimum* (flax). This dicot has undergone a considerable amount of thickening during its one growing season. Note the radially-arranged secondary tracheary elements (1), arising from the vascular cambium (2). The phloem fibres (3) indicate the original position of the protophloem (cf. 10). Cortex (4). (G-Os, LM.)

1 Secondary xylem
2 Vascular cambium
3 Phloem fibres
4 Cortex

373

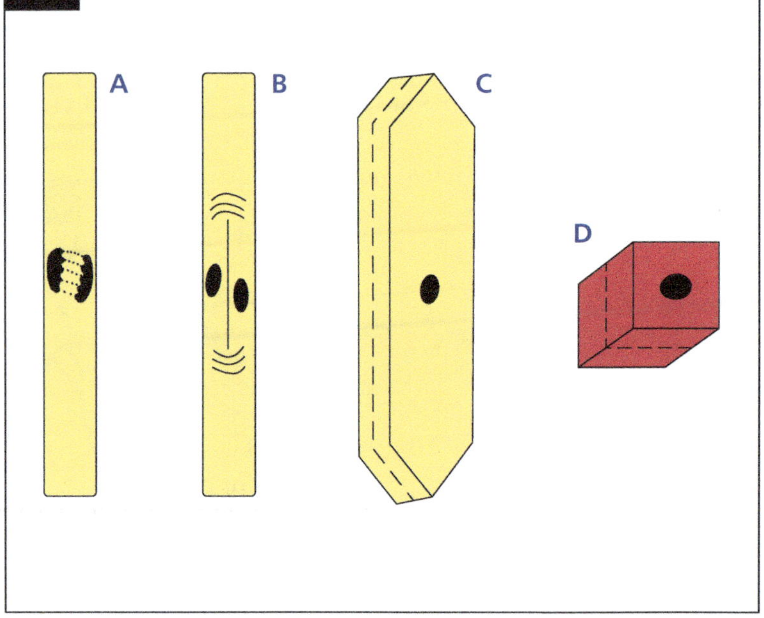

373 Diagrams of the vascular cambium shown in longitudinal views. An isolated fusiform initial is shown in radial view undergoing mitosis (A) and cytokinesis (B), while in C the two daughter cells are viewed from an oblique tangential aspect. D shows a ray initial which has just completed a tangential division.

374 A and B show tangential views of a storied and nonstoried cambium respectively (yellow indicates fusiform and red indicates ray initials).

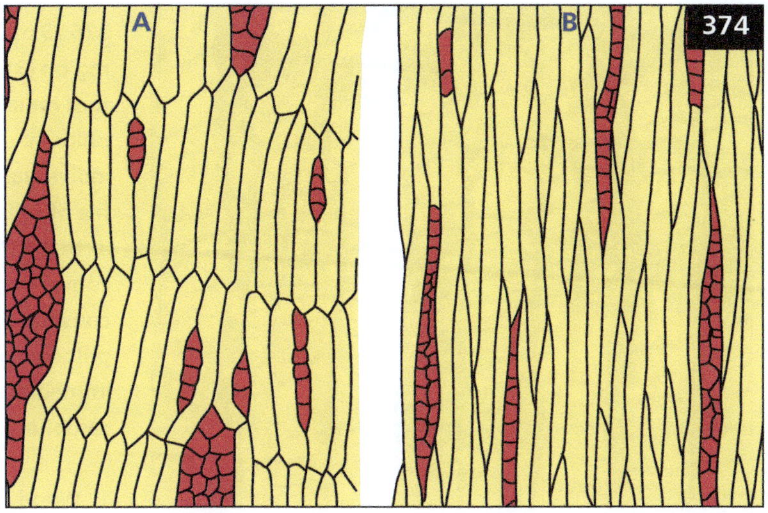

375 Diagrammatic transverse section of the vascular cambium showing its pattern of tangential divisions. In A, division has occurred in the cambial mother cell (yellow) which leads in B to the formation of a phloem initial (blue). Meanwhile, the cambial cell is again dividing and gives rise in C to a xylem initial (red). The phloem initial in C is dividing to form two further initials and a new division (D) is already taking place in the outermost derivative. In the meantime, the cambial cell has formed a further xylem initial towards the inside (D), while the original initial has also divided, so that there are now three xylem initials.

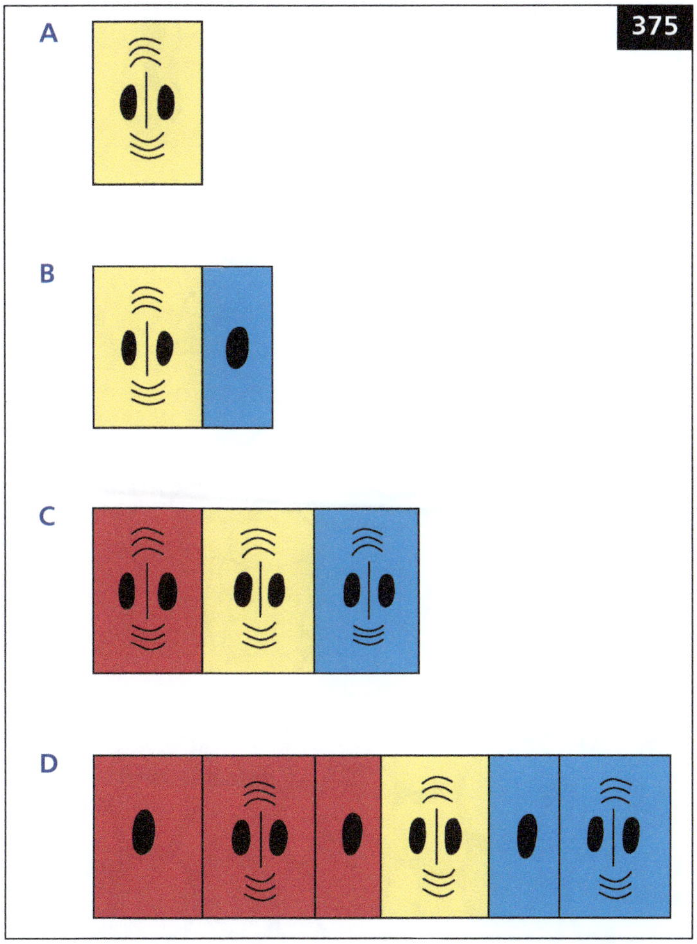

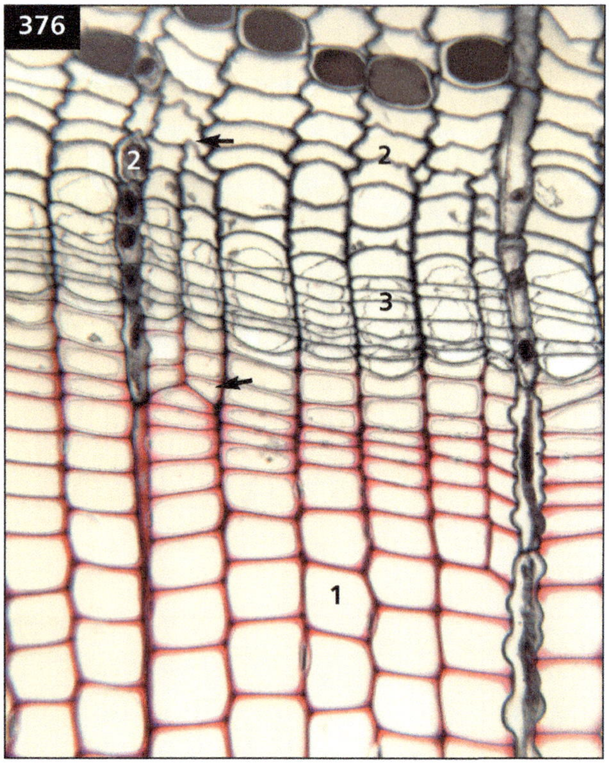

376 TS of wood (1), secondary phloem (2),and cambial region (3) of *Pinus strobus* (pine, a gymnosperm). Note that radial files of cells are continuous from wood to phloem: all cells of each file are produced by an individual fusiform or ray initial. At arrow in xylem, one file of cells becomes two; when the fusiform initial was at that site, it underwent an anticlinal division converting itself to two fusiform initials which consequently produced two files of cells. Arrow in phloem indicates the sieve cells that were produced immediately before and after this change in the cambium. (LM.)

1 Wood
2 Phloem
3 Cambium

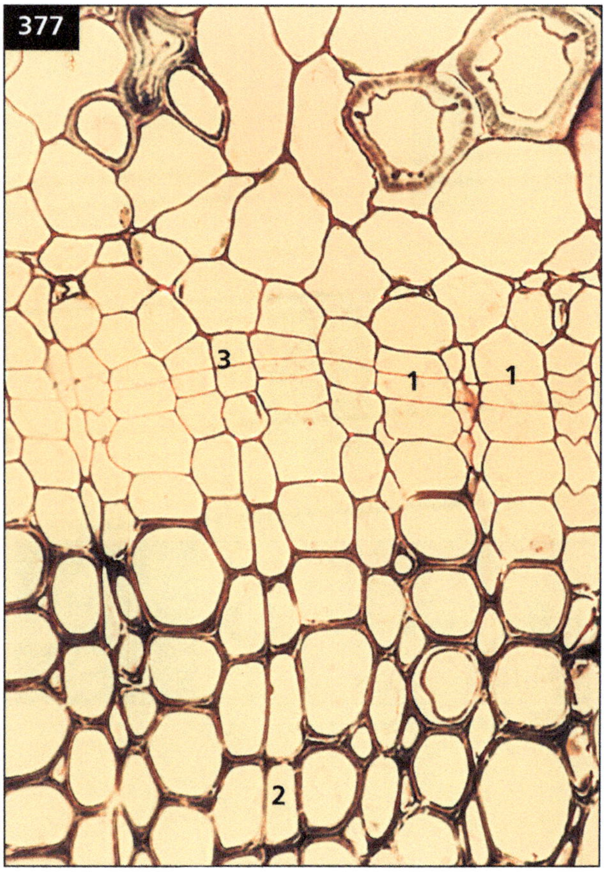

377 TS of the stem of the dicot *Linum usitatissimum* (flax) showing detail of vascular differentiation. The tangentially flattened cells of the cambial zone are well defined and, from its inner face, secondary tracheary elements are differentiating (cf. 372). These undergo radial expansion but little tangential growth so that their origin from specific fusiform initials (1) can be traced. Radial rows of narrower thick-walled ray parenchyma cells also occur (2); these originate from small ray initials (3). The phloem region is very narrow and demarcated externally by phloem fibres. In the phloem the inner conducting elements are probably secondary but their derivation from the vascular cambium is obscure. (G-Os, LM.)

1 Fusiform initials
2 Ray parenchyma cells
3 Ray initials

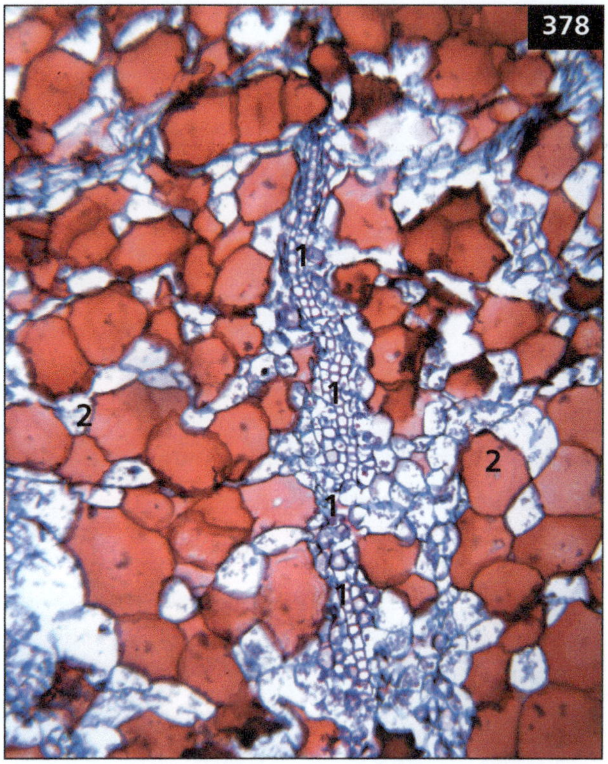

378 *Puna clavarioides*. TS of unusual but not anomalous wood. The vascular cambium has only narrow regions of fusiform initials, producing just a few rows of vessels and tracheids (1), but very wide regions of ray initials producing wide rays (2) filled with mucilage cells (red). (LM.)

1 Vessels and tracheids 2 Rays

379 TS of *Bauhinia* stem. This is a tropical, woody vine that grows in width much more rapidly than in thickness. Its vascular cambium is very active on the two protruding sides, less so elsewhere.

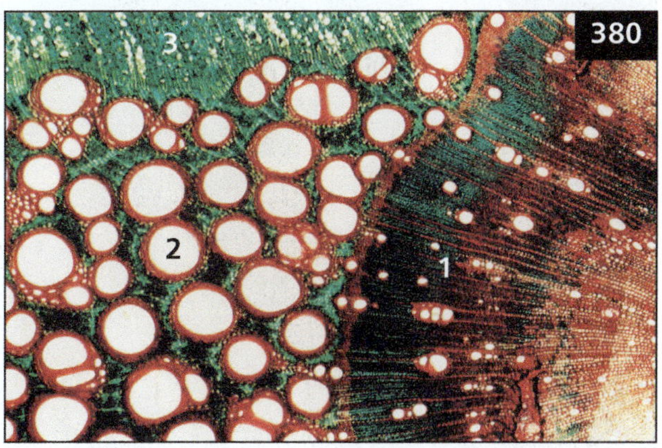

380 TS of the secondary xylem of a young stem of the dicot *Bauhinia*. Note the radial regularity of the early secondary xylem (1) in contrast to the irregularity of the later wood which contains very numerous wide vessels (2). Vascular parenchyma wedges (3). (LM.)

1 Secondary xylem
2 Large vessels
3 Vascular parenchyma wedges

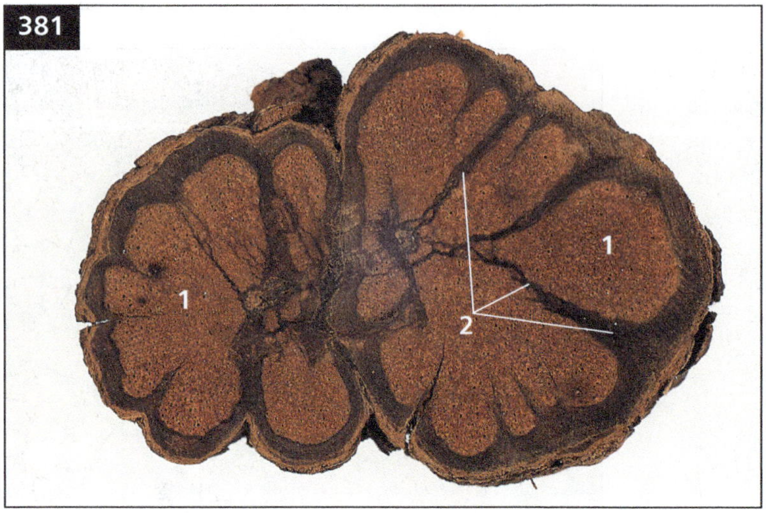

381 TS of the woody liane of the dicot *Bauhinia*. In common with many other lianas, the stem undergoes anomalous secondary thickening. The originally regular xylem becomes split into highly lobed units (1) by the proliferation of the vascular parenchyma. Phloem wedges (2) extend between the xylem lobes.

1 Lobed xylem 2 Phloem wedges

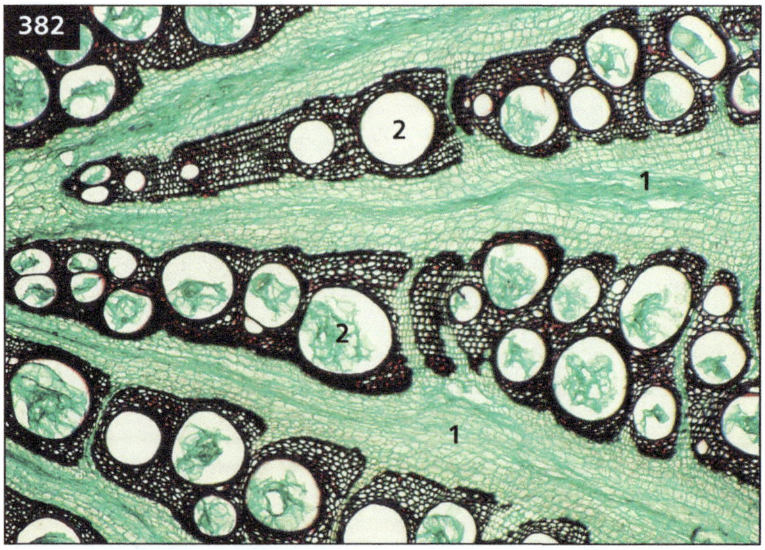

382 TS of the stem of the liana *Tinospora*. Note that the secondary xylem is deeply fissured by the proliferation of the intervening ray parenchyma (1). The xylem, in common with other liana contains many wide vessels (2) embedded in a sclerenchymatous ground tissue. (LM.)

1 Ray parenchyma
2 Large vessels

383 *Artemisia tridentata*. Trunk and branches of a desert sagebrush shrub. Broad strips of its vascular cambium die and stop forming wood, which thus remains thin; associated leaves and buds also die. Other strips of cambium survive, producing bands of thicker wood.

384 TS *Fumana thymifolia* trunk in which some strips of vascular cambium have died (1) while others survived (2) and continued making wood and secondary phloem. Bark (3). (LM.)

1 Points at which vascular cambium has died
2 Adjacent regions where vascular cambium survived
3 Bark

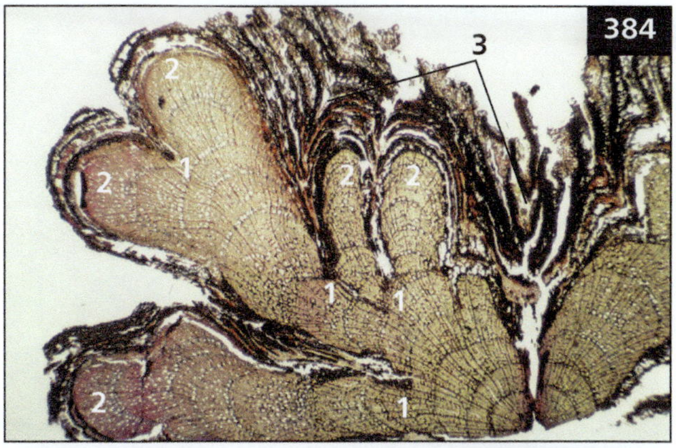

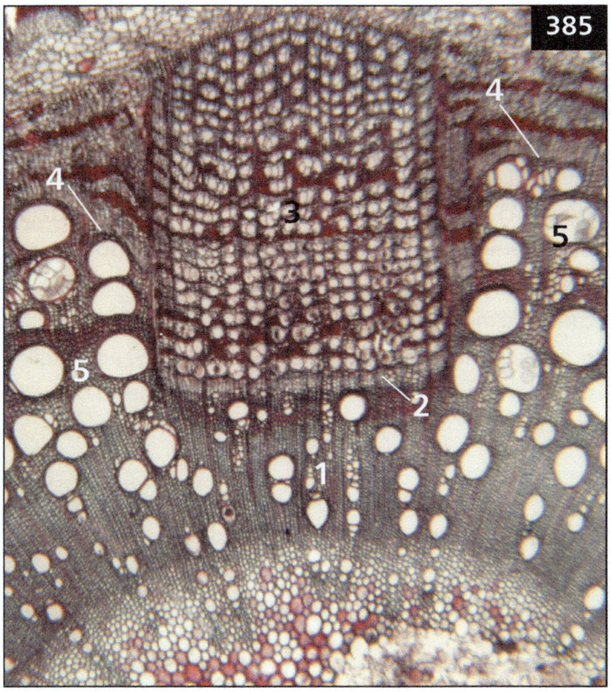

385 TS *Machaerium purpurascens* stem. At first this formed wood (1) with an ordinary bifacial vascular cambium (VC). One strip (2) has converted to being unifacial and producing only secondary phloem (3) and is not being pushed outward because it is not forming wood internally. Other regions of the ordinary VC (4) have continued producing wood (5) and so have been pushed outward, leaving the unifacial VC behind. (LM.)

1 First-formed wood, produced by a completely bifacial vascular cambium
2 Unifacial vascular cambium
3 Secondary phloem produced by unifacial vascular cambium
4 Bifacial vascular cambium
5 Wood produced recently by the bifacial vascular cambium

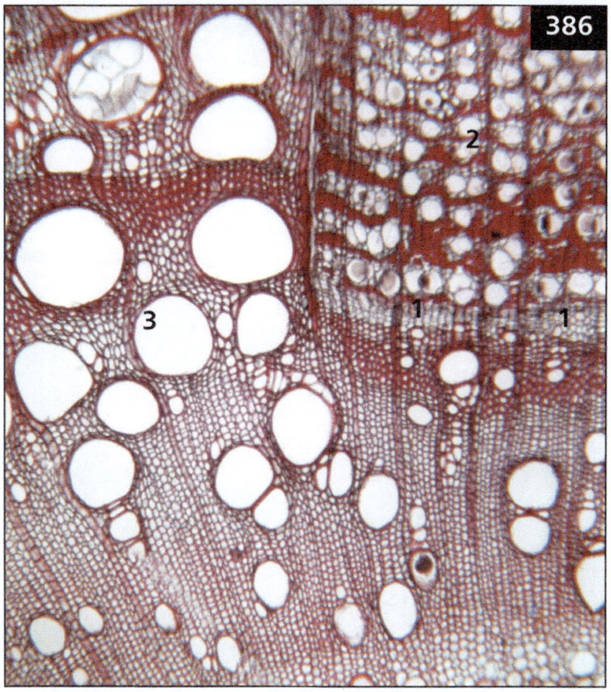

386 TS *Machaerium purpurascens* unifacial vascular cambium (VC, 1) seen at higher magnification than in 385. The secondary phloem (2) is being produced at (1) and is being pushed outward. The wood (3), produced by continued activity of the ordinary bifacial VC, is not moving from where it was produced and consequently the phloem (2) must slide past it. (LM.)

1 Unifacial vascular cambium
2 Secondary phloem
3 Wood produced by bifacial vascular cambium

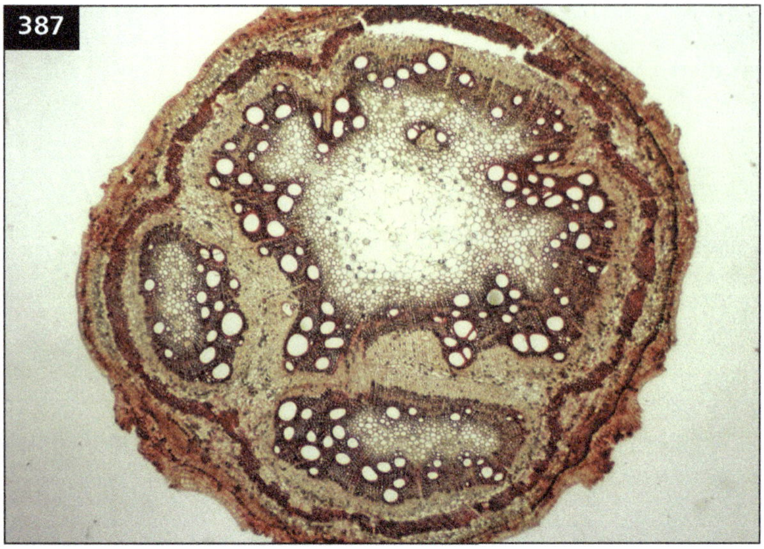

387 TS *Paullinia sorbilis* stem. Three vascular cambia (VC) are present at this point but the number and profiles of the VC are probably different at other points in the stem. (LM.)

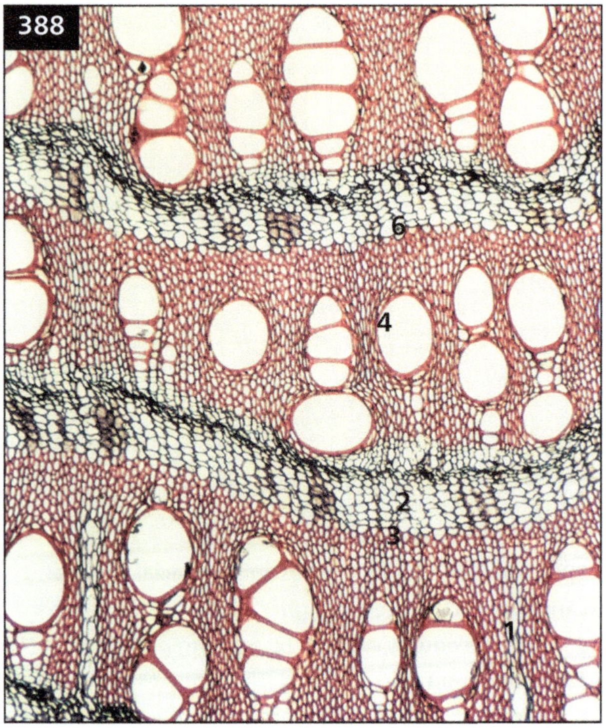

388 TS *Iresine* stem with bands of included phloem. The xylem (1) and phloem (2) were produced by a bifacial, discontinuous vascular cambium (VC) that disorganized when it was located at (3). A new bifacial VC arose in the outermost parenchyma of the phloem (2); it produced the xylem at (4) and phloem at (5) and then disorganized when it was located at (6), and so on. (LM.)

1 & 2 Xylem (1) and phloem (2) produced by vascular cambium that disorganized at 3
4 & 5 Xylem (4) and phloem (5) produced by vascular cambium that disorganized at 6

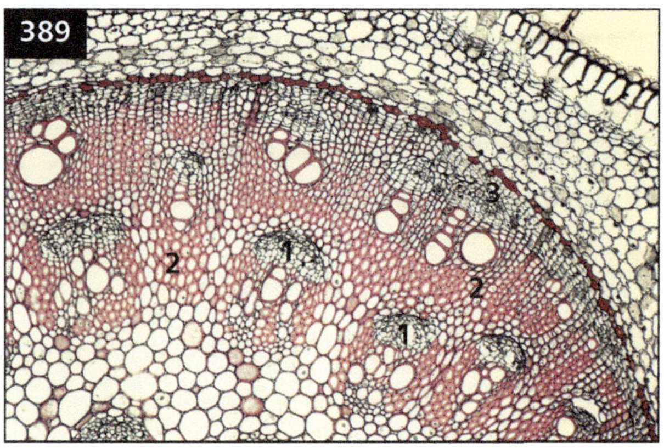

389 TS *Bougainvillea* stem with strips of included phloem (1) surrounded by secondary xylem (2). The vascular cambium (VC) is now located at (3) and most of it will continue to act as an ordinary cambium but strips will disorganize and be replaced by new VC that arises in the outermost phloem parenchyma. (LM.)

1 Included phloem
2 Secondary xylem
3 Vascular cambium

390 Large specimens of the bamboo *Dendrocalamus giganteus*. Note the uniform thickness of the trunks of these monocots and the horizontal scars where the leaf sheaths were attached to the stem.

391 Large specimen of the arborescent monocot *Dracaena draco* (dragon's blood tree). The tips of the branches bear crowded sword-shaped leaves and clusters of flowers. This several hundred-year-old specimen is endemic to the Canary Islands; its much-branched trunk has undergone extensive anomalous secondary thickening and is several metres wide at its base (cf. 392).

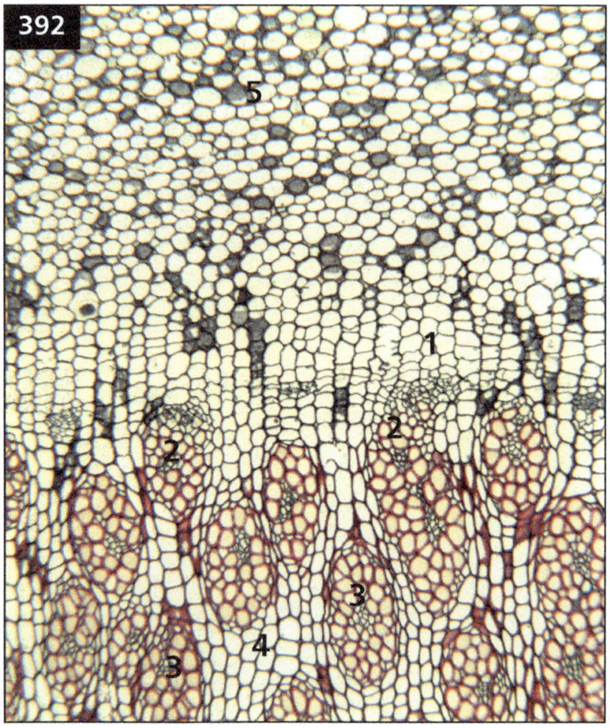

392 *Cordyline terminalis* TS shoot. A cambial zone (1) is producing parenchyma cells to the inner side of this monocot shoot, and some cells subdivide (2) and differentiate into secondary vascular bundles (3); others develop as secondary conjunctive tissue (4). The original cortex is still present (5). (LM.)

1 Cambial zone
2 Newly forming secondary vascular bundles
3 Mature secondary vascular bundles
4 Secondary conjunctive tissue
5 Cortex

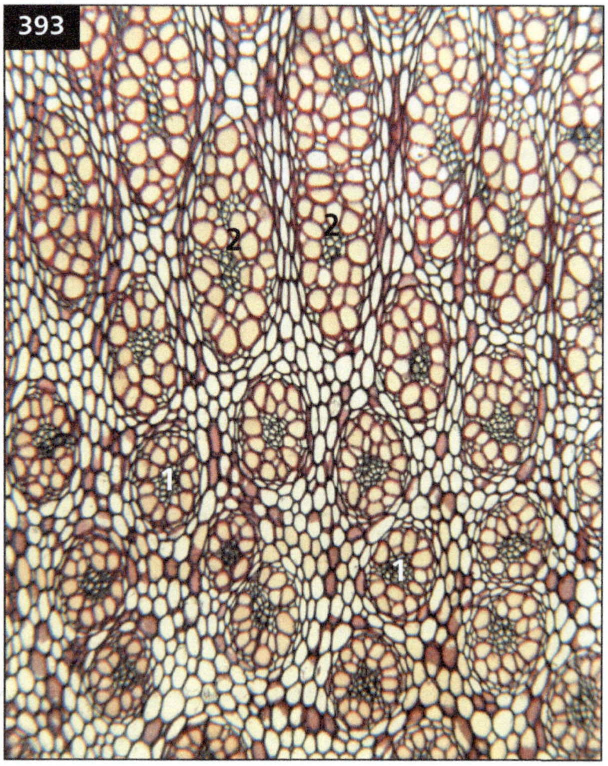

393 *Cordyline terminalis*. TS of inner shoot, showing the primary vascular bundles (1) that constituted the original vascular system formed by the shoot apex, and the secondary vascular bundles (2) that were produced by the monocot cambium. (LM.)

1 Primary vascular bundles
2 Secondary vascular bundles

394 The trunk of the conifer *Sequoiadendron giganteum* (giant redwood). These long-lived trees may grow up to nearly 80 m in height and are protected by a very thick layer of soft cork. This sloughs off unevenly so that a number of growth layers is often visible.

395 Trunk of *Quercus suber* (cork oak) stripped recently of cork. The cork can be peeled away from the trunk of this dicot and harvested (cf. 136, 137), leaving a thin layer of new cork still covering the functional secondary phloem on the trunk. The cork layer on the tree gradually builds up anew and will be harvested in about another 10 years.

396 TS showing the epidermal origin of the cork cambium in the stem of the dicot *Linum usitatissimum* (flax). Note the thin tangential walls (1) which divide the originally single-layered epidermis. Cortex (2), phloem fibres (3). (LM.)

1 Tangential division walls
2 Cortex
3 Phloem fibres

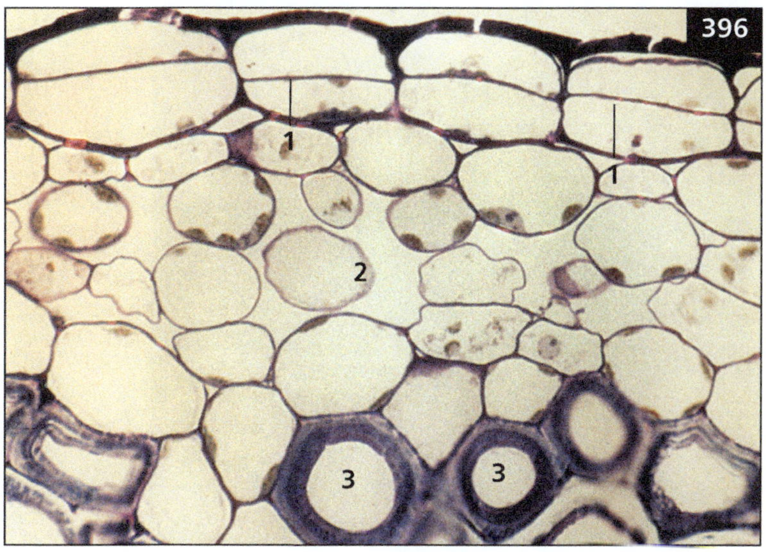

397

397 This trunk of *Cercidium microphyllum* (palo verde) appears green because the cork cambium produces an internal layer of chlorophyllous phelloderm cells. The external cork cells are transparent.

398

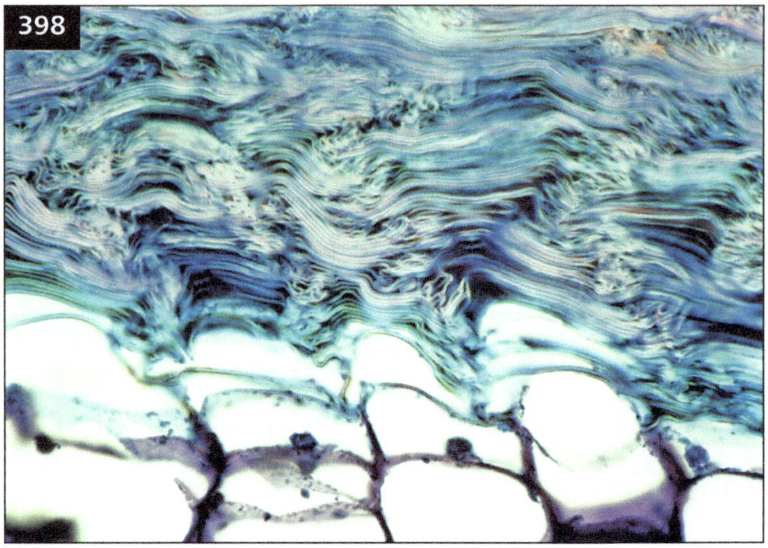

398 TS *Cleistocactus fieldianus* stem showing completely collapsed cork cells; although the layer of cork is only a few millimetres thick, it contains scores of suberized cell walls and is very protective. (LM.)

399

399 Large tree of the dicot *Acer pseudoplatanus* (sycamore) with a hollow trunk. Note how an extensive protective layer of periderm has grown over the exposed margins of the trunk and side branch.

400 Outer surface of the trunk of a large *Cordyline australis* tree. This tree is a monocot with anomalous secondary thickening; note its well-developed layer of protective cork.

401 *Ulmus alata* winged-bark elm. The cork cambia in two opposite strips are greatly more active than in all other regions of the stem.

402 *Chorisa* trunk. The thorns on this tree become more formidable as they are elevated and enlarged by a basal cork cambium.

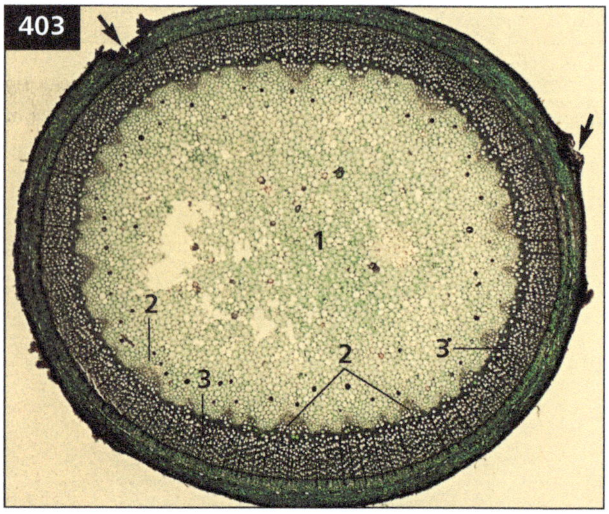

403 TS showing distribution of lenticels (arrows) on a young twig of the dicot *Sambucus nigra* (elder). As secondary thickening proceeds the epidermis is replaced by cork in which numerous lenticels occur. Pith (1), primary xylem (2), secondary xylem (3). (LM.)

1 Pith
2 Primary xylem
3 Secondary xylem

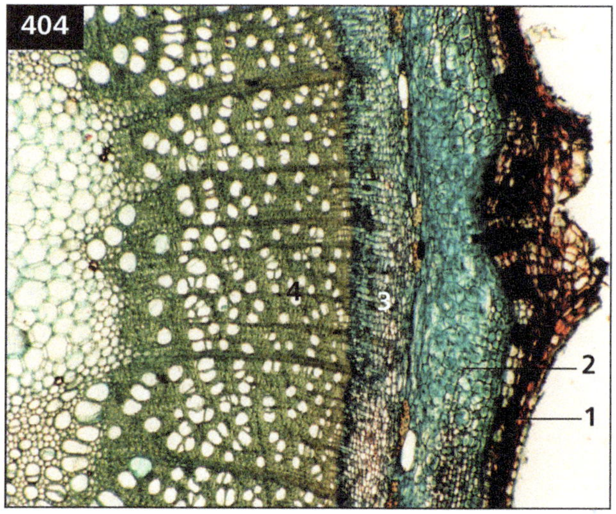

404 TS showing detail of a lenticel from the dicot *Sambucus nigra* (elder) stem. The cork cambium (1) originates in the hypodermis and to the inside a thin layer of secondary cortex (2) has differentiated. Secondary phloem (3) and secondary xylem (4). (LM.)

1 Cork cambium
2 Secondary cortex
3 Secondary phloem
4 Secondary xylem

405 *Ulmus parvifolia* cork cambia develop as irregular patches in the shape of pieces of a jigsaw puzzle. As the bark is shed, it flakes off in pieces that reflect the shape of the cork cambium; the light brown region is newly exposed cork. Dot-like lenticels occur along the intersections between patches.

406 *Prunus serrula* trunk. Most of the bark is so smooth that the extra cells in the elongated lenticels make obvious bumps on the surface. The bright orange cells in the centre of each lenticel (freshly exposed due to new cell production in the interior) are regions especially permeable to oxygen.

CHAPTER 7
The Root

INTRODUCTION

The basic functions of the root system are the uptake of water and mineral nutrients, anchorage and, sometimes, support (**407–410**). Roots are also concerned with the supply of cytokinins and gibberellins to the shoot system and commonly store starch. They sometimes are perennating organs (**430**) and may be modified in various other ways such as into contractile roots, pneumatophores (**411, 412**), prop roots (**413**), aerial absorptive roots (**258**), haustoria (**441, 442**), spines (**414**), and nitrogen-fixing nodules (**415**). Roots also sometimes give rise to adventitious shoots (**145**).

In many dicots the radicle of the embryo (**46, 213**) develops into the vertically growing taproot from which numerous lateral roots originate (**26, 415**). In monocots the radicle is rarely persistent, but instead a fibrous root system develops from the base of the radicle while adventitious roots arise on the lower stem (**416**). The root system of an individual plant may be extensive and it has been calculated that a single mature plant of *Secale cereale* (rye) may form a fibrous root system up to 40 times greater in surface area than the shoot system. Roots can penetrate several metres deep into the soil.

In trees and shrubs the absorptive roots are usually superficially located in the soil but the root system often extends laterally beyond their aerial canopies (**407**). Natural grafting between roots on the same tree (**408, 417**) is common, and also frequently occurs between roots on different individuals of the same species. Grafting involves union of cambia and vascular tissue and this provides a possible route of disease transmission from an infected tree to adjacent specimens.

ANATOMY OF THE MATURE PRIMARY ROOT

In dicots the core of the root is normally occupied by xylem, with commonly three to five protoxylem poles arranged around the central metaxylem (**254, 418, 433, 434**). However, roots with two poles (diarch roots) or more numerous poles (polyarch roots) also occur and sometimes a parenchymatous pith-like dilatated metaxylem is present (**433, 434**). Monocot root tips are much broader than those of dicots and are polyarch with numerous protoxylem poles (**42, 127**); in some palms up to 100 poles occur. A parenchymatous or sclerenchymatous pith-like conjunctive tissue may be present in monocot roots (**127, 257**). In species with aerial prop roots, a much wider pith is often present than in the underground root. In young monocot roots the xylem and phloem are separated by parenchyma but in older polyarch roots this may become sclerified, rendering it difficult to distinguish the boundary of the xylem.

LATERAL AND ADVENTITIOUS ROOTS

Lateral roots develop endogenously (**222**) from the pericyclic parenchyma. They are frequently initiated in the parent root at about the level of the oldest root hairs (**250**). In dicots they arise opposite a protoxylem pole, such that the xylem linkage between lateral and parent root is direct. In monocots the root primordia often originate between the xylem poles.

The initiation of a lateral root involves a number of pericyclic parenchyma cells, which undergo division in varied planes and dedifferentiate to a meristematic state (**419**). Eventually a dome of meristematic cells is formed which begins to protrude into the root cortex (**222**). The lateral root primordium is invested by the endodermis of the parent root. Although the endodermis initially accommodates the growth of the lateral primordium by undergoing anticlinal divisions, ultimately it disintegrates. Before the lateral root emerges it penetrates the overlying cortex and epidermis of the parent root. A well defined root cap is formed and internally the apical organization characteristic of the species is established (**420**). A quiescent centre is initially absent from the root primordium but this develops soon after the emergence of the young root.

Adventitious roots are often present on the stem (**39, 143, 414, 416**), frequently located at the nodes. They usually have an endogenous origin, either from the ray parenchyma or vascular parenchyma (**236**). Roots may also develop in conjunction with adventitious buds on excised leaves (**250, 421**).

In a number of species of tropical and subtropical origin a system of adventitious prop roots is present (**39, 413**). In various species of the aptly named strangler fig (*Ficus*, **417, 422**) the root system develops from a bird-transported seed germinating in the canopy of the host tree. The roots grow downwards on the trunk, fuse around it, and eventually kill the host tree, the trunk of which rots away to leave a hollow fig tree. In tree ferns a dense, tangled mat of adventitious roots invests the trunk and provides significant mechanical support to the shoot (**423**).

SECONDARY GROWTH IN ROOTS

In the majority of dicots some secondary growth occurs (**38, 424**). It is extensive in gymnosperms and woody dicots (**407, 408, 422, 425–427, 435**), and it occurs in the few arborescent monocots (**391**). In the former, the vascular cambium is initiated from parenchyma cells which resume division and first become active between the metaphloem and metaxylem (**254, 427**). These divisions progressively spread laterally over the protoxylem poles to involve the pericyclic cells, so that the vascular cambium completely encloses the xylem core (**38, 427**) and eventually becomes circular (**424–427**). In the secondary xylem wide rays may occur, especially on the radii of the protoxylem poles, but narrower rays are more generally distributed (**425, 426**).

The secondary xylem of angiosperm roots contains wide vessels interspersed between narrower tracheary elements (**424, 426**), but in gymnosperms the absence of vessels provides a more uniform appearance to the xylem (**425**). Growth rings may become apparent in the xylem. The formation of secondary vascular tissue causes an expansion of the core of the root and generally the endodermis, cortex, and epidermis are sloughed off (**427**). However, before this occurs a phellogen arises endogenously within the proliferated pericyclic parenchyma. The vascular cylinder becomes invested and protected by a layer of cork; lenticels allow gas exchange (**426, 428**). In some herbaceous dicots the cork cambium arises superficially (**424**). In arborescent species, roots expand so greatly they exert sufficient pressure to help crack open rocks (**429**), so eventually contributing to the build up of soil.

SUCCULENT ROOTS

Numerous plants have fleshy, succulent roots that store carbohydrates and water (**430**). Root succulence typically is based on parenchymatous wood (**431, 432**) but may be augmented by dilatated metaxylem (**433, 434**); rarely, fleshy secondary phloem accumulates (**435**). Metaxylem dilatation occurs when some immature metaxylem cells (**418**), rather than differentiating into dead tracheary elements, proliferate into a mass of parenchyma that remains alive; the innermost portions of wood rays usually also dilatate (**433**). Unlike pith, dilatated metaxylem has isolated vessels dispersed among its parenchyma cells (**434**). In some succulent roots, sieve tubes make up just a tiny fraction of secondary phloem, the rest consists of broad rays composed of storage parenchyma (**435**). Such phloem may become several millimetres thick. In *Daucus carota* (carrot, **431**), the secondary xylem mainly differentiates as parenchyma rather than conducting elements. In *Ipomoea* the original vascular cambium is replaced by further cambia which form successively deeper in the extensive parenchymatous tissue of the secondary xylem. However, in *Beta vulgaris* (beetroot, **432**) additional cambia form external to the original vascular cambium.

PARASITIC PLANTS

Parasitic plants, which separately evolved on many occasions and occur in 18 dicot families, vary in degree of dependence on the host and extent of their modification. However, all invade their hosts using highly modified roots called haustoria. Hemiparasites perform photosynthesis and have rather ordinary leaves and stems (**436**); holoparasites obtain water, minerals, and bulk carbohydrates from their host, perform no photosynthesis, have little need for leaves, stems, or sunlight, and many are subterranean with extremely modified bodies (**437**).

Hemiparasites

Castilleja (Indian paintbrush; Orobanchaceae) is an herb with almost no modifications; it has green leafy shoots and fibrous roots with inconspicuous attachments to roots of neighbouring grasses. *Phoradendron* and *Viscum* (mistletoes; Loranthaceae and Viscaceae) shoots have a typical dicot anatomy (**436**) but their haustoria penetrate tree branches, infecting the host cambium and wood. The host cambium is stimulated to form more wood, so increasing the surface area of the host/parasite interface. *Ligaria cuneifolia* (**438**) invades the soft cortex of cacti; its haustorium resembles a lumpy spherical callus with no root apical meristem, root cap, or epidermis. It has irregular strands of xylem but no sieve tubes. The host produces many vessels at its interface with the haustorium, but *Ligaria* itself only rarely produces vessels near host tissues (**439**).

Holoparasites

Cuscuta (dodder; Convolvulaceae), one of the least modified holoparasites, is a vine with tiny, achlorophyllous leaves (**440, 441**). Its adventitious roots have threadlike cells able to penetrate individual vessel elements of the host (**442**). In contrast, *Tristerix aphyllus* (Loranthaceae) has almost no ordinary aspects to its vegetative body: a germinating seed invades a cactus host's cortex with only a few root cells, then its embryo dies. The root cells grow as a web of uniseriate strands of cells, lacking all features of a dicot body (**443**). When they encounter host phloem, some cells divide longitudinally, forming multiseriate strands in which a few sieve tube members differentiate but there is no apical meristem, epidermis, cortex, or pith (**444**). Callus-like nodules form below the cactus surface then swell and break the epidermis; adventitious bud apical meristems arise within nodules, then emerge as inflorescences. At this point, *T. aphyllus* has an ordinary floral body but a vegetative body with virtually no ordinary anatomy. Plants of *Rafflesia* (Rafflesiaceae), *Pilostyles* (Apodanthaceae), and many others are similar.

In the Balanophoraceae, infection produces a potato-like 'tuber' of vascular bundles in ground parenchyma with no apical meristem or epidermis (**445**). Host xylem and phloem proliferate outward, into the tuber: the host/parasite interface is outside the host's body (the host and parasite cells can be distinguished because living parasite cells have gigantic nuclei and parasite vessel elements have knobby secondary walls; **446, 447**). Tubers produce endogenous adventitious buds (**448**) that develop into either inflorescences or rhizome-like 'runners'. These lack root hairs, leaves, and axillary buds; their apical meristems lack root caps and leaf primordia. In *Ombrophytum*, runners contain only *Ombrophytum* cells and only initiate new infections, but in *Balanophora*, *Helosis*, and *Langsdorffia*, runners are composites of both parasite and host cells. Even in runners many centimetres long, the central vascular tissues are host cells, outer cortex-like tissues are parasite cells. No part of the vegetative body has any ordinary dicot structure. Near the ends of the runners, adventitious buds (containing only parasite cells) become inflorescences.

ANT-PLANTS

The term 'ant-plant' is reserved for species with unusual structures inhabited by ants. Thus extrafloral nectaries or seed-borne food bodies (elaiosomes) on a plant may entice ants to protect it, clean its surface of fungi and insects, or disperse its seeds, but it is not regarded as an ant-plant if it does not also house ants. Ant-plants are diverse and have evolved in several families of dicots and ferns where various modifications of their hypocotyls, leaf bases, or internodes occur. In some cases, the plant provides only shelter to the ants and in return receives their protection, but other ant-plants are epiphytes inhabiting the nutrient-poor substrate of a host tree bark. Here, their adventitious roots frequently facilitate the absorption of nutrient waste produced by the ants.

Myrmecodia and *Hydnophytum* (Rubiaceae) are epiphytic ant-plants of southeastern Asian rainforests. After a seed germinates, its hypocotyl swells, becoming a lumpy potato-sized 'tuber' filled with hollow chambers (**449, 450**). The chambers are of two types in *Myrmecodia* (with either smooth or warty walls), whereas those of *Hydnophytum* have smooth centres and warty ends. Ants live and tend their young in smooth areas, they deposit waste in the warty regions. The warts are highly cytoplasmic, modified tips of adventitious roots that do not elongate but transfer nutrients from ant waste into the plant tissue. Ants have access by means of naturally formed tunnels that open downward such that rain cannot enter. Several species of these ant plants have numerous minute openings that ventilate the inhabited chambers. As the tuberous hypocotyl continues to enlarge, more interconnected chambers are added; up to 60% of tuber volume consists of hollow chambers.

Similar chambers are formed by several ant-plant ferns; *Lecanopteris* has flat, crustose rhizomes with hollow centres, in *Myrmecophila* they are more swollen, and *Solanopteris* ('potato fern') has lumpy rhizomes. In all three taxa, the chambers form as part of ordinary plant development. However, natural entrances do not occur but instead the ants chew an opening and then live in some chambers while using others for waste. Nutrients are absorbed by minute, villus-like projections from the chamber walls of *Lecanopteris*, but ordinary adventitious roots arise near the openings of *Solanopteris* and then grow into the chambers, absorbing the waste nutrients. These ferns do not provide nectar or food bodies, but *Solanopteris* is most heavily colonized and flourishes best when growing in trees that have abundant nectaries.

Three well-known 'food and shelter for protection' symbioses exist in species of *Piper* (Piperaceae), *Cecropia* (Moraceae), and *Acacia cornigera* (bull-thorn acacia, Fabaceae). Ants occupy hollow internodes in the first two taxa, either using naturally occurring chambers or chewing away living pith cells. The stipules of *A. cornigera* develop as enormous thorns but the ants chew entrance holes and hollow them out before the thorn's fibres harden (**451**). All three genera are remarkably specialized in their ability to feed ants. *Cecropia* produces Mullerian bodies that store glycogen (more typical of animals) rather than starch. *A. cornigera* produces foliar nectaries that supply carbohydrates while their leaflet tips develop into protein- and lipid-rich Beltian bodies (**452**). *Piper* forms single-celled, opalescent food bodies but only if the partner ant (*Pheidole bicornis*) is actually occupying the plant.

POLYMORPHIC ROOT SYSTEMS IN MANGROVES

Mangroves such as *Avicennia* (**453**), *Brugueira*, *Rhizophora* (**413**), and *Sonneratia* are dicot trees that grow extensively and luxuriantly on the saltwater coastal and delta swamps of many subtropical/tropical regions, but also sometimes occur in more temperate locations. They belong to several families and have numerous species. Their root systems are profusely developed, very extensive, and show various modifications related to their saline hypoxic environment (**411–413, 454, 455**). In some mangroves sodium is actively pumped out of their roots while subsequently salt may be concentrated in their bark and leaves which are continuously shed; in *Avicennia* and *Sonneratia* leaf-borne glands accumulate salt before excreting it onto the leaf surface.

Many mangrove species develop large adventitious prop or stilt roots. These radiate outwards from the lower trunk (**413**) and larger branches, helping to stabilize the tree against the buffeting of tidal waters. Profusely covered with lenticels, the prop roots grow down into the mud for 20–30 cm and then branch to form thick cable-like roots (**455**), which grow horizontally away from the tree. Smaller anchoring roots may grow even deeper into the mud. In mangroves which lack well-developed prop roots, such as *Avicennia marina* (**453**), cable roots arise directly from the base of the trunk (**454**). A network of more superficial absorptive roots is also often developed from the cable roots. In addition, many mangroves develop pneumatophores (snorkel-like roots, **455**, **456**) that are negatively gravitropic and grow vertically upward. A pneumatophore is abundantly covered with lenticels and internally a wide aerenchymatous cortex (supported by numerous sclereids) helps aerate the underground root system; a pneumatophore actually feels spongy and up to 50% of its internal volume is gas space.

MYCORRHIZAE AND ROOT NODULES

In many plants, fine hairs develop just behind the tips of growing roots (**250**), greatly increasing the root's absorptive surface. However, root hairs are frequently sparse in trees and may be absent from some gymnosperms. Nevertheless, most woody plants and many herbs develop accessory root/micro-organism symbiotic systems of nutrient supply and absorption. These either take the form of root nodules (**415**, **457–459**) or endo- and ectomycorrhizae (**460–463**); in *Alnus rubra* (red alder) and some legumes both nodules and mycorrhizae can occur on the same root system. Some species form either type of mycorrhizal association with fungi and occasionally, as in *Eucalyptus*, *Juniperus* (juniper), and *Salix* (willow), both types may be present on the same individual.

Root nodules are inhabited by symbiotic micro-organisms (**465**) that fix or transform atmospheric nitrogen into nitrogenous compounds utilizable by the plant. *Rhizobium* bacteria occur in the nodules of numerous legumes (**415**), such as *Acacia*, *Glycine*, and *Trifolium*, as well as other taxa such as members of the Ulmaceae (elm family). The actinomycete *Frankia* (**465**) inhabits nodules in various species of *Alnus* (**457**, **458**) and other members of the Betulaceae (birch family) and *Comptonia peregrina* (bog-myrtle in the Myricaceae, **459**). The vesicular stage of *Frankia* in the root nodules (**465**, **466**) fixes free soil nitrogen; thus the plant receives nitrogenous compounds while the micro-organism obtains sugars. In *Alnus* the nodules may live up to 10 years and sometimes grow to tennis ball size. The individual lobes (**464**) of a compound nodule are modified root tips which sometimes grow out into normal roots (**459**). Nitrogen-fixing cyanobacteria of the genus *Nostoc* are present in surface roots of the cycad *Macrozamia riedlei*.

Symbiotic endomycorrhizal associations involve the fine-roots of a wide variety of trees. Examples include *Acer* (maple, sycamore), *Fraxinus* (ash), *Juglans* (walnut), *Magnolia*, *Olea* (olive), and species of conifers such as *Juniperus* (juniper), *Sequoia sempervirens* (coastal redwood), and *Sequoiadendron giganteum* (giant sequoia). Their root tips normally develop some root hairs and usually there is little external evidence of the associated fungus. Within the root, however, the fungi penetrate the cell walls of the outer cortex and produce highly branched arbuscules (**462**) but they remain separated from the root protoplast by the plasmalemma. The thread-like fungal hyphae ramify through the soil and provide the tree with additional access to essential inorganic ions (especially phosphorous) and water located in soil some distance away from the nearest young roots.

Ectomycorrhizal associations occur between fungi and the roots of most conifers and about 90% of northern hemisphere temperate broadleaf trees. Southern hemisphere taxa such as *Eucalyptus*, *Nothofagus* (southern beech), and some tropical dipterocarps also form ectomycorrhizae. Infected roots typically lack root hairs, are distorted, and may be coral-like (**461**). The fungus is visible externally on the root as a smooth sheath (**463**), while internally its hyphae ramify within the intercellular spaces of the outer cortical cells; they do not penetrate the root's cell walls (**460**). In late summer/autumn the fruiting bodies of many common temperate mushrooms and toadstools represent the aerial extension of various ectomycorrhizal fungi (**467**).

407 Root system of a mature tree of the dicot *Fraxinus excelsior* (ash). This was growing on a steep, eroded slope and shows the richly branched, superficial roots which extend laterally some metres from the trunk.

408 Detail of the superficial root system of the dicot *Fagus sylvatica* (beech). This was a large tree and the roots growing in the lower side of a steep slope showed frequent grafting; this bracing provides additional stability for the trunk and aerial canopy.

409 Buttress roots of the large rain forest tree *Parkia javanica*. This specimen had several prominent buttresses at the base of its trunk which extended laterally several metres and were a metre or so high at their origin from the trunk. Buttresses occur on a number of rain forest tree species and may reach up to several metres high. They help to stabilize large trunks and their aerial canopies; the buttresses have numerous lenticels and are also thought to aid in aeration of the underground root system.

410 Specimen of the dicot *Ficus benghalensis* (banyan tree) showing a massively developed prop root system; these roots develop adventitiously from the branches, grow downwards into the soil, and develop into prominent supporting pillars.

411 Pneumatophores of *Avicennia nitida* (mangrove) growing in saline estuarine mud. The negatively gravitropic roots of this dicot develop from long horizontal roots which extend from the base of the tree trunk and grow in the oxygen-depleted mud. The breathing roots are protected by a cork covering but contain many lenticels which allow aeration of the horizontal root system.

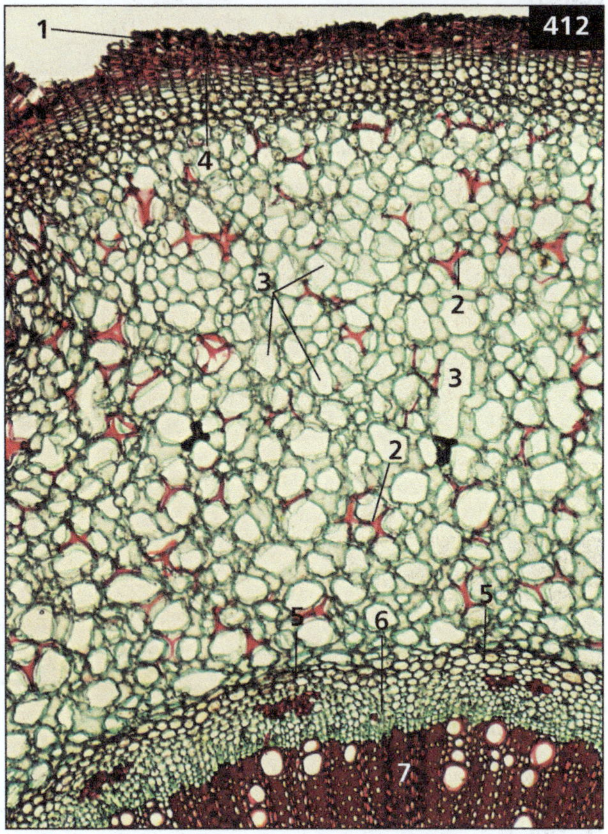

412 TS of a small pneumatophore of the dicot *Avicennia nitida* (mangrove). This negatively gravitropic root is covered by cork (1) containing numerous lenticels. Cortical sclereids (2) support the parenchyma cells which are separated by extensive intercellular spaces (3) providing aeration to the submerged horizontal roots. Cork cambium (4), endodermis (5), secondary phloem (6), secondary xylem (7). (LM.)

1 Cork
2 Cortical sclereids
3 Intercellular spaces
4 Cork cambium
5 Endodermis
6 Secondary phloem
7 Secondary xylem

413 *Rhizophora mangle* (spider mangrove) growing in saline estuarine mud. These dicots show a dense tangle of adventitious stilt roots which help to stabilize the tree and transport water and nutrients to its trunk. Although covered by bark, numerous lenticels in the stilts allow aeration of the roots growing in the oxygen-depleted mud. In the foreground several seedlings, which germinated in the fruits while still attached to the tree, have fallen into the mud and started to grow.

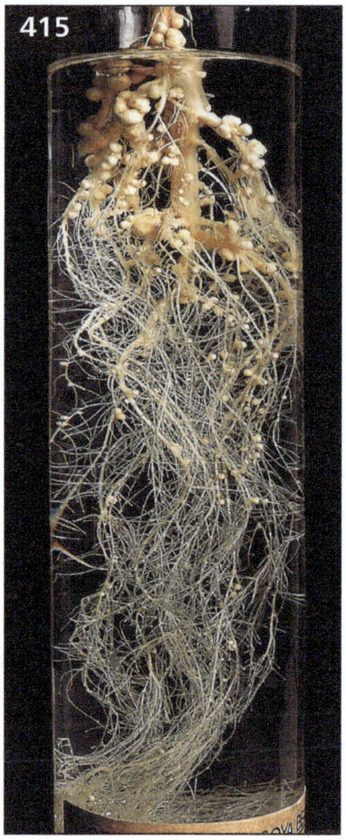

415 Root system of the dicot *Glycine soja* (soybean) showing nodules. This plant was grown in water culture containing balanced mineral salts, except for the absence of nitrates or ammonium salts. Nevertheless, the symbiotic bacteria (*Rhizobium*) present in the root nodules were able to fix free nitrogen so that the soybean plant grew vigorously. Legumes commonly develop root nodules which, on their decay, enrich depleted soils.

414 Long spines on the trunk of the palm *Crysophila nana* (a monocot). These modified adventitious roots are of limited growth and the apices and root caps are replaced by sclerenchymatous points.

416 Adventitious roots at the base of a palm trunk. This unidentified specimen was growing in the rainforest in Queensland, Australia.

417 Roots of strangler fig (*Ficus microcarpa*, a dicot) investing a rainforest tree. These roots are growing down its trunk and have developed from a seed germinating in the droppings of a bird in the canopy of the 'host' tree. The woody roots frequently graft together and prevent the 'host' tree trunk from expanding, eventually cutting off the food supply so that it dies (cf. 422).

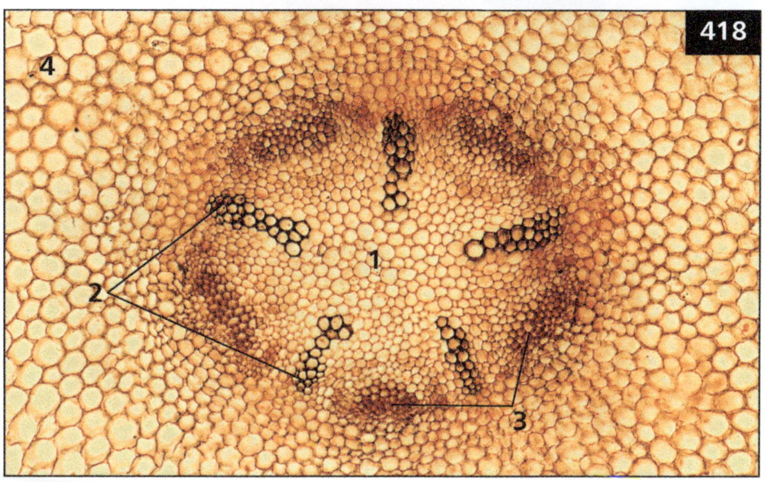

418 TS of the primary root of the dicot *Vicia faba* (broad bean). Note the extensive immature metaxylem (1), five arcs of xylem with the protoxylem (2) situated centrifugally, and the strands of phloem (3) alternating with the xylem. The endodermis is not yet clearly defined and no vascular cambium is apparent. Cortex (4). (LM.)

1 Potential metaxylem
2 Protoxylem
3 Phloem
4 Cortex

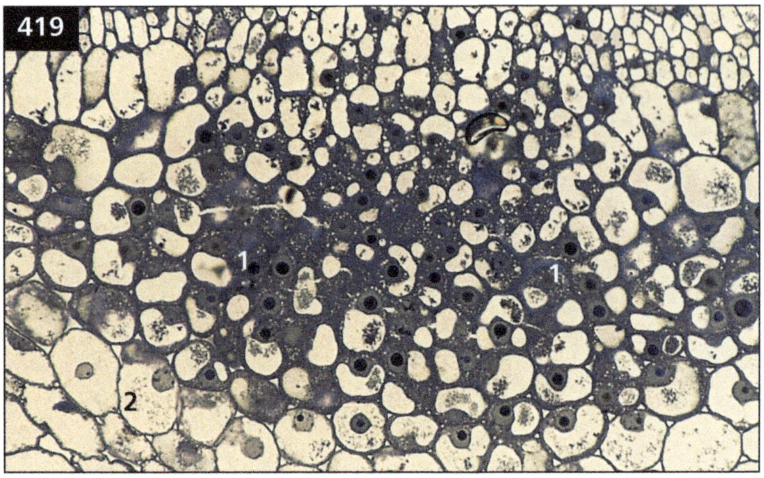

419 LS of a lateral root primordium of the dicot *Pisum sativum* (pea). This has formed by the division and dedifferentiation of the pericyclic parenchyma of the parent root. The densely cytoplasmic cells of the primordium (1) bulge out into the parenchyma of the parent root cortex (2). (G-Os, LM.)

1 Root primordium
2 Root cortex

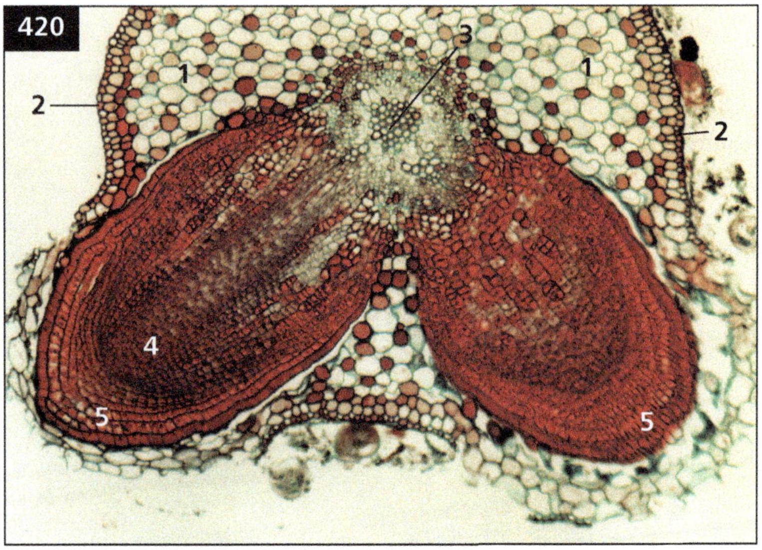

420 LS of lateral root primordia in the dicot *Salix* (willow). These originated opposite protoxylem poles from the pericyclic parenchyma of the parent root (cf. 222) and have now penetrated its cortex (1), with the root tips distorting the normally circular epidermis (2). The vascular cylinder of one root primordium (left) links with the vascular system of the parent root (3). Root apical meristem (4), root cap (5). (LM.)

1 Cortex
2 Epidermis
3 Vascular cylinder
4 Root apical meristem
5 Root cap

421 *De novo* root and bud production on leaf cuttings of the monocot *Sansevieria trifasciata* (mother-in-law's tongue). The basal end of each was inserted in moist compost and all show adventitious roots at the basal ends. Three cuttings have also developed large leafy buds connected by short rhizomes to the parent leaf segments.

422 Hollow 'trunk' of a strangler fig tree (*Ficus microcarpa*). This actually represents the fused roots of an epiphytic seedling which germinated in the crown of a rainforest tree and grew down its trunk to the soil. Meanwhile the 'host' tree's vascular system was cut off by the investing roots (cf. 417) of the fig, and the tree trunk eventually died and rotted away leaving the 'trunk' of the fig in its place.

423 TS of the old trunk of *Cyathea* (tree fern). Note in this rainforest fern the thick mantle of adventitious fibrous roots (1) which arise from and invest the petiolar bases (2). The latter persist after the leaves wither; they surround a complex sclerenchymatous vascular core (3), but secondary growth is absent. The leaf bases and mantle of roots help support the stem which may be several metres high and bears evergreen leaves several metres long.

1 Fibrous roots
2 Petiolar bases
3 Vascular core

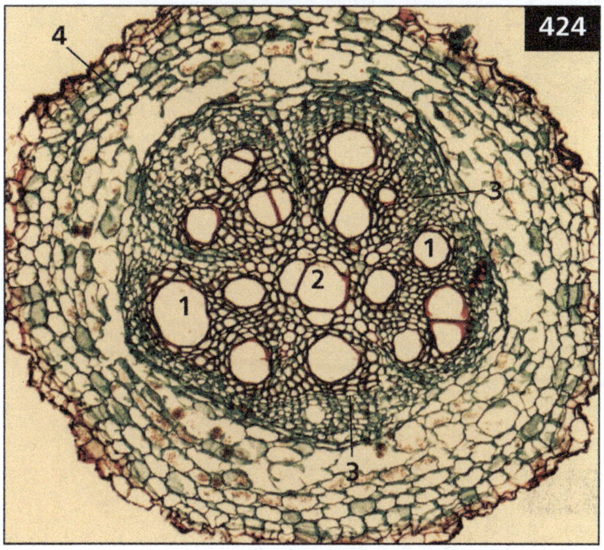

424 TS showing early secondary growth in the fleshy root of the dicot *Ipomoea batatas* (sweet potato). In the centre a cylinder of secondary xylem with prominent vessels (1) invests the core of primary xylem (2). At a later stage the vascular cambium (3) produces large quantities of xylem parenchyma leading to the formation of a root tuber. Note the cortical origin of the cork cambium (4). (LM.)

1 Secondary xylem
2 Primary xylem
3 Vascular cambium
4 Cork cambium

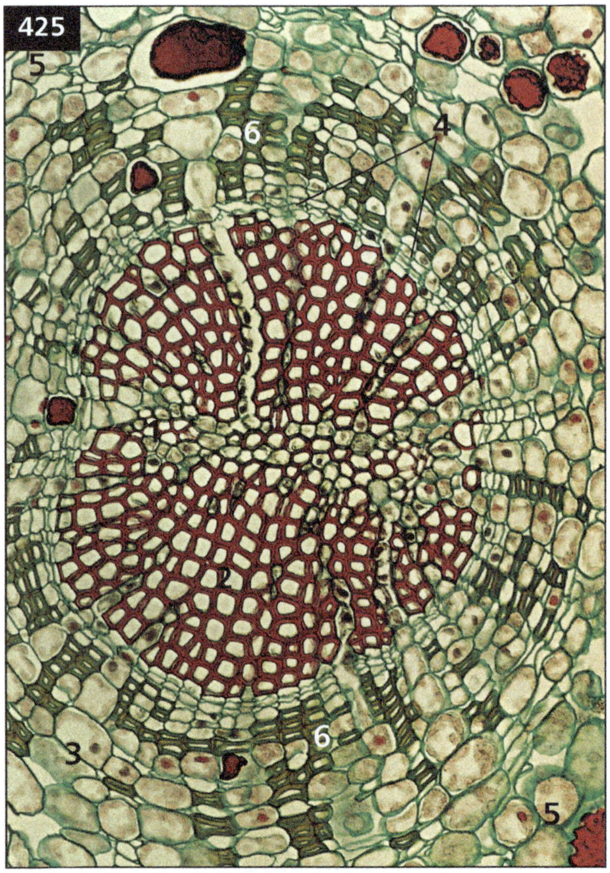

425 TS showing early secondary thickening of the young root of the gymnosperm *Ginkgo biloba*. The primary xylem is diarch and the narrower protoxylem elements (1) contrast with the somewhat wider but uniform tracheids of the secondary xylem (2). Both this and the secondary phloem (3) show radial cell lineages traceable to their origin in the vascular cambium (4). Cortex (5), phloem fibres (6). (LM.)

1 Protoxylem elements
2 Secondary xylem
3 Secondary phloem
4 Vascular cambium
5 Cortex
6 Phloem fibres

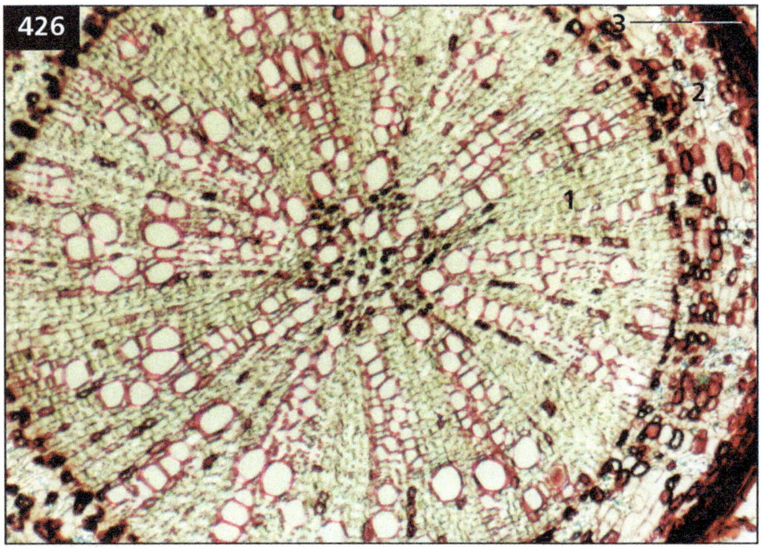

426 TS of a young root of the dicot *Tilia cordata* (lime). A broad cylinder of secondary xylem (1, with many wide vessels) is surrounded by a narrow layer of secondary phloem (2) which contains numerous fibres. A thick layer of cork (3) covers the root surface. (Polarized LM.)

1 Secondary xylem
2 Secondary phloem
3 Cork

427

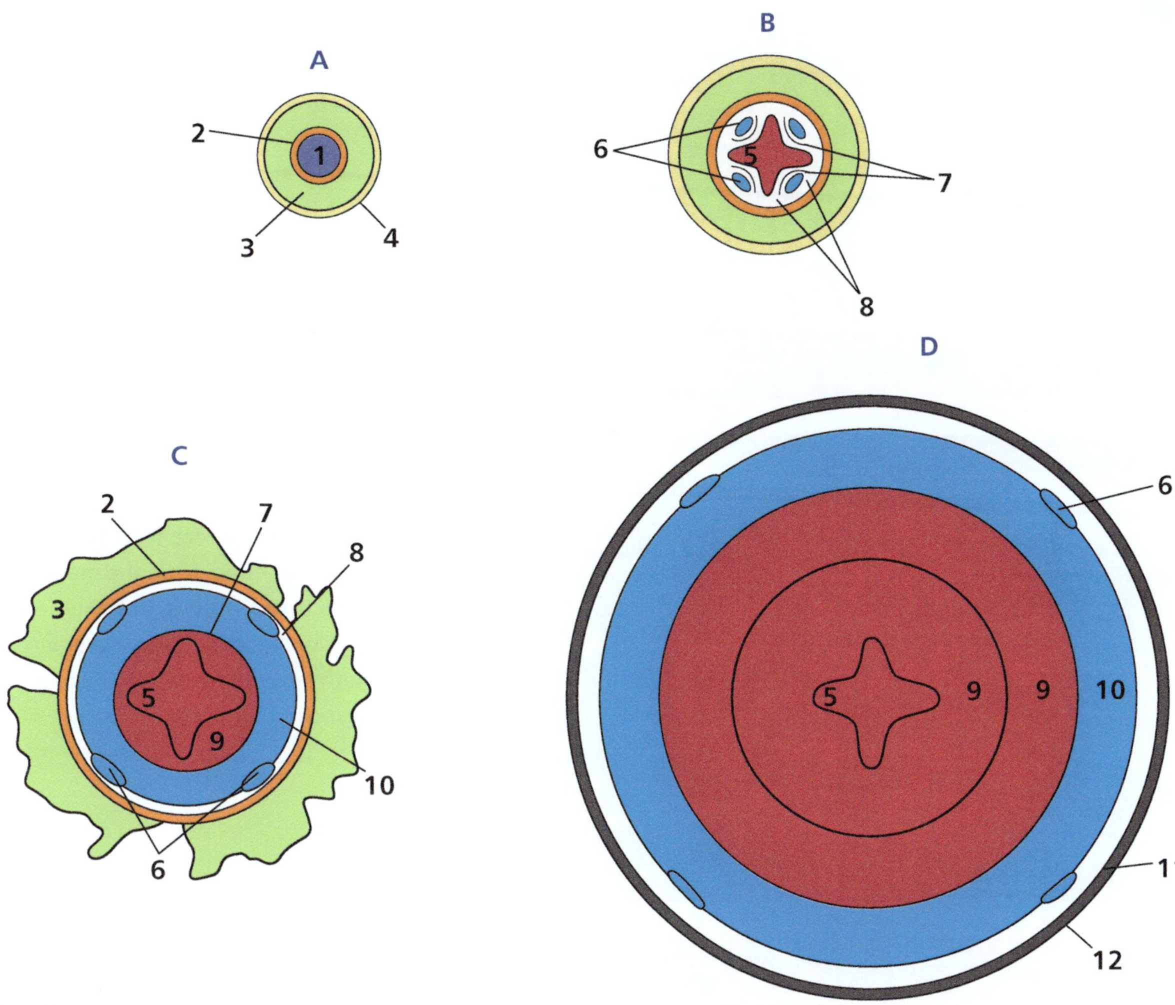

427 Diagrammatic development of the root in a woody dicot as seen in TS. A: Just behind the root apex a central cylinder of procambium (1) is separated by the single-layered endodermis (2) from the cortex (3) and epidermis (4). B: Beyond the root hair zone the mature primary root shows a tetrarch xylem (5) separated from the phloem strands (6) by narrow arcs of vascular cambium. (7). C: The vascular cambium has differentiated laterally from the pericycle (8) and now encloses the xylem. The vascular cambium has given rise to a layer of secondary xylem (9) internally and secondary phloem (10) externally. Primary phloem strands are visible on its outer margin while the primary xylem remains intact internally. The epidermis has ruptured and the cortex and endodermis are being sloughed off. D: The root shows further secondary thickening with two growth rings in the secondary xylem. These are not evident in the secondary phloem although the crushed remains of the primary phloem are still evident. The endodermis and external tissues have sloughed off and a cork cambium (11) has arisen in the pericycle to form a protective layer of cork (12).

1 Procambium
2 Endodermis
3 Cortex
4 Epidermis
5 Tetrarch primary xylem
6 Primary phloem strands
7 Vascular cambium
8 Pericycle
9 Secondary xylem
10 Secondary phloem
11 Cork cambium
12 Cork layer

428 Large lateral root of the dicot tree *Quercus petraea* (oak) exposed by the flood waters at the side of Loch Lomond, Scotland. Note the prominent lenticels visible in the outer cork.

429 Fine roots from a large tree of the dicot *Fraxinus excelsior* (ash) growing in the cracks of a rock face. These roots have grown down from the main branch roots in the soil above, and their expansion by secondary thickening growth helps to further break open the exposed rock face.

430 Plant of *Lophophora williamsii* (peyote) with succulent root (1) merging uniformly with succulent shoot (2). Water, starch, and alkaloids are stored in both stem and root parenchyma. Only the blue-grey portion of shoot is aerial, all the rest is subterranean.

1 Succulent root   2 Succulent shoot

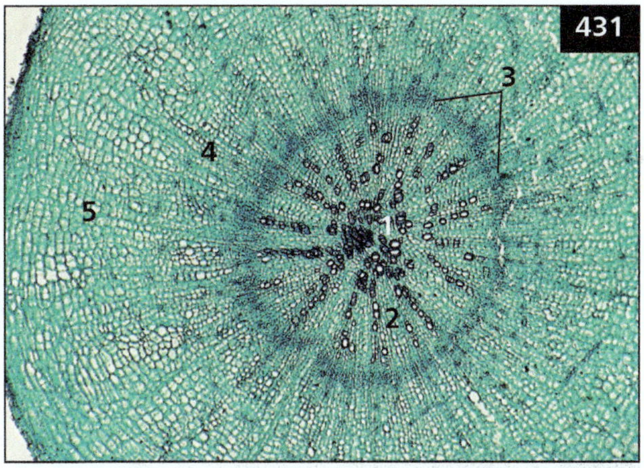

431 TS of the fleshy root of the dicot *Daucus carota* (carrot). The narrow core of primary xylem (1) is surrounded by an extensive but mainly parenchymatous secondary xylem (2) in which a few tracheary elements occur. The well-defined vascular cambium (3) also produces centrifugally a largely parenchymatous secondary phloem (4). Cortex (5). (LM.)

1 Primary xylem
2 Secondary xylem
3 Vascular cambium
4 Secondary phloem
5 Cortex

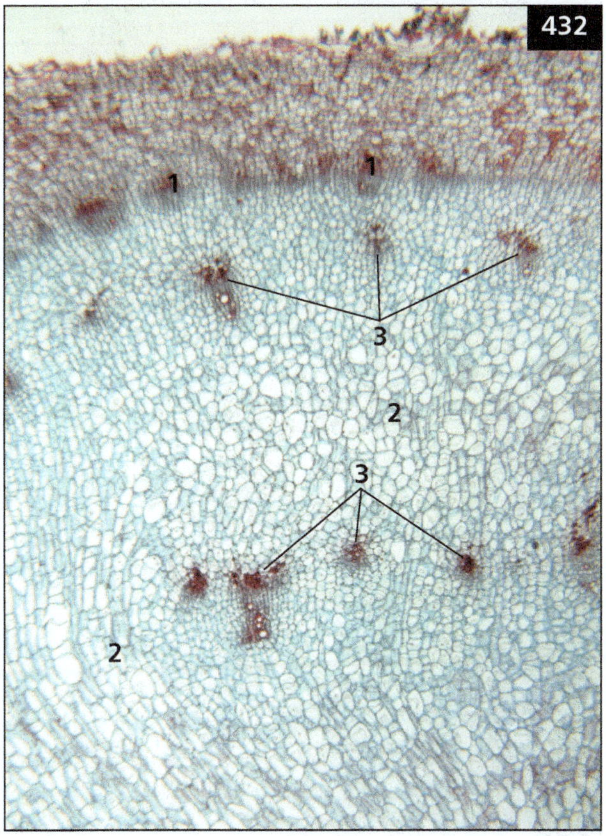

432 TS of fleshy root of *Beta vulgaris* (beet), showing multiple, simultaneous vascular cambia. The original vascular cambium (1) is outmost and produces extremely parenchymatous wood (2) with just a few clusters of vessels (3). New vascular cambia arise around the vessel clusters and produce even more parenchyma. (LM.)

1 Vascular cambium
2 Parenchymatous wood
3 Clusters of vessels

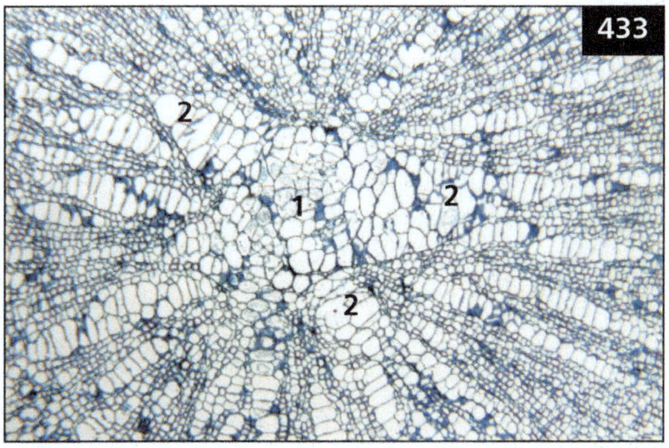

433 TS of centre of succulent woody root of *Thelocactus rinconensis*. What appears to be pith is actually dilatated metaxylem (1) and the dilatated innermost portions of rays of secondary xylem (2). Even the nondilatated portions of wood have a high percentage of storage parenchyma rather than fibres or vessels. (LM.)

1 Dilatated metaxylem
2 Dilatated xylem rays

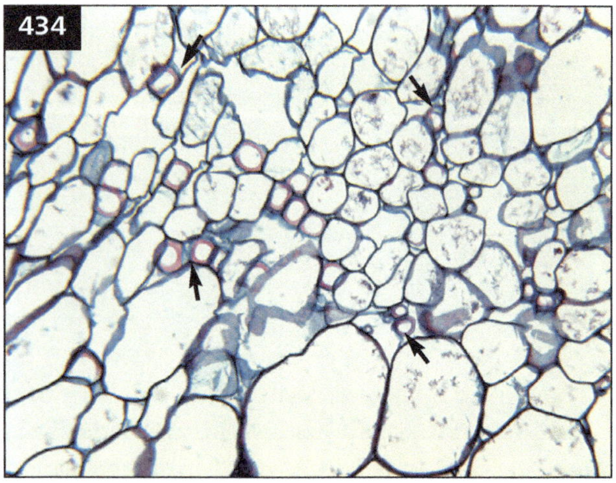

434 TS of centre of succulent woody root of the cactus *Encephalartos strobiliformis*. The presence of vessels (arrows) among the parenchyma cells indicates that this is dilatated metaxylem, not pith. Pith can contain medullary bundles, but not isolated vessels. (LM.)

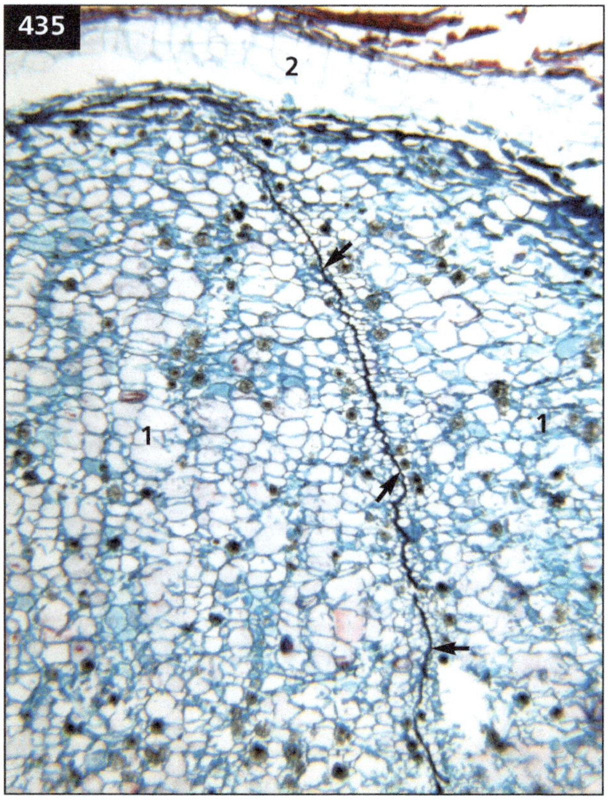

435 TS of outermost portion of succulent root of *Maihueniopsis darwinii*, showing very thick secondary phloem. Narrow black streak (arrows) consists of collapsed sieve tube members in axial system; all the rest (1) is two broad rays consisting of storage parenchyma, druses, and a few mucilage cells (pink). Bark is present (2). (LM.)

1 Phloem rays
2 Bark

436 *Phoradendron tomentosum*, a hemiparasite, is attached to its host plant by a haustorium, a modified root. The shoot has chlorophyll, is photosynthetic, and has relatively ordinary leaves and stems. Its host plant has abscised its leaves for winter, and water loss through the parasite is a significant danger to the host.

437 Excavated plant of *Corynea crassa*, a holoparasite. It spends most of its life as a subterranean mass (lower part of image) but thick stalks elevate the closed inflorescences above ground, where they would have opened and exposed the flowers to pollinators.

438 *Ligaria cuneifolia* is a hemiparasite that attacks cacti (*Corryocactus brevistylus* here). The softness of the host allows the distribution of the parasite's invasive haustorium to be studied easily. *Ligaria* attacks the cactus's cortical bundles (cf. 333).

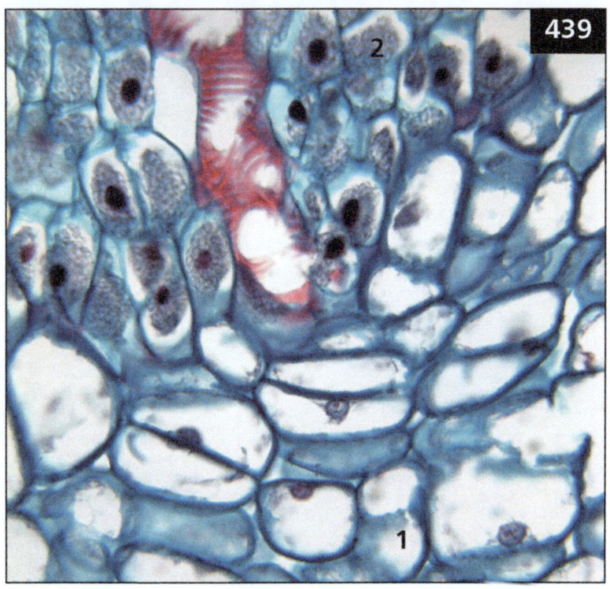

439 TS of cactus host (1) and *Ligaria haustorium* (2); *Ligaria* nuclei stain dark red whereas those of the host are pink with a red nucleolus. *Ligaria* has formed a vessel that runs to its interface with the host, but there is no host xylem to be tapped. (LM.)

1 Host cortex

2 Cells of parasite

440 *Cuscuta* is a holoparasitic dicot vine with tiny achlorophyllous leaves and long internodes. An individual plant can spread across several different host bushes, parasitizing each with adventitious roots modified to be haustoria.

441 Close-up of a *Cuscuta* vine (orange) twining around a host leaf. The numerous pegs are adventitious roots/haustoria that are attacking the leaf's epidermis (cf. 442). By wrapping around the host, *Cuscuta* braces itself such that the haustoria penetrate rather than merely pushing the *Cuscuta* stem away from the host.

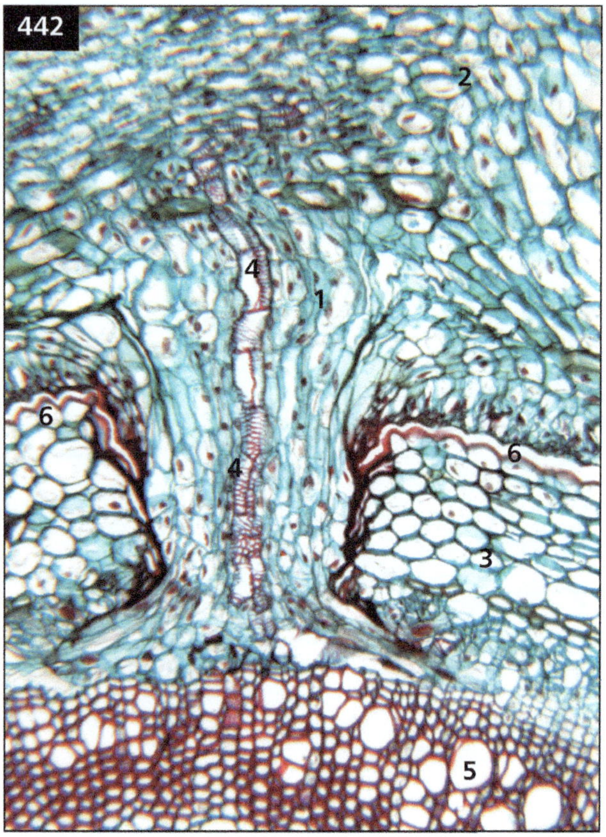

442 RLS of *Cuscuta* haustorium (1), LS of *Cuscuta* stem (2), TS of host stem (3). A set of *Cuscuta* vessel elements (4) has differentiated within the haustorium, which has reached the host wood (5) and is now spreading laterally between host wood and phloem. Epidermis and cuticle (6) of host are present except where broken by the haustorium. (LM.)

1 *Cuscuta* haustorium
2 *Cuscuta* stem
3 Host stem
4 *Cuscuta* vessel elements in haustorium
5 Host wood
6 Epidermis and cuticle of host

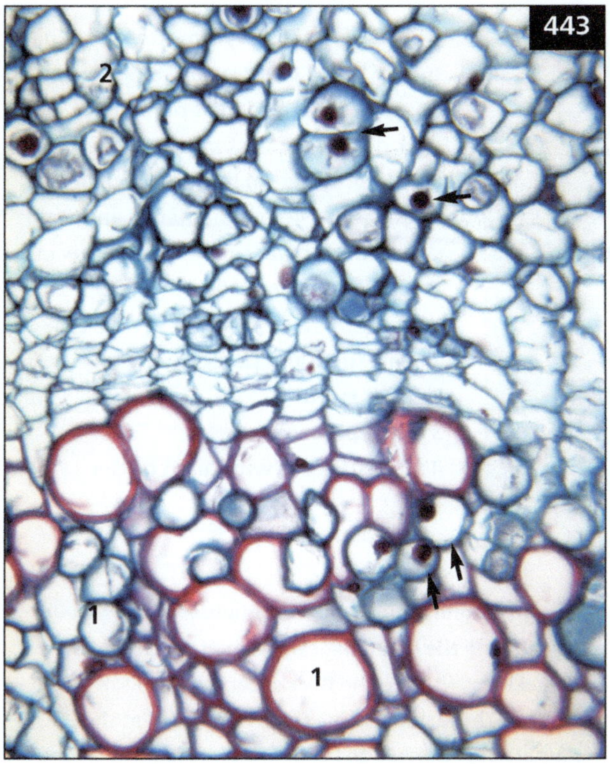

443 TS of wood (1), vascular cambium and secondary phloem (2) of host *Trichocereus chilensis* invaded by uniseriate strands (arrows) of *Tristerix aphyllus*. Like those of *Ligaria*, *Tristerix* nuclei stain very dark red. These uniseriate strands constitute the entire body of the *T. aphyllus* at this stage of its life. (LM.)

1 Host wood

2 Host secondary phloem

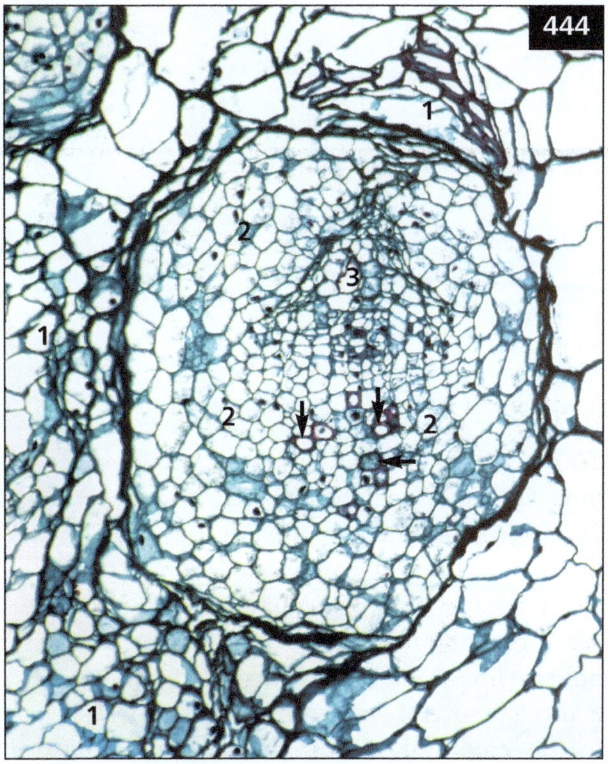

444 TS secondary phloem (1) of host *Trichocereus chilensis* occupied by a multiseriate strand of *Tristerix aphyllus* (2). The *T. aphyllus* has formed idioblastic vessel elements (arrows) and phloem (3) but has no epidermis, cortex, or pith. (LM.)

1 Host phloem
2 Multiseriate strand of *Tristerix aphyllus*
3 Phloem of *T. aphyllus*

445 Excavated plant of *Ombrophytum subterraneum*. Foreground is a young inflorescence (1) that has just emerged from the tuber (2) by tearing the overlying tuber tissues (the volva, 3). The inflorescence has many thick branches, each with a white scale at its end. Background is mature infructescence, covered in tiny pink fruits.

1 Young inflorescence
2 Tuber
3 Volva

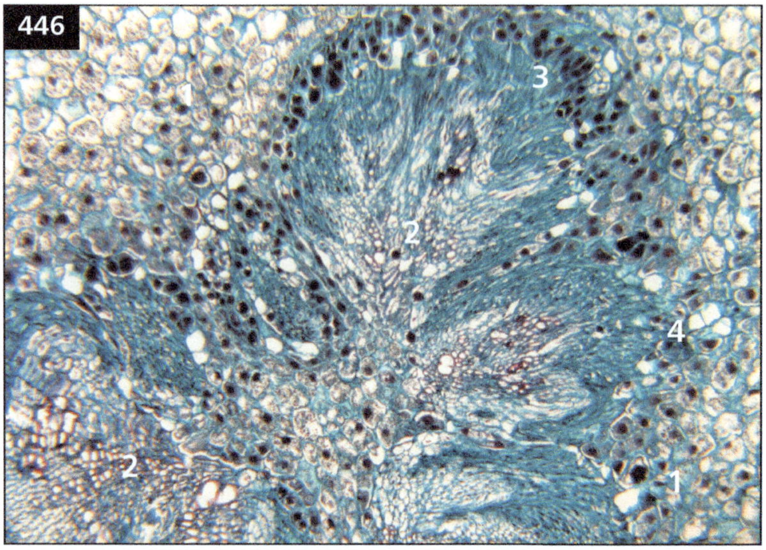

446 TS of *Ombrophytum subterraneum* tuber (1) at a point where it is invaded by host xylem (2) and host phloem (3). Very dark dots (4) at host/parasite interface are *Ombrophytum* nuclei, which are huge (about 40 μm diameter) and stain dark red. Host vascular tissues grow into the parasite's body. (LM.)

1 Parasite tuber parenchyma
2 Host xylem
3 Host phloem
4 Parasite cells with giant nuclei

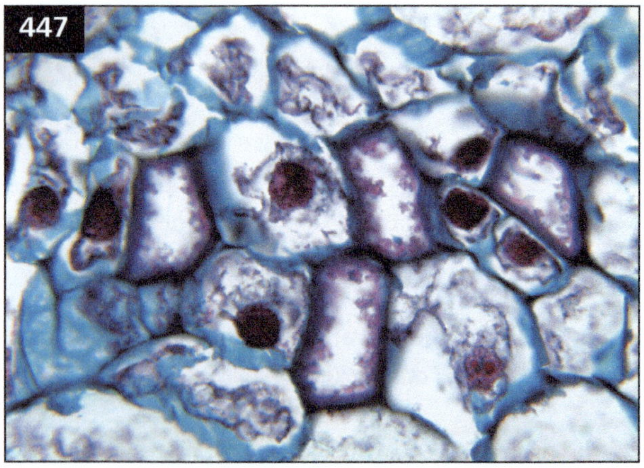

447 High magnification of *Ombrophytum subterraneum* cells showing the giant, densely stained nuclei and the unusual ingrowths on the thick secondary walls of vessel elements. These two features allow every cell at the host/parasite interface to be identifiable. (LM.)

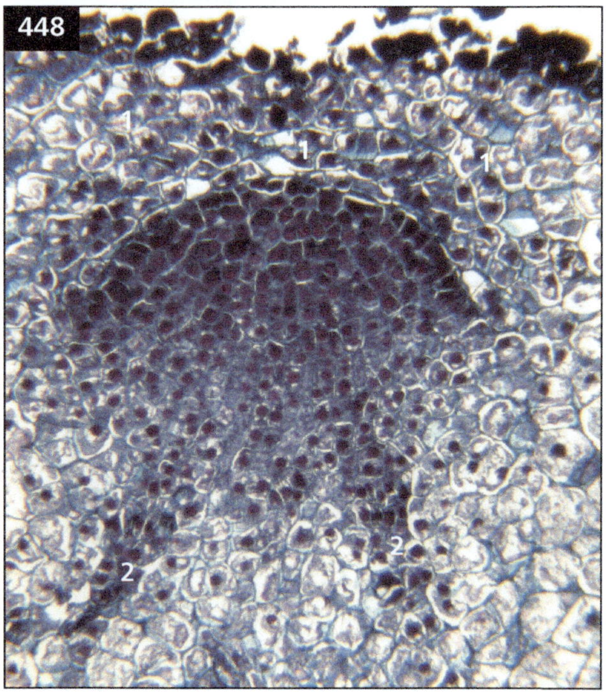

448 TS of *Ombrophytum subterraneum* tuber, LS of adventitious inflorescence bud that has formed within the tuber. Overlying tuber cells (1) will initially expand as the inflorescence grows, forming the volva, but they will ultimately be torn apart. Surprisingly, vascular bundles (2) of the inflorescence approach but do not make contact with vascular bundles of the tuber. (LM.)

1 Tuber cells that will become volva
2 Vascular bundles of inflorescence

449 Hypocotyl of young plant of *Hydnophytum formicarium*, starting to enlarge (note that shoot base is narrow). A chamber and an opening are already present, allowing ants ready access to living spaces.

450 Section through greatly enlarged hypocotyl of an older plant of *Myrmecodia tuberosa*, filled with chambers and occupied by ants. (Image courtesy of Dr. C. Puff; Puff C, Chayamarit K, Chamchumroon V [2005]. *Rubiaceae of Thailand. A Pictorial Guide to Indigenous and Cultivated Genera*. The Forest Herbarium, National Park, Wildlife and Plant Conservation Department, Bangkok, Thailand.)

451 Thorn-like stipules of *Acacia cornigera* must be hollowed out by ants before their fibrous mesophyll sclerifies.

452 Beltian bodies on leaflet tips of *Acacia cornigera* are harvested and eaten by the ants. Beltian bodies store glycogen, a glucose polymer that is more typical of animals than plants.

453 *Avicennia marina* dicot mangrove tree with its roots submerged by brackish water at high tide of the Paramata river near Sydney, Australia.

454 *Avicennia marina* trees showing clusters of small adventitious roots (incipient prop roots, 1) hanging from branches. Note also the dense mat of pneumatophores (2) and numerous *Avicennia* seedlings (3) growing amongst them.

1 Adventitious roots
2 Pneumatophores
3 *Avicennia* seedlings

455 Specimen of *Avicennia marina* mangrove tree with its vertical breathing roots and horizontal cable roots (exposed by water erosion) at low tide on the coast of Queensland, Australia.

456 Knee or breathing roots (pneumatophores) of the dicot mangrove *Brugiera gymnorhiza* exposed at low tide in a mangrove swamp on the coast of Queensland, Australia.

457 Large compound nodule in the root of the dicot *Alnus glutinosa* (alder). This is composed of numerous closely crowded individual nodules which are infected with cells of the symbiotic actinomycete *Frankia*, which are able to fix free nitrogen in the soil atmosphere.

458

458 *Alnus glutinosa* (alder) roots with two large nodules (inhabited by the nitrogen-fixing actinomycete *Frankia*) exposed by water erosion of tree's roots on the banks of Loch Lomond, Scotland.

459

459 Root nodule of the nitrogen-fixing dicot *Comptonia peregrina* which has been grown for a month in liquid culture. Note the numerous root tips; most are stunted but some are growing out into normal lateral roots.

460

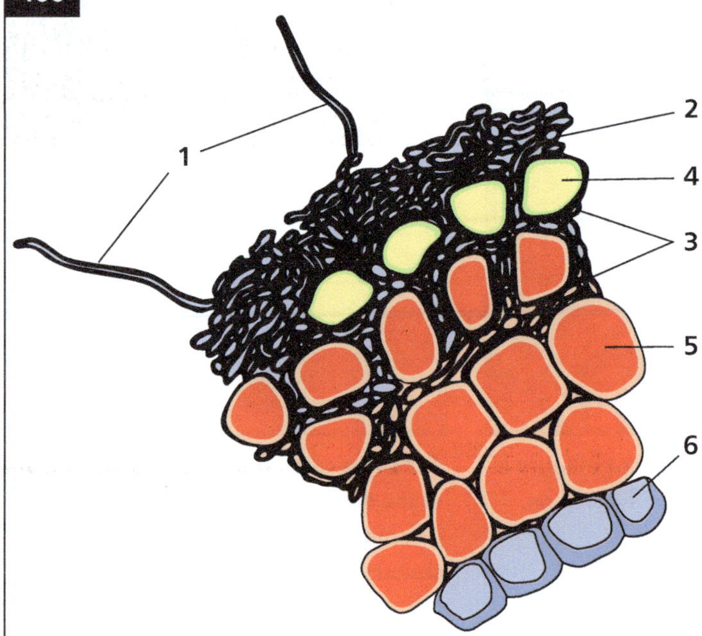

460 Diagrammatic TS of the cortex of a dicot root invested by an ectomycorrhizal fungus. Fungal hyphae in soil (1), investing hyphal sheath (2), fungal hyphae penetrating cortical intercellular spaces (3), root epidermis (4), inner cortical cells (5), root endodermis (6).

1 Fungal hyphae
2 Hyphal sheath
3 Hyphae penetrating intercellular spaces
4 Root epidermis
5 Inner cortical cells
6 Root endodermis

461 Diagram of coralloid ectomycorrhizal fine roots of the conifer *Pinus* (pine). Main root (1), distorted coralloid root branches (2), investing fungal sheath (3).

1 Main root
2 Distorted coralloid root branches
3 Investing fungal sheath

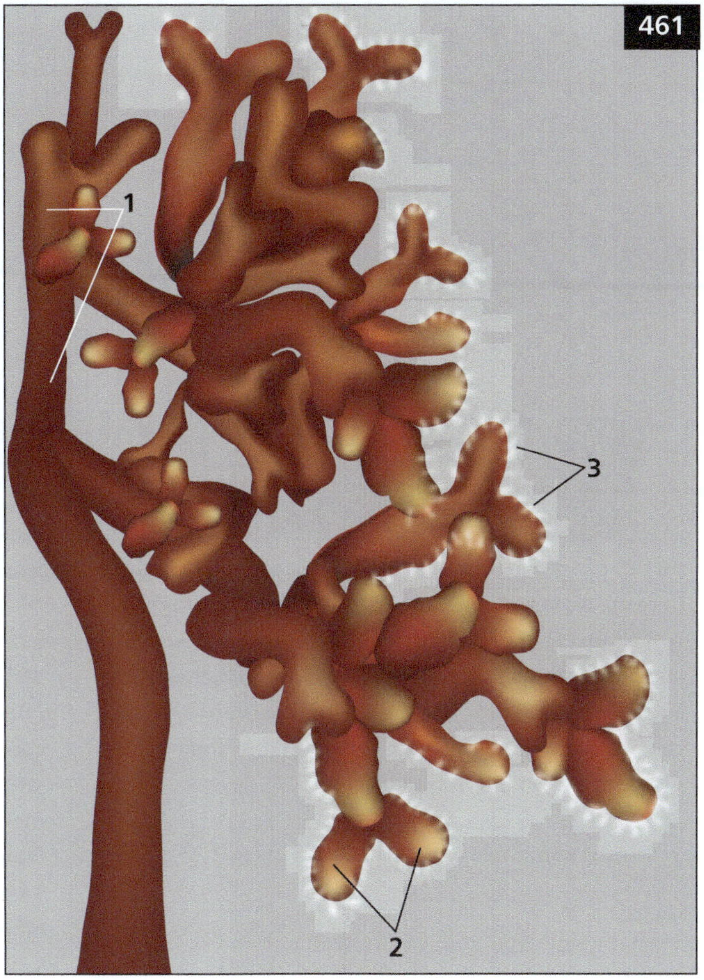

462 Diagram showing a section through the cortex of a fine tree root infected by an endomycorrhizal fungus. Fungal hyphae in soil (1), fungal arbuscle in the inner cortex (2), fungal vesicle (3), fungal hypha penetrating cortical cell wall (4), uninfected cortical cell (5), host root endodermis (6), host epidermis (7).

1 Fungal hyphae
2 Fungal arbuscle
3 Fungal vesicle
4 Hyphae penetrating cortical cell wall
5 Uninfected cortical cell
6 Host root endodermis
7 Host epidermis

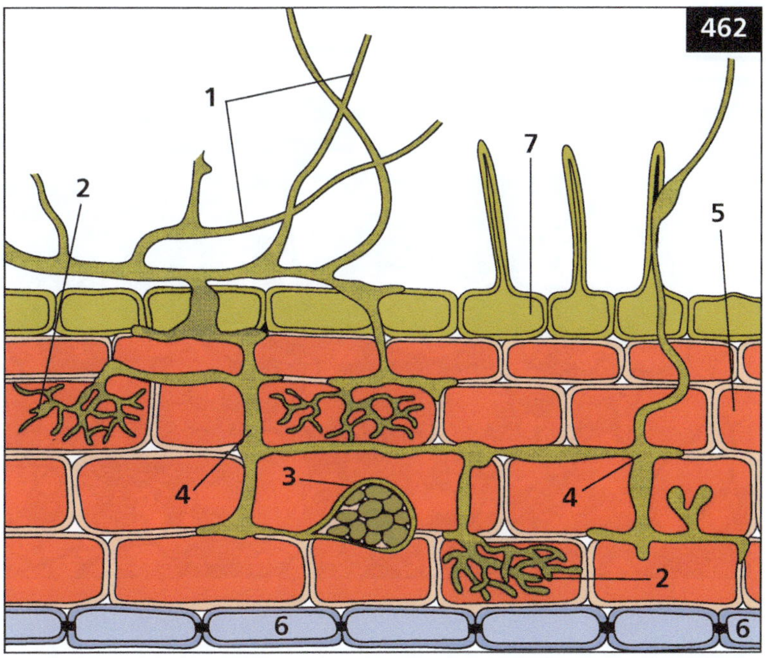

463

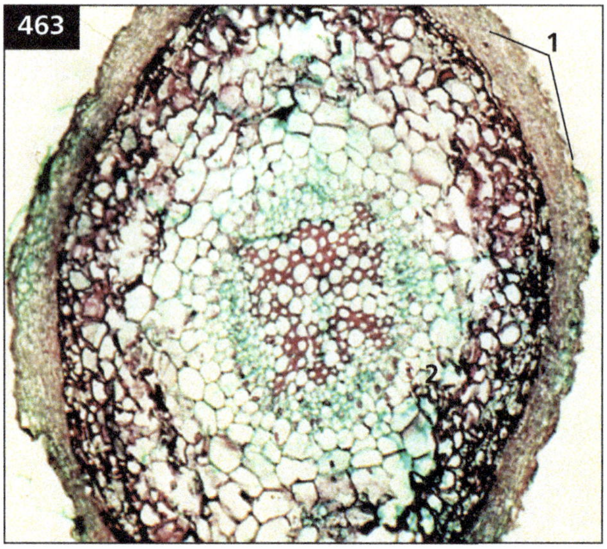

463 TS of a mycorrhizal root of the dicot *Fagus sylvatica* (beech). Note the investing mat of symbiotic fungal mycelium (1); the hyphae penetrate into the surrounding soil and absorb phosphates which are then supplied to the root. They also penetrate between the root epidermal cells into the cortex (2), but their distribution remains apoplastic. (LM.)

1 Fungal mycelium
2 Root cortex

464

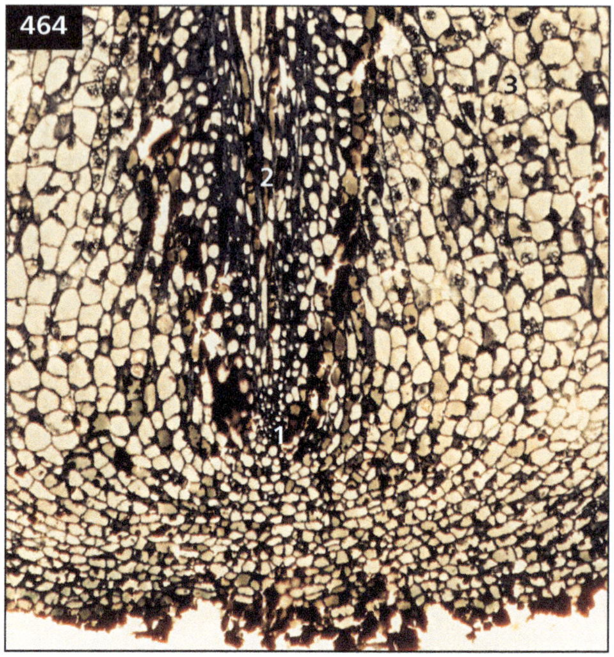

464 LS of a root nodule of the dicot *Alnus glutinosa* (alder). This single nodule was part of a much larger compound nodule (cf. 457) and represents a swollen lateral root of determinate growth. Note the apical meristem (1), vascular cylinder (2), and cortex (3). No *Frankia* is evident in this section but the cortical tissue basal and lateral to this area showed early stages of infection (cf. 465). (LM.)

1 Root apical meristem
2 Vascular cylinder
3 Cortex

465

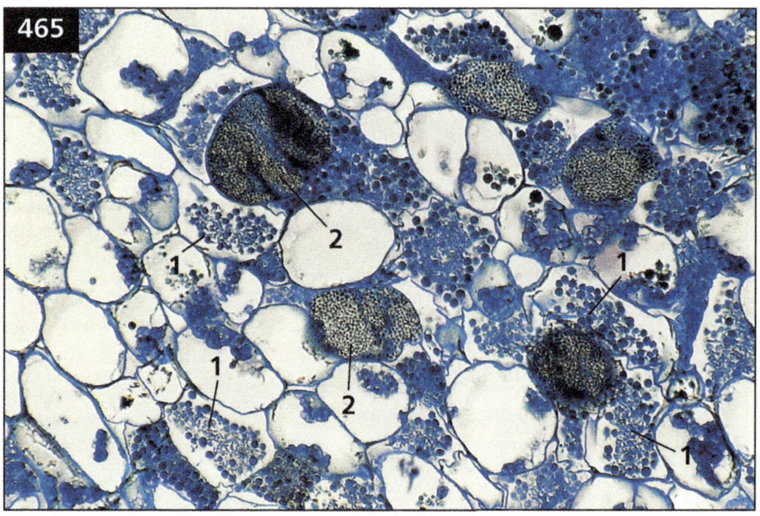

465 Cortical tissue of *Alnus glutinosa* (alder) root nodule showing the distribution of hyphae of the symbiotic actinomycete *Frankia*. This section shows host cells containing numerous endophytic stages of the actinomycete. In many cortex cells (1) the fine hyphal network bears swollen vesicles (which fix nitrogen), while other cells (2) contain the sporangia. (G-Os, LM.)

1 Hyphae with swollen vesicles
2 Host cells containing sporangia

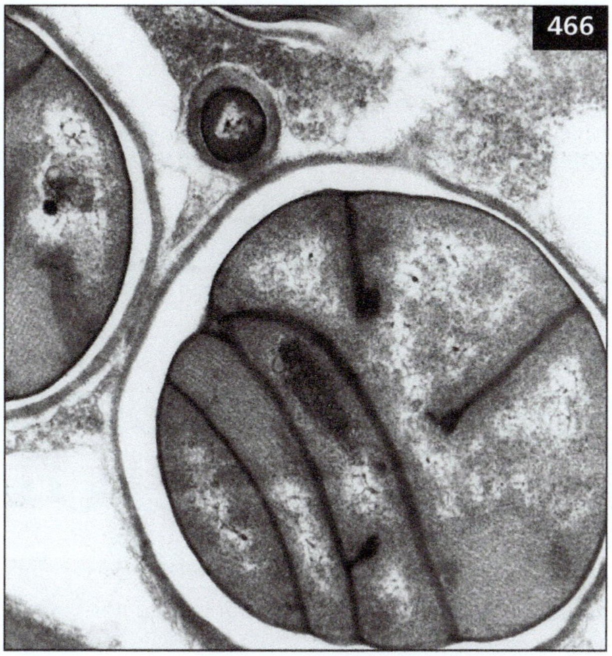

466 TEM of a vesicle of the actinomycete *Frankia* (in which nitrogen is fixed) growing symbiotically in a cortical cell of a root nodule of the dicot tree *Alnus glutinosa* (alder).

467 Aerial fruiting body of *Amanita muscaria* (fly agaric) which is a common ectomycorrhizal fungus associated with the root systems of various species of the dicot *Betula* (birch). (Copyright T.N. Tait.)

CHAPTER 8

Plant Reproduction

ASEXUAL (VEGETATIVE) REPRODUCTION

Vegetative reproduction in nature

In numerous land plants, ranging from bryophytes to many vascular taxa, vegetative reproduction occurs naturally as an adjunct to, or sometimes a replacement for, the sexual process (**468–470**). Considerable areas of land may become colonized by clonal individuals and in northwest America *Populus tremuloides* (quaking aspen) rarely reproduces by seed but extensive forests have arisen from adventitious root buds (**471**).

Many plants show modifications related to both perennation and vegetative reproduction. Rhizomes, which are long-lived stems with numerous adventitious roots and often bearing scale leaves, grow more or less horizontally beneath the soil surface; their axillary or terminal buds give rise to aerial shoots (**469, 470, 472**). In *Pteridium aquilinum* (bracken) the rhizome internodes may reach 1 m or more in length and lie deep beneath the soil surface. Rhizomes are generally thick and fleshy and in the palm *Nypa fruticans* can reach up to 50 cm in diameter; however, some may be only a few millimetres thick, as in various grasses.

Rosette plants often develop thin runners or stolons, with long internodes (**473**). These stems creep along the soil surface, rooting at the nodes where new plantlets eventually develop due to the decay of the intervening tissues. Stem tubers are typically underground swollen shoots arising from axillary stolons on the aerial plant as in *Solanum tuberosum* (**338, 474**), now domesticated as the edible potato, but with several hundred wild relatives still distributed from the southwestern USA to the Andes in South America.

A perennating taproot typically incorporates some hypocotyl tissue, with dormant buds, where it was originally attached to the aerial shoot as in *Daucus carota* (carrot, **475**). Both bulbs and corms represent vertically growing subterranean shoots. However, a bulb has a very condensed stem bearing many leaves packed with food reserves (**476**), whereas the greatly swollen stem of a corm stores the food for the growth of the next season's aerial shoot (**477**). In numerous taxa, the aerial stems, inflorescences, and leaves bear various vegetative bulbils and small tubers (**27A**) which easily break off from the parent, form roots, and establish themselves as new plants.

In many herbaceous and woody plants, including some conifers such as *Sequoia sempervirens*, adventitious buds develop on their roots (**145**). In various tree species such as *S. sempervirens*, *Tilia*, *Ulmus*, *Populus*, and *Liquidambar*, extensive clonal stands (**471**) remain linked together via their root systems. In addition, adventitious roots may grow from the shoots and branches (layering) of shrubs and trees when immersed in water or touching the soil. A semi-independent stem/trunk results (**478**); this may remain connected with the parent or may eventually become separated by decay.

A number of dicot trees such as *Aesculus* and *Tilia* show bud sprouting (often profuse) from the old trunk (**239**). This is often due to the reactivation of dormant axillary buds buried in the bark but sometimes, as in cauliflorous trees, the buds are adventitious. Following coppicing, pollarding, and other pruning procedures, new shoots often appear near the cut surfaces of the trunk or branch (**479**); these may be adventitious or axillary in origin. In some cases the buds develop *de novo* from the cut surface by dedifferentiation of the vascular cambium or, as in several tropical taxa, the xylem parenchyma.

If the plumule (seedling shoot) of *Linum usitatissimum* (flax) is badly damaged, within several days numerous replacement adventitious bud primordia (**237**, **238**) develop on the hypocotyl. These originate from epidermal cells but the activity spreads to the adjacent cortical tissue; vascular strands then differentiate to link the larger buds to the hypocotyl vascular system and several buds grow into replacement shoots.

In various conifers the sexually produced proembryo divides vegetatively into several embryos (polyembryony). In species of *Pinus* the fertilized egg divides to form a four tiered, 16-celled proembryo. Filamentous derivatives of the lower tier of cells eventually separate vertically into four embryos (**480**), but normally only one embryo develops to maturity per seed. In some angiosperms such as *Citrus* species and *Funkia ovata*, the embryo may develop normally from a fertilized haploid egg but further embryos form asexually (apomictically) from the nucellus or other diploid components of the ovule. In taxa where microspecies are common (as in *Taraxacum*, **481**, **482**; *Hieracium*, *Alchemilla*, and *Rubus*) the diploid megaspore does not undergo reduction division (meiosis) but instead divides mitotically to form a diploid egg. In such cases the embryos develop from an unfertilized diploid egg and/or diploid ovular tissue.

Artificial vegetative propagation

Cuttings from various seed plants are widely employed in horticulture. In stem cuttings axillary buds are already present but adventitious roots usually form *de novo* (**143**); however, root primordia already occur on the intact stem in taxa such as *Salix* and *Glechoma* (**236**). In leaf cuttings, both roots and buds develop adventitiously (**421**, **483**), while new buds arise on root cuttings (**145**, **484**).

In an excised organ, wound healing occurs in the vicinity of the cut surface. Damaged cells die and the adjacent cells become suberized and a cork layer sometimes forms. The lumens of tracheary elements become blocked by gums and tyloses intruding from xylem parenchyma cells while the underlying, undamaged tissues proliferate to form a superficial parenchymatous callus. Adventitious roots then develop endogenously (but usually not directly from the callused surface) near the basal end of the stem or leaf cuttings (**143**) and penetrate the epidermis or periderm laterally. In herbaceous stems adventitious roots frequently arise by the dedifferentiation of parenchyma lying between the vascular bundles, but can also be initiated from the phloem parenchyma or pericycle. In woody cuttings roots commonly originate in parenchyma of the secondary phloem or in ray parenchyma, but may also develop from the vascular cambium, pith, and lenticels. In some difficult-to-root cuttings, the sclerenchymatous cylinder external to the vascular tissues apparently impedes lateral root growth but instead these may emerge from the basal callus.

Excised leaves from more than 300 flowering plant taxa, mainly dicot species of tropical/sub-tropical origin, spontaneously develop adventitious roots and buds when grown as cuttings. These include many decorative houseplants (**483**) such as *Begonia*, *Peperomia*, *Saintpaulia* (African violet), *Streptocarpus* (African primrose), and *Sansevieria* (mother-in-law's tongue, **421**). Adventitious roots and shoots typically form at the basal end of the cut leaf. Roots are initiated endogenously from dedifferentiated parenchyma cells associated with the leaf veins and usually appear before the adventitious buds. The latter arise from epidermal or mesophyll cells near the leaf mid-rib, prominent veins, or the petiole (**144**). In *Peperomia* and *Sedum* cuttings a callus first develops at the excised surface of the leaf, and new roots and buds develop from meristemoids arising within it.

In various species with fleshy roots, such as *Taraxacum*, *Crambe*, and *Cichorium*, new growth occurs from the callus formed at the excised surfaces of the cutting; regeneration is markedly polar with buds soon forming at the surface nearest the parent shoot system (**484**) and roots later at the other extremity. In species such as *Armoracia rusticana* (horseradish), both buds and roots develop from the cork cambium in a nonpolar way.

The advent of *in vitro* techniques, employing appropriate hormonal and cultural regimes, has vastly extended the range of plants in which propagation from excised leaves stems, floral parts, tissues, and isolated cells/protoplasts is possible (**485**). This occurs either via the induction of separate buds and roots in the culture or by the formation of a somatic (asexual) embryo. In the latter case a plumule and one or more cotyledons lie at one end of the embryo while the radicle terminates the other pole (**486**). Immature pollen grains (haploid microspores) of many plant taxa can also be cultured *in vitro* and stimulated to develop either directly into embryos or undergo an initial callus phase from which adventitious organogenesis/embryogenesis later occurs. In most species the generative nucleus of the pollen grain (see below) remains inactive, but the larger vegetative nucleus divides to give a haploid embryoid. However, prior to the development of their investing cell walls, the haploid nuclei sometimes fuse spontaneously resulting in embryos which are diploid or of higher ploidy.

SEXUAL REPRODUCTION IN SEEDLESS VASCULAR PLANTS

The great majority of ferns is homosporous, with numerous haploid spores being produced by meiosis in the sporangia borne on the leaves of the diploid plant (sporophyte, **14**). When the spores are shed they develop into small, but free-living, green gametophytes (**487**) on which the gametes are later formed. Fusion of the motile sperm with the egg results in a diploid embryo; this re-establishes the sporophyte generation which is soon independent of the gametophyte. The present day herbaceous fern ally *Huperzia selago* (**16**) has a very ancient lineage; it is a survivor of an otherwise long extinct plant group of heterosporous tree lycopods represented by the very common fossil tree *Lepidodendron* of the mid-Devonian to Carboniferous periods. Like these fossil ancestors, *Huperzia selago* is heterosporous: the megasporangium usually produces four haploid, thick-walled megaspores while the microsporangium develops numerous thin-walled haploid microspores (**16**). The advent of heterospory ca. 400 million years ago was followed by the evolution of pollen and seeds. The Carboniferous seed *Lagenostoma ovoides* (**488**, which actually grew on the seed-fern *Medullosa*) is a common fossil example. It is considered that the megasporangium was the precursor of the naked ovule, with only one member of the spore tetrad (formed by meiosis from a diploid megaspore mother cell) developing to maturity (**45A**). Microspores evolved into pollen grains.

SEXUAL REPRODUCTION IN SEED PLANTS

Reproductive phases in seed plants

All seed plants have a juvenile phase during which they cannot sexually reproduce followed by an adult reproductive phase; the shoot morphology of the two phases is usually similar. For example, the twigs and leaves of an oak sapling look like those of a mature tree; similarly, in most cacti the shoot of the adult (**489**) has no special characters that distinguish it from a juvenile incapable of flowering. In some genera there are indeed great differences, with the adult flowering phase of the cactus occurring as a terminal or lateral cephalium. A juvenile *Melocactus* grows for many years with a single shoot apical meristem producing an unbranched, globose vegetative shoot. However, after the phase transition, all further tissues produced by this same meristem develop with an adult shoot morphology and the phyllotaxy switches from several vertical ribs to tight helices. The surface is thickly covered with spines and trichomes; stomata and cortical chlorenchyma (**490**) are absent and the epidermis becomes a cork cambium. The adult wood is parenchymatous in contrast to the juvenile which has some fibres (**491**, **492**). In the cephalium the flower buds develop beneath an armour of spines and trichomes but then elongate and open above this covering. Terminal cephalia are also present in other genera such as *Discocactus* and *Backebergia* (**493**).

Cacti with terminal cephalia produce no new photosynthetic tissue in the adult shoot and as the plant ages, so does its chlorenchyma. Other genera have solved this problem with lateral cephalia: when such a plant 'goes through puberty',' only one sector of an apical meristem undergoes the transition to the adult phase while the other sectors continue developing a juvenile morphology (**494, 495**). From the same shoot apex some cells develop into a juvenile epidermis with stomata, while other cells from another sector of the same apical tunica layer immediately form a cork cambium. Internally some cells develop as a broad chlorophyllous cortex while others develop as a narrow, nonchlorophyllous cortex (**496–498**). On the cephalium ribs the axillary buds produce flowers (and an abundance of long spines and trichomes) but the buds on juvenile ribs never do. For the rest of the shoot's lifetime its apical meristem produces some tissues that flower and others that photosynthesize. Lateral cephalia occur in *Cephalocereus*, *Coleocephalocereus*, *Espostoa*, *Facheiroa*, and *Micranthocereus*.

The cactus *Browningia candelaris* is unique: it grows as a juvenile unbranched, spiny column until 2 m tall, but then produces many almost spineless adult branches that are able to flower (**233, 499**).

Tissues and organs of sexual reproduction in seed plants

All seed plants are heterosporous and, unlike the lower vascular plants, do not have a free-living gametophyte generation. In the gymnosperms a single haploid megaspore matures to form the embryo sac within the naked female ovule (**11, 45**). This megaspore is formed by the meiosis of a solitary megaspore mother cell located in the diploid nucellus (megasporangium). Only one of the four resultant spores matures while the others abort (**45**). Both the embryo sac and nucellus are enclosed within an integument and these structures collectively constitute the ovule which is retained on the tree. The haploid megaspore divides repeatedly to form a mass of gametophytic tissue within the enlargened embryo sac, and eventually one or more eggs become demarcated within this tissue close to the micropyle (**500**). Following pollination of the ovule by a microspore, the egg is fertilized by a sperm and the diploid embryo develops. The testa of the mature seed is formed from the modified integument of the ovule, while copious food reserves for the future seedling are contained within the tissue of the female gametophyte surrounding the embryo (**45, 500**).

In flowering plants the ovule is enclosed within an ovary (**45**) in contrast to the exposed ovule in gymnosperms. The female gametophyte within the embryo sac is typically reduced to eight cells and one of these, at the micropylar end of the embryo sac, represents the egg. In contrast to gymnosperms double fertilization occurs in angiosperms: one sperm fertilizes the egg to form the diploid embryo (**45**) while the other sperm fertilizes the two central polar nuclei to form the triploid, nutritive endosperm (**45**) which is initially coenocytic.

General features of flowers

Most flowers contain both male and female sex organs (**1, 44**) but some flowers are unisexual (**501**) and may occur on the same plant or separate individuals. The flowers of *Magnolia* and *Rhododendron* illustrate the general features of a basal angiosperm and an eudicot, respectively. In *Magnolia* the upper half of the elongate receptacle (floral axis) is covered by a large number of spirally arranged and separate female organs (carpels) which collectively constitute the apocarpous gynoecium. Below the carpels numerous spirally arranged male organs (stamens forming the androecium) are inserted onto the receptacle by short filaments. These are terminated by large anthers containing the pollen (microspores). At the base of the receptacle there are usually nine large, petaloid, perianth members while the flower itself terminates a short pedicel or flower stalk (**32**). In the flower of *Rhododendron*, by contrast, the components are in multiples of five and all show a considerable degree of fusion. The perianth consists of a small green calyx with five basally fused sepal lobes, the corolla is composed of five fused red petals. In the androecium there are 10 free stamens, their filaments (of uneven lengths) terminated by pores through which numerous pollen tetrads are extruded in fine viscous threads. The gynoecium shows a single stigma terminating a long style with a single ovary at its base; however, this syncarpous gynoecium consists of five carpels which are completely fused together.

In contrast, the monocot flower of *Lilium* (**1**) has six petaloid perianth members arranged in two alternating whorls. These encircle six stamens, while the syncarpous gynoecium consists of three fused carpels. The three compartments of the ovary are clearly visible on the fruits developing after the flowers wither. The ovary in *Lilium* is terminated by a long, slender style which is tipped by a slightly swollen stigma. As in *Magnolia* and *Rhododendron*, the ovary is superior since its point of insertion lies above the rest of the floral parts. In other flowers the ovary may be inferior (**502, 503**). In the early development of most flowers the organ primordia arise in centripetal sequence so that the perianth is initiated first and the gynoecium last.

Perianth

In some flowers the perianth members are similar in appearance (**1, 44**) but in many others the perianth is differentiated into the outer sepals, constituting the calyx, and the inner petals, forming the corolla (**2, 504, 505**). There are also some flowers (usually wind-pollinated) in which the perianth is vestigial (**157, 506**). The calyx consists of several sepals that are frequently green and leaf-like (**267, 532**) but their mesophyll is not usually differentiated into palisade and spongy layers. Besides their protective role in investing the unopened flower bud, sepals are sometimes brightly coloured (**504**) and attractive to pollinators. In *Taraxacum* the hairy pappus (calyx, **482, 507**) of each floret assists in the wind dispersal of the numerous small fruits.

The corolla in most flowers consists of a number of petals which are variously modified and attract pollinators (**504, 505, 508–510**) but the corolla is greatly reduced or absent in flowers where the pollen is wind-distributed (**506**). In many bee-pollinated flowers the epidermal cells of the petals contain ultraviolet-absorbing flavonoid pigments which are visible to the bee and act as guides to the nectaries (**511, 512**). Betacyanins in the epidermal vacuoles of the petals and carotenoids in the chromoplasts often cause bright colouring which attracts pollinators. Animal-pollinated flowers often have nectaries (**157, 504**).

Petals are usually ephemeral structures with a thin cuticle and few stomata, while the mesophyll is usually nonphotosynthetic. In actinomorphic (radially symmetrical) flowers perianth members are all of similar size and distributed regularly around the receptacle (**1, 504, 508, 511**). However, many flowers are bilaterally symmetrical (zygomorphic, **267, 505**). As with the other floral organs, the petals are often fused (**513, 514**) although sometimes their tips remain free.

Stamens (androecium)

Each stamen is terminated by an anther which produces the pollen grains (**515–517**). The anther is borne on a filament which is either inserted directly onto the receptacle or onto the corolla. The filaments are sometimes of unequal length (**516**), and may be very short (**506**), while in orchids filaments are absent. In some flowers the filaments are fused to form a tube round the style (**518**) and in the Asteraceae the anthers are fused (**514**). A vascular bundle runs along the filament (**519**) and supplies water and nutrients to the anther. The anther is commonly bilobed (**516**) with two cylindrical pollen sacs present in each lobe (**514, 520**). The ripe anthers usually split longitudinally to release their pollen (**515, 520**), but in some taxa, for example in *Rhododendron*, *Erica*, and *Calluna* (Ericaceae, heathers), the pollen is released from apical pores (**517**).

In the young anther (**513**) several layers of hypodermal cells divide periclinally within each pollen sac to give rise to a central mass of sporogenous cells and the investing parietal tissue (**514**). The latter differentiates into the hypodermal endothecium, an intermediate parenchymatous layer and the tapetum which surrounds the sporogenous tissue (**521**). The endothecial walls normally develop numerous bands of thickenings on the anticlinal and inner periclinal walls (**522**) which may be lignified. Most anthers dehisce along specialized longitudinal tracts of epidermal cells (stomia) which are underlain by unthickened endothecial cells. In species such as *Phaseolus vulgaris* (bean), self-pollination occurs in the unopened flower (cleistogamy). The grains germinate *in situ* and the pollen tubes penetrate the anther wall to reach the stigma.

The tapetum plays a vital part in the nutrition of the developing sporogenous tissue, and consists of a layer of densely cytoplasmic cells (**521**) which are commonly bi- or multinucleate (**523**). In many anthers these cells remain intact but in species with an invasive (amoeboid) tapetum their walls degenerate (**524**). As a result a coenocytic periplasmodium is formed from their combined protoplasts, which directly invests the developing sporogenous cells. Within each anther the pollen mother cells typically undergo synchronous meiotic divisions (**523, 524**).

In many dicots the four immature haploid pollen grains (microspores) formed by meiosis of the pollen mother cell (microsporogenesis) lie in a tetrahedron (**524**); hence each grain has three inner faces (**525**) in contact with the other grains. However, an isobilateral segmentation pattern is more frequent in monocots. The pollen grain is initially uninucleate (**526**) but the first mitotic division, which often occurs in the anther, divides the maturing grain into a large vegetative and a smaller generative cell. In *Drosera* the pollen grains remain united as tetrads, but in the majority of flowering plants the four grains separate as the investing callosic wall of the pollen mother cell dissolves. In many orchids, asclepiads, and *Rhododendron* the pollen adheres together in a sticky mass termed a pollinium (**517**).

The outer wall of the mature pollen grain (exine) is extremely complex (**526–528**) and comprises an inner nexine (composed of sporopollenin, which is very resistant to decay) and an outer sexine. The latter is permeated by sporopollenin and is composed of fused rods (baculae). The exine pattern differs from species to species (**527, 528**) and is of great taxonomic significance. The inner region of the pollen grain wall is cellulosic and is termed the intine.

Carpels (gynoecium)

In many flowers the carpels are fused to each other (syncarpy, **508, 514, 515**); they remain separate (apocarpy) in only a few taxa (**44**). Various degrees of fusion exist: the ovaries may be united but the styles remain separate (**515**) or only the stigmas remain free (**529, 530**). While most floral parts are generally ephemeral, the ovary and the ovules it contains continue growth after fertilization and develop into the fruit and seeds (**531, 532**).

The stigma is the pollen receptor (**45, 508**); in wind-pollinated flowers the stigmas are often feather-like, while in other plants the stigmatic epidermis is either papillose or hairy (**530**). The majority of carpels possess a style (**515, 516**) but in some species this is very short. In syncarpous species in which the styles are also fused, there is usually one main longitudinal vascular bundle per stylar component (**514, 533**). The centre of the style consists of transmitting tissue (**533**) and, after the pollen grains germinate (**534**), the pollen tubes grow downwards through this tissue towards the ovary, absorbing nutrients from this tissue en route (**535**).

The ovules are attached to a thickened region of the ovary wall termed the placenta. The carpel is generally interpreted as a folded and modified leaflike organ with its abaxial surface outermost; its margins are normally fused and typically two longitudinal placentae lie internally (adaxially), close to the fused leaf margins (**536**). In syncarpous ovaries axile placentation is common, with the margins of the carpels fused at the centre of the ovary (**13, 531**). However, the placentae may also be located at the outer margin of the ovary (parietal placentation) or basally (**502**). Marginal placentation is common in apocarpous ovaries (**536, 537**).

Each ovule is connected at its base (chalaza) to the placenta via the funiculus (**45, 538, 539**). The ovule is invested by two (or sometimes one) thin integuments enclosing the nucellus (megasporangium). At the apical end of the ovule a narrow channel, the micropyle, penetrates the integuments, exposing the surface of the nucellus. The orientation of the ovule relative to the funiculus is variable but commonly the funiculus is bent through 180 degrees (anatropous ovule, **539**). The ovary walls and placentae are usually richly vascularized (**536**), and a number of smaller veins branch into the funiculi (**539**) but rarely extend to the nucellus.

Early in development of the ovule a megaspore mother cell becomes demarcated at the micropylar end of the nucellus (**540**) and undergoes meiosis. In the commonest situation as in *Polygonum* (monosporic type) a row of four haploid megaspores is formed. Only the most interior (chalazal) cell develops further while the outer three megaspores degenerate (**541**). The single functional megaspore enlarges greatly within the expanding nucellus, and its nucleus undergoes three rounds of division to give the eight nuclei of the mature megagametophyte (**539, 541**). Because cell walls are formed only after all nuclei are present in flowering plants, the megagametophyte is called an embryo sac (**45**).

Other patterns of megaspore mother cell division also occur. In the bisporic type, one of the two derivatives of the first meiotic division undergoes further divisions and gives rise to the eight nuclei of the embryo sac while the other derivative degenerates. In tetrasporic development, the megaspore mother cell (**540**) undergoes meiosis but all four nuclei are confined within the common cytoplasm of the embryo sac. These nuclei then undergo a variable number of mitoses, so that the mature embryo sac often contains more than eight nuclei. In the semi-mature embryo sac of *Lilium* (**542**) one haploid nucleus becomes located at the micropylar pole while the other three nuclei at the chalazal end fuse to give a triploid nucleus. Subsequent mitoses and rearrangement of the nuclei lead to three haploid nuclei being situated at the micropylar end, one haploid plus one triploid nucleus in the centre of the embryo sac, and three triploid nuclei at the chalazal end.

The eight nuclei and associated cytoplasm of a developing embryo sac become separated by thin walls and are located in specific regions. Three antipodal cells are located at the chalazal end of the sac, two synergids and a median egg cell lie at the micropylar end, while the central region contains two polar nuclei within the endosperm mother cell. The egg cell is normally larger than the synergids; its nucleus and most cytoplasm usually lie towards the chalazal pole and here the wall is scantily developed. In the synergids, the cytoplasm is concentrated at their micropylar poles and their walls may be modified into a wall labyrinth (filiform apparatus) similar to that present in transfer cells (**113**). The synergids apparently transport nutrients to the egg from the nucellus and may form absorptive haustoria within this tissue. In most species, pollen tubes enter the embryo sac through one of the synergids. The antipodal cells commonly degenerate before fertilization of the egg, but they may persist and also develop haustoria.

Fertilization

When a compatible pollen grain is deposited on the stigma a pollen tube grows out through one of the germination pores in the pollen grain wall (**528**) where the exine is sparsely developed. The cytoplasm and the nuclei migrate into the pollen tube which, after penetrating the stigma, grows downwards in the stylar transmission tissue (**533, 535**). In the anther the generative nucleus of the pollen grain may already have divided into two sperm nuclei; otherwise, this occurs within the pollen tube. The nuclei and cytoplasm concentrate at the tip of the elongating pollen tube; behind this the tube is highly vacuolate and is often sealed off by plugs of callose. The pollen tube frequently enters the embryo sac via the micropyle but in rare cases may directly penetrate the integuments. Once entry to the embryo sac has been gained, a pore forms near the tip of the pollen tube and the two sperm are liberated. One sperm nucleus fuses with the egg nucleus and the other fertilizes the centrally located polar nuclei. In most flowering plants the cytoplasm surrounding the sperm nuclei is not transmitted at fertilization and therefore inheritance of chloroplasts and other organelles is generally via the female line.

Development of the seed

The fertilization of the haploid egg by a haploid sperm gives rise to a diploid zygote which subsequently divides repeatedly and develops in a highly organized manner into the embryo (**46, 213**). The mature embryo commonly undergoes a period of dormancy within the protective seed coat (testa) which develops from the integuments of the enlargened ovule. The embryo itself may be packed with food reserves (**104**), and in albuminous seeds such as those of monocots, further food is stored in the mass of endosperm surrounding the embryo (**549**). In exalbuminous seeds such as beans and peanuts, the endosperm is consumed during seed development. In the Caryophyllales little endosperm is formed but the nucellus develops into a nutritive perisperm.

Embryo development

Following entry of the sperm nucleus through the wall-free region of the egg cell, a wall is formed in this area of the zygote. The diploid nucleus then undergoes mitosis and usually a transverse wall divides the zygote into basal and terminal cells. The basal cell divides mainly transversely to form a suspensor (**46**); this pushes the terminal cell away from the micropyle and into the endosperm which is developing (**52, 543, 544**) within the expanding embryo sac. In legumes the suspensor cells often show highly polyploid, amoeboid, nuclei (**545**) and apparently synthesize growth substances of importance for the development of the embryo. The terminal cell undergoes divisions in various planes to form a globular proembryo (**544**) and soon an outer layer of anticlinally dividing protoderm cells becomes established.

In dicots the enlarging globular proembryo becomes transformed into a heart-shaped structure as paired cotyledons at its chalazal end develop. The embryo now elongates and differentiation of the procambium and ground tissues occurs (**546**). The radicle apex becomes demarcated at the micropylar pole of the embryo and merges into the hypocotyl above (**547**). Meanwhile, at the other end of the hypocotyl and between the cotyledons, a bulge representing the plumule becomes apparent. In the mature seed the plumular apex may either remain small (**547**) or is larger and has already given rise to its first foliage leaves (**213, 548, 549**), while the radicle shows a root cap and apex. Depending upon the architecture of the embryo sac, the embryo may be straight or variously curved. In albuminous seeds abundant endosperm is present and the cotyledons are thin and leaf-like, in contrast to their swollen appearance in nonendospermous seeds (**547, 548, 550**).

In monocots, the early development of the embryo parallels that in dicots, but only a single lateral cotyledon is formed (**32**). Monocot seeds are commonly albuminous and the single cotyledon acts as a digestive organ: in *Cocos nucifera* (coconut) the cotyledon enlarges greatly, fills the centre of the coconut then digests the white coconut 'meat' (cellular endosperm, **551**) and transports the nutrients to the embryo through the vascular bundles of the cotyledon. In grasses the endosperm is digested by the scutellum, which is sometimes regarded as a modified cotyledon. During early germination the plumule of grasses is protected by a cylindrical leaf-like coleoptile while the radicle is initially ensheathed by the coleorhiza (**552**).

In some flowering plants the mature embryo shows little morphological differentiation, consisting of little more than a few parenchyma cells and their seeds are minute. Examples are orchids, bromeliads, and many parasitic plants.

Endosperm

In the majority of angiosperms two haploid polar nuclei occur in the embryo sac and their fusion with one of the sperm nuclei leads to the development of the triploid primary endosperm cell (**45**). In some species multiple polar nuclei occur so that the resulting endosperm is polyploid. In a few species fertilization of the polar nuclei does not occur and in others division of the primary endosperm nucleus ceases very early. In many albuminous seeds (**549**) the nucellus and the inner integument degenerate as the endosperm develops in the maturing seed.

In most flowering plants the initial divisions of the primary endosperm nucleus are not followed by cytokinesis and a coenocytic nuclear endosperm develops (**52, 543**). However, subsequent free-wall formation leads to the cellularization of the endosperm (**538**). These walls resemble those in some *in vitro* cultured tissues and, in contrast to normal cell plate development, neither microtubules nor Golgi bodies are apparently involved in their growth. The random wall formation results in some endosperm cells being multinucleate. In the mature fruit of *Cocos nucifera* (**551, 553, 554**) the watery milk of the enclosed seed represents the remnants of the coenocytic cytoplasm, while the white flesh results from its cellularization.

Seed coat

The integuments of the ovule (**539**) develop into the seed coat (testa) after fertilization (**547, 555**). The testa usually contains a hardened protective layer (**551**) which may develop in the outer integument (**556, 557**) or from the inner integument. In the legumes the inner integument eventually degenerates but the outer epidermis of the remaining integument differentiates into a palisade layer of sclereids (**557**). At the hilum two layers of sclereids occur, with the outer forming from funicular tissue (**556**). A group of tracheids, which is probably concerned with water uptake during germination, also develops in this region. In cereals and grasses the grain (**552, 558,**

559) is not a seed but rather a fruit (caryopsis), since the thin ovary wall is fused to the integument of the single ovule.

The fruit

Following fertilization the ovary develops into the fruit which encloses and protects the seeds (**531**, **532**, **537**, **554**, **555**). Some plants, however, form parthenocarpic fruits without fertilization of the ovules, as in *Musa* (banana) and *Ananas comosus* (pineapple) (**244**). A great diversity of fruits occurs but their typology will not be considered in detail here. Most fruits are termed simple since they are derived from a single ovary, which may be apocarpous (**44**, **537**) or syncarpous (**531**, **532**, **560**). Aggregate fruits are formed from several carpels of an individual flower as in *Fragaria vesca* (strawberry, **561**) while multiple fruits are formed from a number of flowers such as *A. comosus*.

In addition to tissues directly derived from the ovary wall, adjacent accessory tissue may also contribute to the fruit body. In *Fragaria vesca* (**561**) the swollen receptacle forms the centre of the fruit, while in *A. comosus* the inflorescence scales and swollen floral axis participate in the fruit. The closely-crowded flowers of *Banksia* (**562**) develop into dehiscent woody fruits embedded in the swollen inflorescence axis (**563**) in which some secondary thickening occurs. In fruits developed from an inferior ovary (**531**, **532**, **564**), an outer layer of receptacle or perianth tissue is joined to the ovary wall. In *Malus sylvestris* (apple) the flesh represents the swollen floral tube while the core is formed from the ovary (**565**).

The changes which occur in the fruit during maturation vary considerably in different plants. In grasses the ovary contains only a single seed (**559**) which at maturity is relatively small and the maturation changes in the ovary wall are mainly effected by vacuolation growth and cell sclerification. But in large-seeded or fleshy fruits (**565**) active cell division accompanies growth and in *Persea americana* (avocado, **555**) division occurs throughout its growth. In large and heavy fruits the original delicate flower stalk may become woody as further vascular tissue differentiates within it to meet the increased requirements of water and foodstuffs for the developing fruit.

Fruit wall

The pericarp of fleshy fruits is often three-layered, with the exocarp comprising the skin, the mesocarp is represented by the flesh, and the endocarp comprises the stone cells surrounding the seed (**566**). However, accessory tissues often contribute to the flesh as in *Malus sylvestris* (apple, **565**) and most cacti (**564**).

The cuticle-covered outer epidermis of the ovary wall is usually persistent in the fruit (**567**, **568**) but the stomata are usually nonfunctional. Cork sometimes develops in various fruits and the patchy appearance of certain apple varieties is due to localized cork formation. The inner epidermis of a fruit adjacent to the loculus (**567**) is rarely cutinized. The sclerenchymatous endocarp of some fruits (**566**) originates from the inner epidermis and the adjacent ground tissue of the ovary wall. In many fleshy fruits the inner epidermis becomes secretory and in citrus fruits the flesh is formed by proliferating juice sacs which extend into the carpel locules (**568**).

The ground tissue of the ovary wall develops variously. In fruit walls which are dry at maturity much sclerenchyma develops (**551**, **554**, **569**, **570**). In soft fruits (**555**, **560**, **561**, **565**) the ground tissue undergoes extensive vacuolation and predominantly consists of parenchyma penetrated by vascular strands linked to the pedicel. In the developing fruit the parenchyma often shows maturation changes; for example, the softening flesh of *Pyrus communis* (pear) and *Persea americana* results from breakdown of the middle lamellae and the degeneration of the walls of this tissue.

Seed release from dry fruits

In plants where the fruit is not dispersed by animals (**563**, **569**, **570**) its principal function is seed protection. Indehiscent fruits, which are dry at maturity, often have thin walls which can be easily penetrated by the germinating embryo (**571**). Dehiscent fruits frequently have thick walls and break open to release the seeds (**572–574**). In legumes the pod dehisces longitudinally along the suture of the carpel margins and also along the median vein (**537**, **575**). In many legumes a single layer of exocarpic fibres lies with their long axes more-or-less at right angles to that of the pod axis, while the endocarpic fibres run parallel to its axis. As the pericarp dries, the exo- and endocarp contract in cross directions and the pod eventually ruptures (**575**). In some fruits (**570**) the dehiscence zone is not related to carpellary sutures or vascular bundles and the fruit may break open by pores (**572**).

Forage fruits and seed dispersal

In dehiscent fruits the release and dispersal of the seed occurs while the fruit is still attached to the plant. Fleshy fruits are generally attractive as food for animals and the enclosed seeds are often protected from the animal's chewing and digestive juices by a sclerenchymatous endocarp (**555, 566**), or thick-walled or mucilaginous seed coats. The seeds are later distributed in the animal's droppings. Small fleshy fruits (**561**), which are often nonaromatic, are mainly eaten by birds which have good colour vision but poor sense of smell. Larger aromatic fruits (**565, 576**) are usually eaten by mammals, who have a well-developed sense of smell. Fleshy fruits also often fall to the ground where they are either eaten or rot and release the seeds.

Passive dispersal of fruits and seeds

Many fruits develop barbs or have sticky surfaces, and so become attached to passing animals and are passively distributed. Dispersal of fruits and seeds by the wind is also common (**571, 574, 577**) but these rarely travel great distances. It has been calculated that even in the small and light fruits of *Taraxacum* (**482**) only about 1% of the fruits are dispersed further than 10 km. Some species have explosive fruits (**575, 578**) with the ejected seeds being thrown up to several metres. Water dispersal is much less common but the large fruits of *Cocos nucifera* (coconut) and *Lodoicea maldavica* (sea coconut) are dispersed in the ocean currents with their enclosed seeds protected by a thick and sclerified endocarp. The fruits of mangroves such as *Avicennia* and *Rhizophora* have a single seed which germinates *in situ*. In *Rhizophora* the radicles grows up to 50 cm over a number months while the fruits are still attached to the tree (**579**). The precociously germinated mangrove seedlings eventually fall off into the water or mud and, once firmly rooted, develop into further mangrove forest (**413, 580**).

468 *Conocephalum conicum* clone. This widespread and large liverwort commonly spreads densely in damp and shaded conditions, on flat rocks and walls.

469 *Pteridium aquilinum* (bracken) newly emerging in spring after dying back to its rhizomes in late autumn. Note the dead remains of last year's foliage. This very common fern occurs throughout much of the temperate and sub-tropical world and spreads clonally via a complex system of rhizomes lying up to 1 m beneath the soil surface. The foliage is very toxic to grazing animals and contains a potent mix of carcinogens.

470 *Acorus calamus* (sweet flag). The aerial sword-shaped leaves of this monocot commonly occur in dense, large patches of boggy ground. It spreads clonally from its persistent large rhizomes which normally grow underground, but here (in early spring) have been exposed by the running waters of a drainage ditch on Iona, west Scotland.

471 This grove of *Populus tremuloides* (quaking aspen) actually may be just one or two plants. The roots produce adventitious shoot buds which develop into what appears to be ordinary trees, although all are clones of the original seedling and are interconnected by a single root system.

472 Rough hillside pasture in Scotland showing the common and pernicious clonal weed *Pteridium aquilinum* (bracken) with its newly emerging fronds in spring. This fern is very toxic to grazing sheep and catttle but is extremely difficult to eradicate because it propagates vegetatively from perennial rhizomes which may lie up to 1 m beneath the soil surface.

473 *Ranunculus repens* (creeping buttercup). In late spring the runners of this widespread and common perennial dicot grow rapidly over damp, uncultivated or waste ground surfaces. Adventitious roots form at the nodes and small plantlets develop; later the intervening thin connecting runner stems decay, so the plant spreads clonally.

474 *Solanum tuberosum* var. 'Desire' (potato). The fat underground tubers of this dicot are packed with starch. The tubers develop from stem-borne stolons, the buds of which swell into tubers underground; later the stolons decay and buds on the tubers sprout (cf. 338), forming aerial green plants that are all clones of the original plant.

475 *Daucus carota* (carrot). This common dicot vegetable is a biennial: the 'rootstock' (actually a swollen primary root/hypocotyl) is harvested in its first season after being sown from seed. If left in the ground in colder climates, the aerial shoot would die down in autumn. Later, as shown in this Figure, the aerial shoots can re-grow from buds on the terminal hypocotyl.

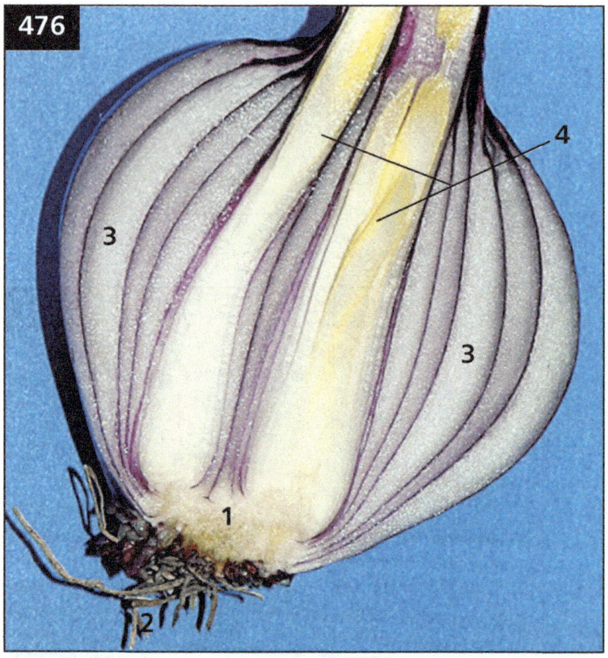

476 *Allium cepa* (red onion). This widespread vegetable, with numerous varieties and related species, is actually a biennial but the underground bulb is grown from seed and harvested as an annual. The bulb has been sectioned lengthwise to reveal its small stem (1) from which numerous adventitious roots (2) originate basally. The bulk of the bulb is composed of swollen leaf bases (3) in which the food reserves, synthesized in the aerial plant, are stored for the buds (4) to grow out in the following year.

1 Basal stem
2 Adventitious roots
3 Swollen leaf bases
4 Bud

477 LS corm of *Crocus*. This decorative monocot perrenates from an underground condensed stem that becomes packed with food reserves (mainly starch, here stained blue-black with iodine) derived from the photosynthetic activity of the aerial foliage. The starch will provide nutrients for the following year's new leaves and flowers, which will form from the dormant bud (1). Scaly remnants of leaves (2).

1 Dormant bud
2 Remnants of leaves

478 *Fagus sylvatica* (beech). Main branch (1) of a mature dicot tree has rooted (2) and further divided into branches (3, 4) which are developing semi-independently. Although the image is somewhat distorted (and therefore over emphasizes the differences), in reality the diameters of both branches (3, 4) are much wider than that of the main branch (1) near its origin from the tree trunk.

1 Main branch
2 Site of rooting
3, 4 Branches beyond the point of rooting

479 *Populus* (poplar) dicot tree stump. Note the profuse sprouting of adventitious buds from the vascular cambium (1) and also the secondary phloem and bark (2). Some buds originating from the bark may represent previously dormant, buried axillaries.

1 Vascular cambium
2 Secondary phloem and bark

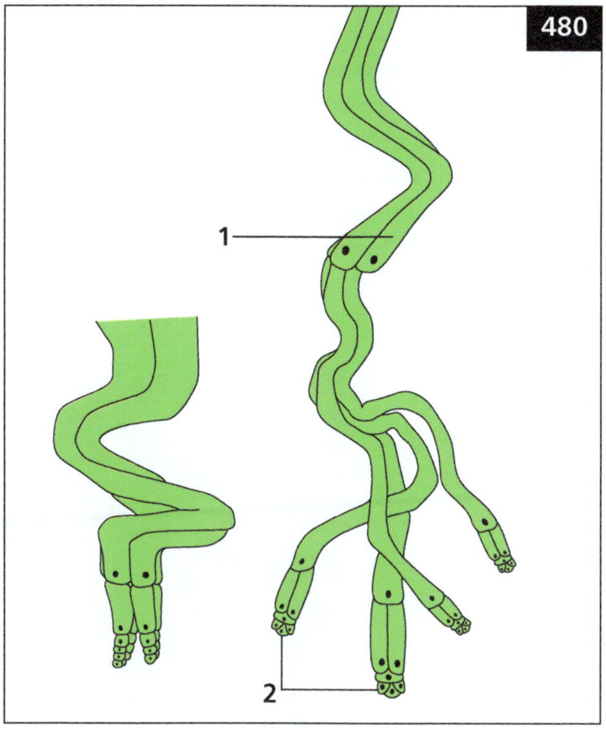

480 Diagram showing early embryology in *Pinus* (pine). In many species of pine and other conifers the original sexually produced proembryo terminates a file of suspensor cells. These divide longitudinally, so producing four potential embryos but generally only one survives in the mature seed.

1 Suspensor
2 Embryo

481 *Taraxacum officinale* (dandelion). The inflorescence ('flower') of this common herbaceous dicot bears numerous florets. Despite being visited by numerous insects, these do not normally set seed sexually after pollination but rather develop apomictic (asexual) embryos. The parachute-like seeds (cf. 482) are distributed widely by wind.

482 *Taraxacum officinale* (dandelion). Ripe inflorescence head show numerous apomictically formed barbed achenes (1), each bearing a terminal parachute-like pappus (2) developed from a modified calyx.

1 Achenes 2 Pappus

483 Potted-up specimens (all regenerated from *in vitro* grown leaf explants) of several species/cultivars of common dicot houseplants *Begonia*, *Streptocarpus*, *Peperomia*.

484 Excised root of *Crambe maritima* (seakale, a dicot). After several weeks culture in a moist and warm environment, adventitious buds have proliferated from the cambial region at the end of the root nearest the original aerial shoot.

485 *Crambe maritima* (seakale) tissue culture grown on a nutrient agar medium (1) *in vitro*: the culture bears a number of organized structures (2), some of which further develop into asexual embryos when sub-cultured (cf. 486).

1 Agar medium
2 Masses of organized cells

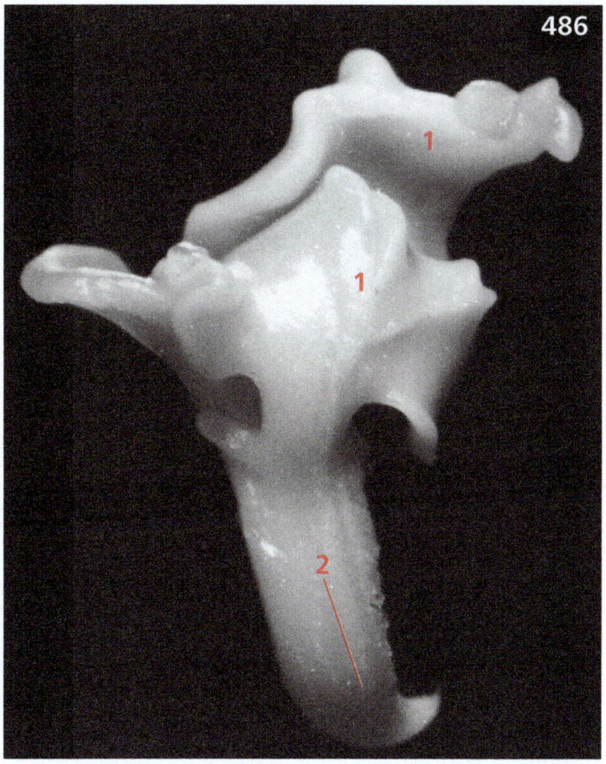

486 *Crambe maritima* asexual embryo grown in tissue culture. Cotyledons (1), radicle (2).

1 Cotyledons
2 Radicle

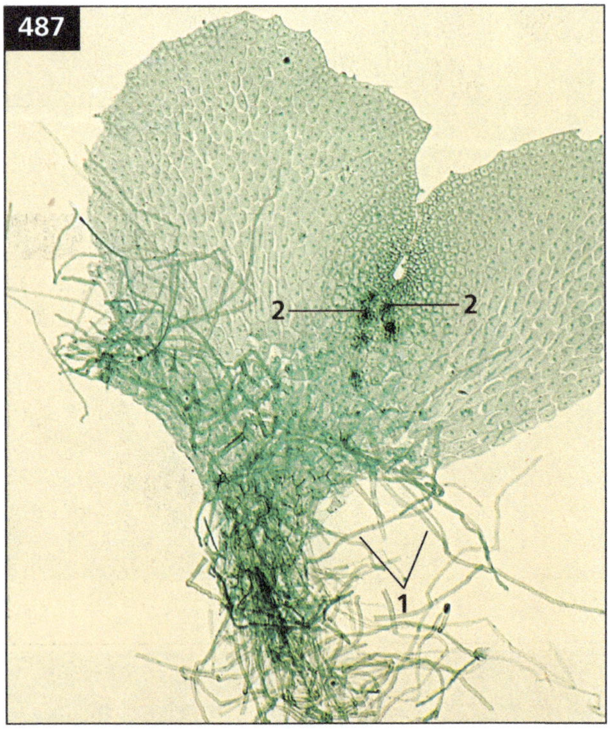

487 Lower surface view of a fern prothallus. When the haploid spores are released from the sporangia on the fern frond (cf. 14) they germinate to form small dorsiventrally flattened gametophytes adhering to the soil by numerous rhizoids (1). The female reproductive organs (2) develop near the notch in the gametophyte and the eggs are fertilized by flagellate sperms; the zygote is diploid and the embryo rapidly develops into a young fern plant (sporophyte) independent of the short-lived gametophyte. (LM.)

1 Rhizoids
2 Female reproductive organs

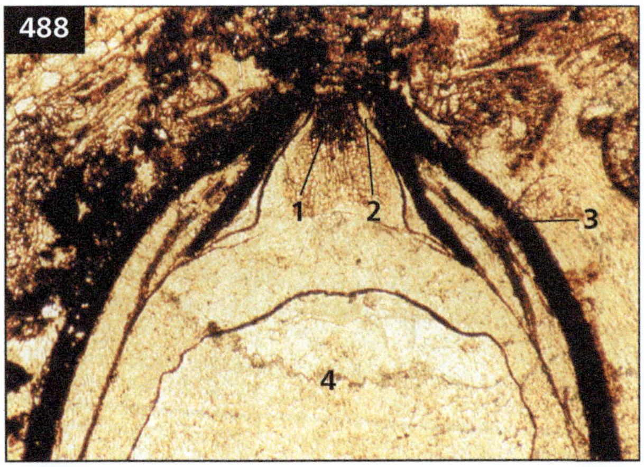

488 *Lagenostoma ovoides*. LS of a fossil seed cut in section of Carboniferous rock. Nucellar beak (1), pollen chamber (2), integument (3), female gametophyte (4).

1 Nucellar beak
2 Pollen chamber
3 Integument
4 Female gametophyte

489 The adult shoot of *Rathbunia alamosensis* with flower buds produced by axillary meristems situated in the spine clusters. As in most cacti, the vegetative shoot of the adult has no special characters that distinguish it from a juvenile incapable of flowering.

490 Mature plant of *Melocactus* showing one monopodial shoot produced by a single apical meristem. The juvenile (1) and adult (2) portions of this cactus shoot differ strongly: the juvenile body is broad, green, photosynthetic, and has few spines while the cephalium, bearing one fruit (arrow), is so densely covered with spines that photosynthesis is impossible.

1 Juvenile portion of shoot
2 Adult portion of shoot (cephalium)

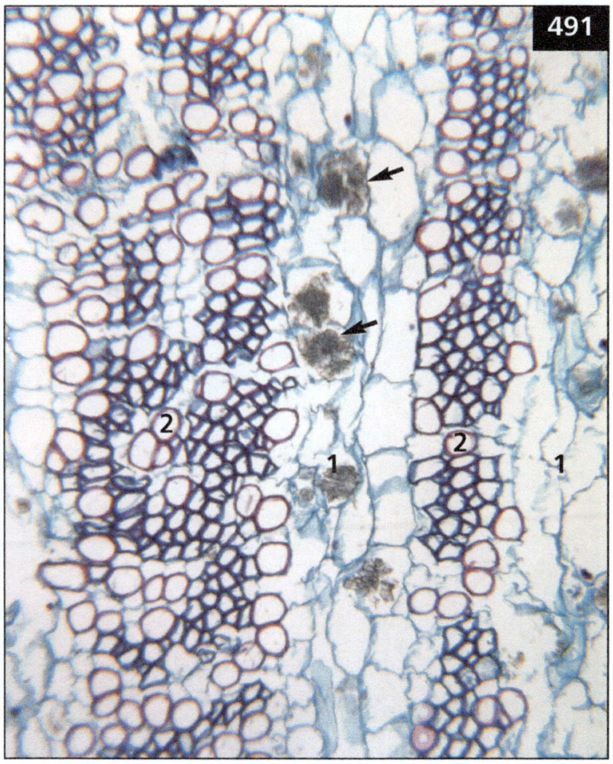

491 TS of juvenile body of *Melocactus intortus*. The juvenile wood in this cactus has parenchymatous rays (1) with druses (arrows) and narrow axial regions (2) with a matrix of xylary fibres, vessels, and paratracheal parenchyma. (LM.)

1 Rays in wood
2 Axial regions of wood with fibres and vessels

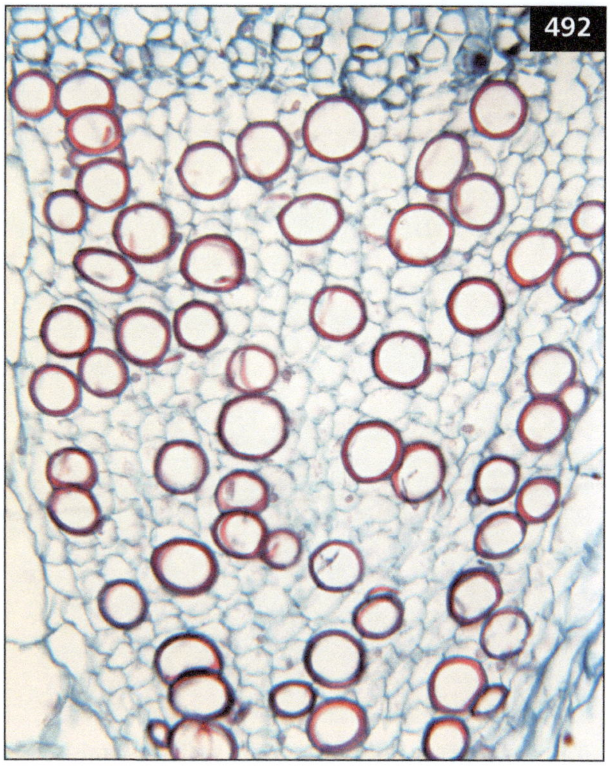

492 TS of adult (cephalium) body of *Melocactus intortus*. The adult wood of this cactus differs from the juvenile (cf. 491) in having no xylary fibres; instead the wood matrix consists of parenchyma with abundant vessels. (Broad parenchymatous rays also occur but are not shown.) (LM.)

493

493 The cactus *Backebergia (Pachycereus) militaris* has a protracted juvenile phase (green shoots) and large terminal adult cephalia (dark and gold regions). Each juvenile branch abscises its cephalium after it is several years old, then several axillary buds at the top of the juvenile branch produce more juvenile photosynthetic tissue, and later further cephalia are formed.

494

494 Plant of *Espostoa melanostele* with a lateral cephalium. The plant grew as a juvenile (1) until several metres tall, then one side underwent phase transition and became adult (2), the other three sides continue to grow as juvenile phase. Flowers are produced only from the lateral cephalium. Monopodial growth, all tissues are produced by a single shoot apical meristem.

1 Juvenile portion of shoot
2 Adult portion of shoot (cephalium)

495

495 Shoots of *Micranthocereus estevesii*, with lateral cephalia.

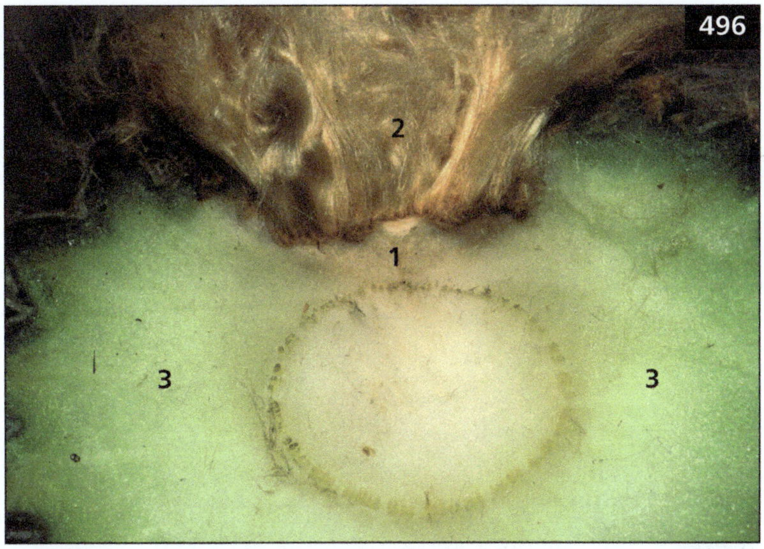

496 TS through *Espostoa melanostele* at midlevel of lateral cephalium. Cortex is thinner below the cephalium (1), causing it to be sunken. Adult cortex has no ribs and no chlorophyll; flower buds are protected by a mass of dead trichomes (2). Wood and pith are slightly thinner on the adult side as compared to the three juvenile sides (3).

1 Adult cortex
2 Cephalium trichomes
3 Juvenile cortex

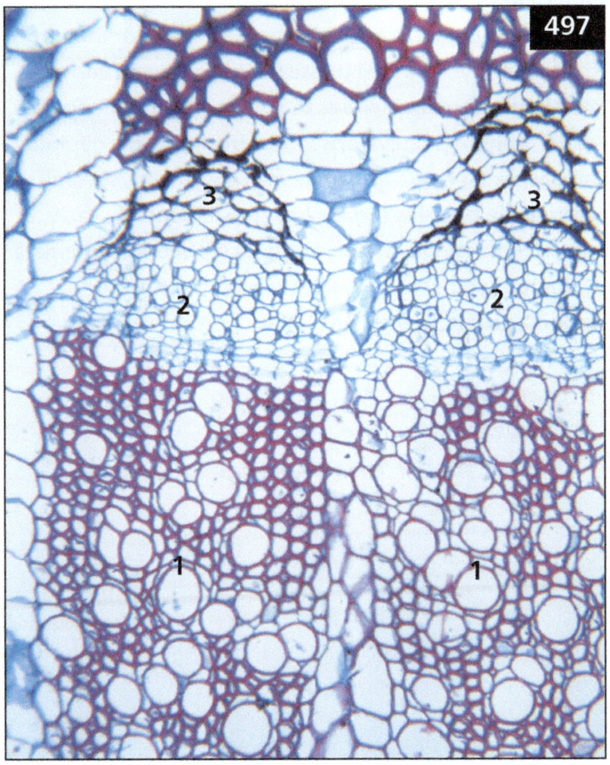

497 TS of wood (1) and secondary phloem (2) of juvenile portion of *Espostoa lanata*, collected at the level of 496 and 498. Juvenile wood has a solid matrix of xylem fibres, and even some ray cells are lignified. Only a little collapsed phloem has accumulated (3). (LM.)

1 Wood of juvenile portion
2 Secondary phloem of juvenile portion
3 Collapsed phloem of juvenile portion

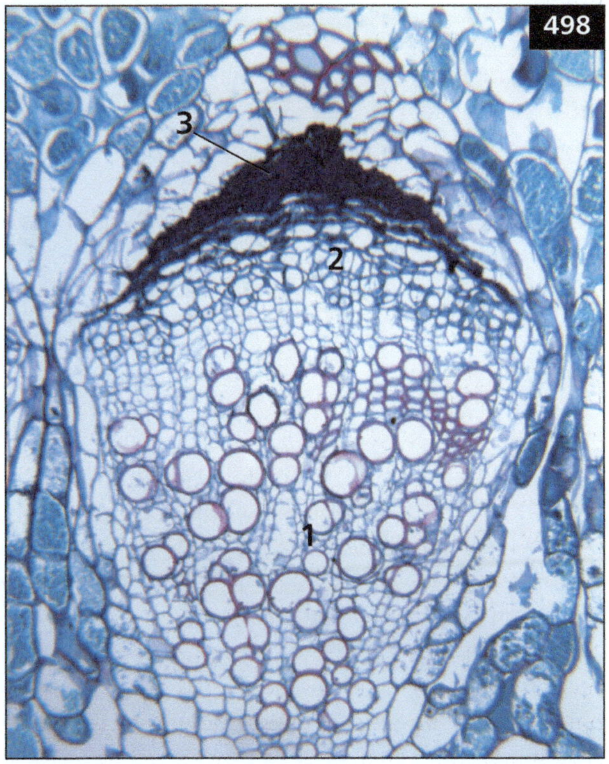

498 TS of wood (1) and secondary phloem (2) of adult portion of *Espostoa lanata*, collected at the level of 496 and 497. Adult wood is much less plentiful and has a matrix of parenchyma with only a few fibres. A great deal of phloem has been produced and collapsed (3).

1 Wood of adult portion
2 Secondary phloem of adult portion
3 Collapsed phloem of adult portion

499 Mature plant of *Browningia candelaris*. The trunk is the juvenile portion; it is spiny, thick, unbranched, and determinate. Once it stops growing, numerous narrow branches are produced: these are the adult portion and produce flowers (arrows) but few spines.

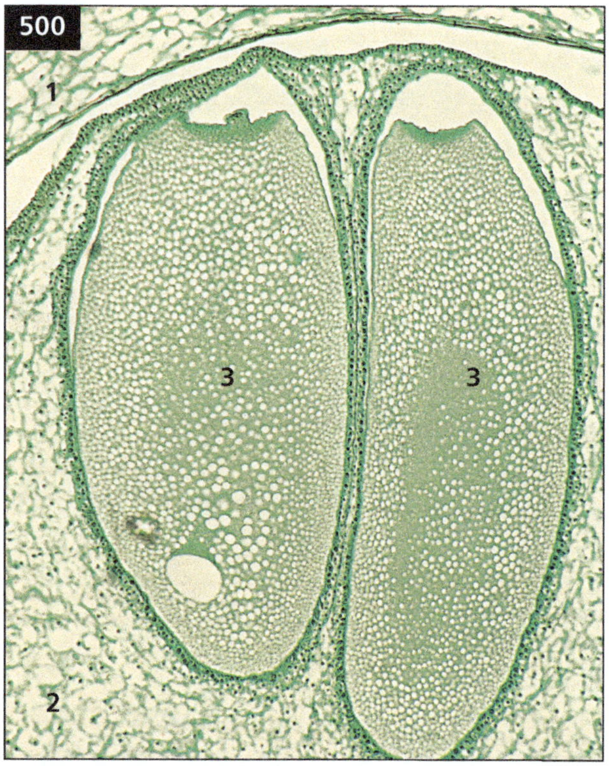

500 LS of the cycad ovule of *Zamia* showing a pair of large archegonia. In seed plants the megasporangium (nucellus, 1) is invested by sterile integuments and the whole structure comprises the ovule (cf. 11). The female gametophyte (2) develops *in situ* from a single haploid megaspore (cf. 45) and, at its micropylar end, several vestigial female sex organs develop. Essentially each consists of a single massive egg (3) embedded in female gametophytic tissue. Fertilization is effected by multiflagellate sperm. (LM.)

1 Nucellus
2 Female gametophyte
3 Egg

501 Unisexual flowers of the dicot *Begonia sempervirens*. Both sexes occur in the same plant: the irregular (zygomorphic) male flower shows two large and two small perianth members and a central cluster of yellow stamens (1). In the regular actinomorphic female flower five or more perianth members surround an inferior ovary terminated by convoluted stigmas (2).

1 Stamens
2 Stigmas

502 LS of an immature floret of the dicot *Helianthus* (sunflower). The numerous florets are condensed on a capitulum (1) to make the large sunflower head. The floret has a central inferior ovary with a single basal ovule (2). Anthers (3), bract (4), sepal (5), petal (6). (LM.)

1 Capitulum
2 Basal ovule
3 Anthers
4 Bract
5 Sepal
6 Petal

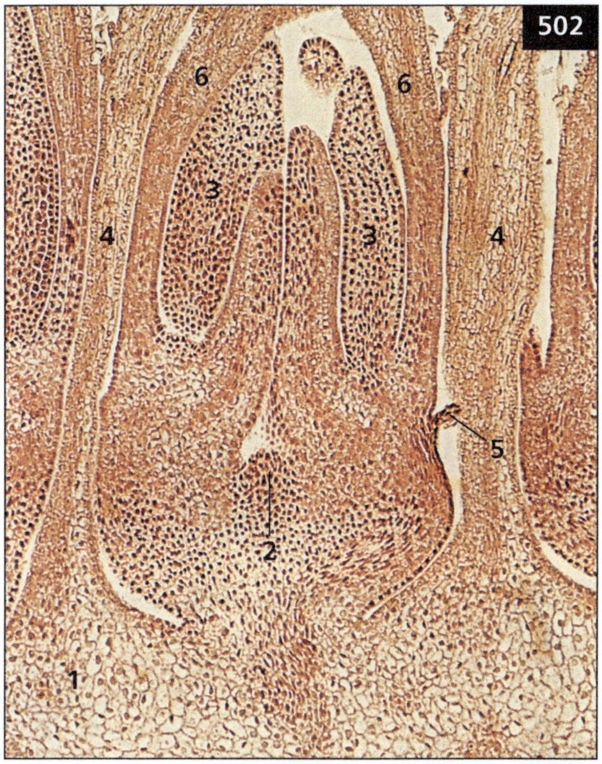

503 Cactus flowers, such as this *Echinocereus coccineus*, are borne at the end of short, green floral shoots (1); the shoot grows upward and surrounds the ovary (large arrow), which consequently is located not only below the perianth and stamens, but also below a considerable amount of vegetative tissue. The spine clusters (small arrows) are axillary buds, which indicate that this green organ is a shoot, not a pedicel. Floral shoot (1), petals (2), stigmas (3).

1 Floral shoot
2 Petals
3 Stigmas

504

504 Dicotyledonous actinomorphic flowers of *Aquilegia* (columbine). Each is composed of five orange-red petaloid sepals (1) alternating with five yellow petals (2) that terminate in orange-red nectar-secreting spurs (3). In the centre of each flower numerous stamens and five free carpels occur. Long-tongued bumble bees visit the flowers for nectar and pollen.

1 Petaloid sepals
2 Petals
3 Nectar-secreting spur

505

505 Zygomorphic flowers of the dicot *Cytisus scoparius* (broom). In the flower bud the tubular calyx (1) does not enclose the petals, but the large standard petal (2) surrounds the lateral wings (3) and abaxial pair of adherent keel petals (4). The stamens and single carpel enclosed within the keel would be exposed by large visiting bees.

1 Calyx
2 Standard petal
3 Lateral wings
4 Keel petals

506

506 Male and female inflorescences of the dicot *Alnus glutinosa* (alder). The male catkin (1) bears numerous groups of three florets borne on short branches from the inflorescence axis. Each floret is reduced to two vestigial perianth members and four stamens with freely exposed anthers. The as yet unopened female catkins (2) also bear groups of three florets; each has a bicarpellary ovary with two long, curved styles pollinated by wind-borne grains.

1 Male catkins
2 Female catkins

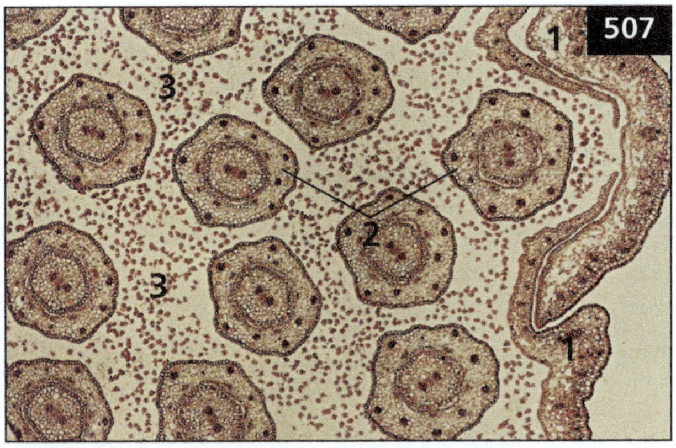

507 TS of an immature capitulum of the dicot *Taraxacum officinale* (dandelion). Numerous florets (cf. 502) are surrounded by large leafy bracts (1) borne on the margin of the capitulum. Each floret has a tubular corolla (2), while the numerous external hairs (3) represent the calyx. (LM.)

1 Leafy bracts
2 Corolla
3 Calyx (pappus)

508 Large regular flower of the dicot *Papaver* (poppy). The four crumpled petals (1) show bluish-black pollen shed onto their adaxial surfaces. The copious pollen from the massed anthers (2) attracts insects which settle on the wide stigmatic disc (3) and thus effect cross-pollination.

1 Petals
2 Anthers
3 Stigmatic disc

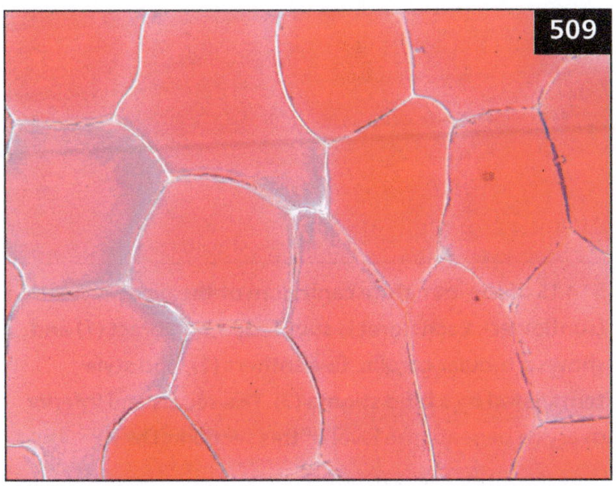

509 The pigments in these fresh, unfixed petal epidermis cells of *Lilium* (lily) are located in the central vacuole and thus appear to occupy the entire cell; however, a nucleus and a thin layer of cytoplasm are present in each, adjacent to the cell wall (cf. 510). (LM.)

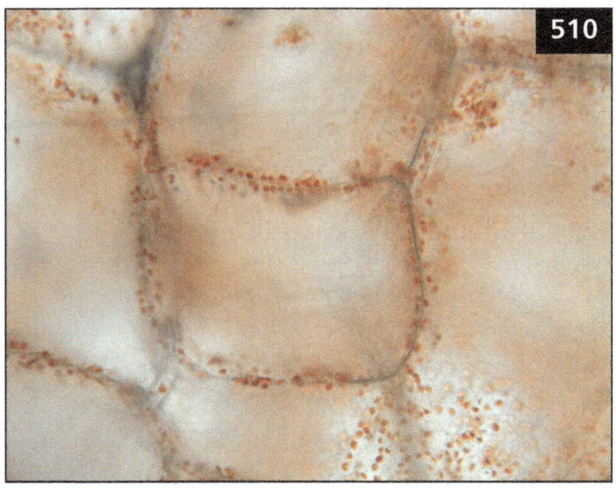

510 The pigments in these fresh, unfixed epidermis cells of *Capsicum annuum* (red pepper) occur in chromoplasts located in a thin layer of cytoplasm between the central vacuole and the cell wall. Although the chromoplasts appear sparse, the tissue has a brilliant orange colour (cf. 509). (LM.)

511, 512 Flowers of the dicot *Ranunculus* (buttercup) under normal (511) and UV illumination (512). Note the regular arrangement of the five petals (1), the abundant stamens (2), and the numerous free, superior, carpels (3). In 512, the nectar guides on the adaxial surfaces of the petals are revealed; these are visible to bees and hover flies and apparently guide them to the nectaries at the base of the petals. Pollen is shed onto the petals and is easily picked up by insects. (Copyright of T. Norman Tait.)

1 Petals
2 Stamens
3 Carpels

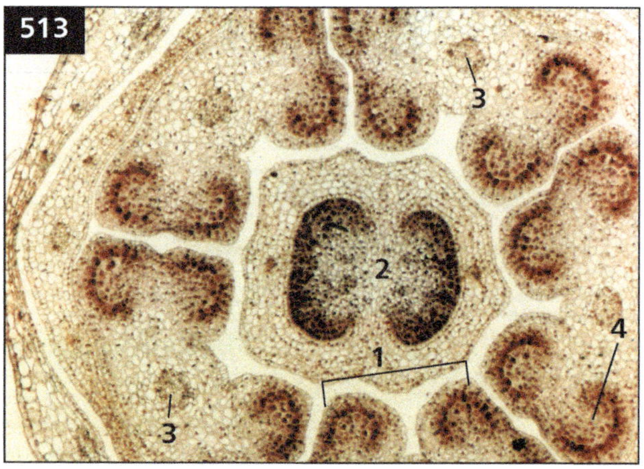

513 TS of the actinomorphic flower bud of *Solanum tuberosum* (potato). This dicot has five anthers (1) and a bilocular, superior ovary (2). Connective vascular bundle (3), pollen sac (4). (LM.)

1 Anthers
2 Superior ovary
3 Vascular bundle
4 Pollen sac

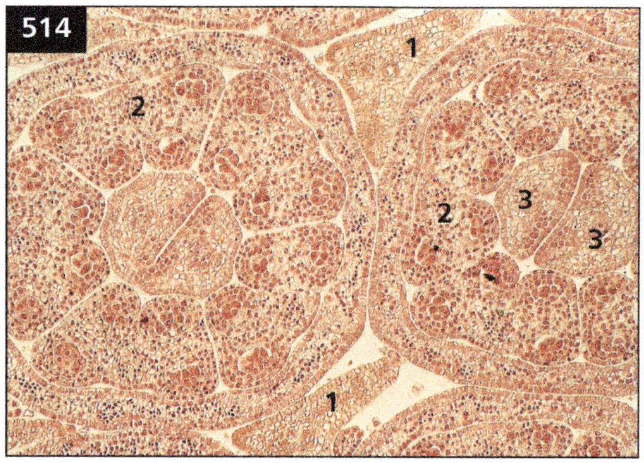

514 TS of the immature capitulum of the dicot *Helianthus* (sunflower). Each floret is subtended by a bract (1) and shows a tubular corolla, five anthers (2), and style terminated by a bifid stigma (3). The sepals and inferior ovary lie beneath the level of this section. (LM.)

1 Bract
2 Anthers
3 Bifid stigma

515 Close-up of the reproductive organs of the flower of the monocot *Crocus*. The anthers (1) have dehisced longitudinally and reveal the massed pollen grains, while the long style (2) branches into three expanded stigmas (3). Long-tongued bees and butterflies visit the flower for the nectar secreted at the base of the perianth and brush against the copious pollen.

1 Anther 2 Style 3 Stigmas

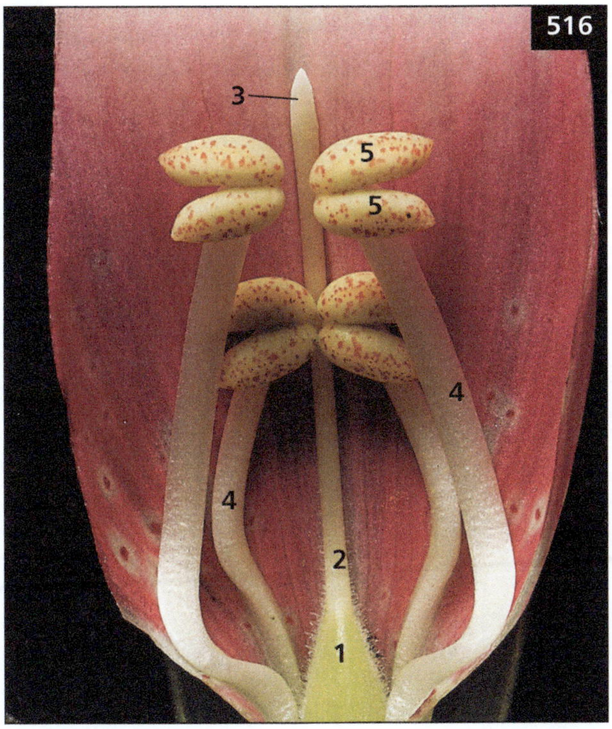

516 Close-up of the flower of the dicot *Digitalis purpurea* (foxglove). The irregular tubular corolla (cf. 267) has been split open to reveal the superior ovary (1) with a long style (2) and terminal stigma (3). There are four stamens with their filaments (4, two short and two long) joined to the corolla tube. Each bears a two-lobed anther (5). Pollination is by bumble-bees visiting for the nectar secreted at the base of the ovary.

1 Ovary
2 Style
3 Stigma
4 Filaments
5 Anthers

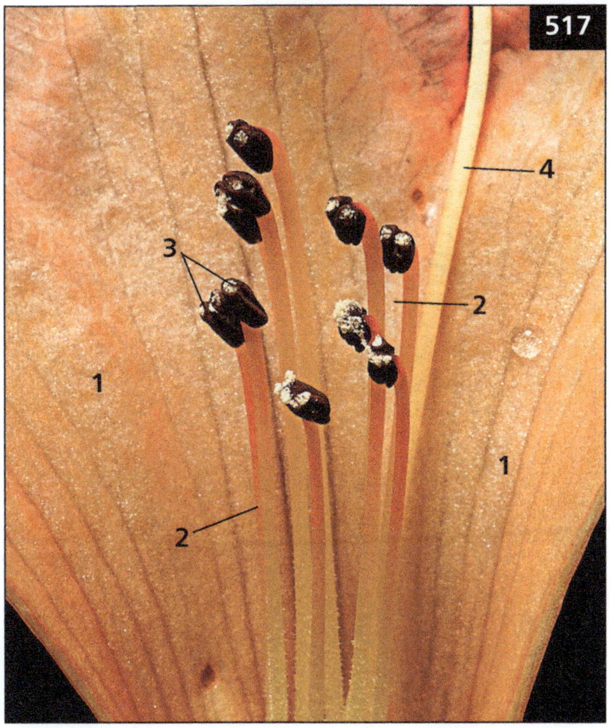

517 Close-up of the flower of the dicot *Rhododendron*. The corolla tube (1), has been split open to reveal the ten stamens, the filaments (2) of which are of varying lengths. The two-lobed anthers exude their pollen from terminal pores (3); the pollen tetrads adhere to each other, forming white masses that stick to bumble-bees visiting for nectar secreted at the base of the corolla. Style (4).

1 Corolla tube
2 Filaments
3 Terminal pores
4 Style

518 *Hibiscus*. Dicot flower showing the staminal tube (1) formed from fused filaments which, due to several branchings, bear numerous anthers (2). The tube encloses the style, the protruding tip of which bears five swollen stigmas (3).

1 Staminal tube
2 Anthers
3 Stigmas

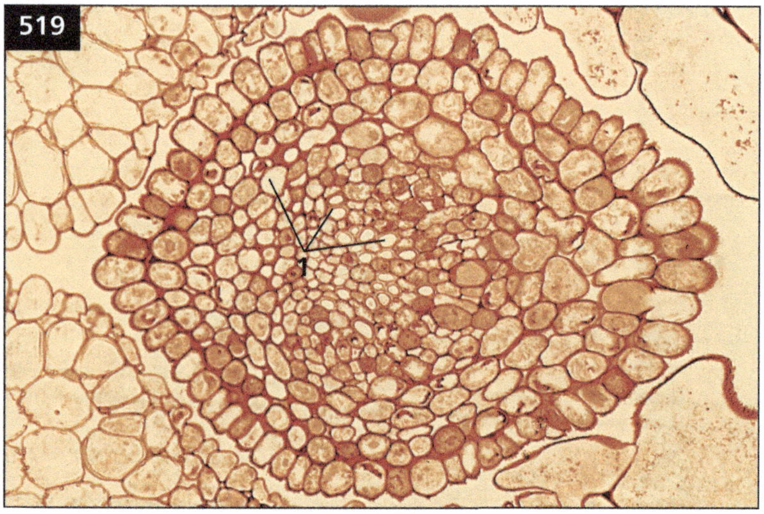

519 TS of the filament of an immature anther of the dicot *Sinapis*. Note its central vascular strand which transports nutrients to the developing pollen. Tracheary elements (1). (G-Os, LM.)

1 Tracheary elements

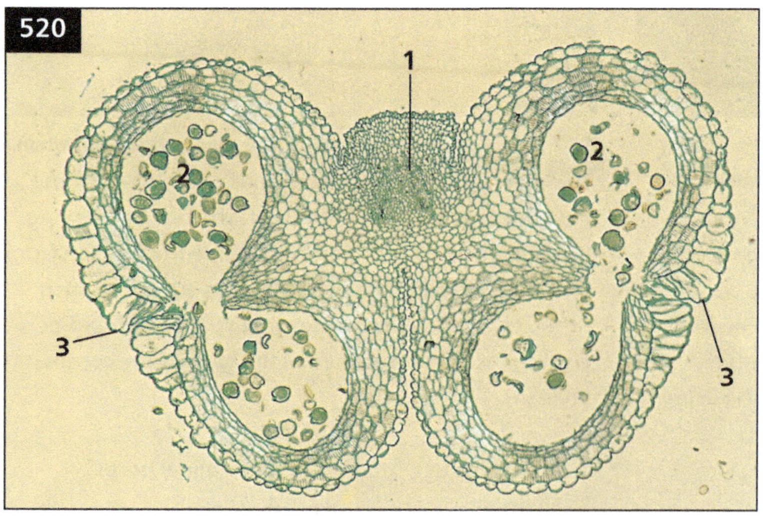

520 TS of the dehisced anther of the monocot *Lilium* (lily). The two anther lobes are joined by the connective tissue containing a vascular bundle (1). The wall between the two pollen sacs has broken down and the pollen grains (2) have been released by the dehiscence of the anther walls at the stomia (3). The hypodermal walls (except at the stomia) are elaborately thickened (cf. 522). (LM.)

1 Vascular bundle
2 Pollen grains
3 Stomium

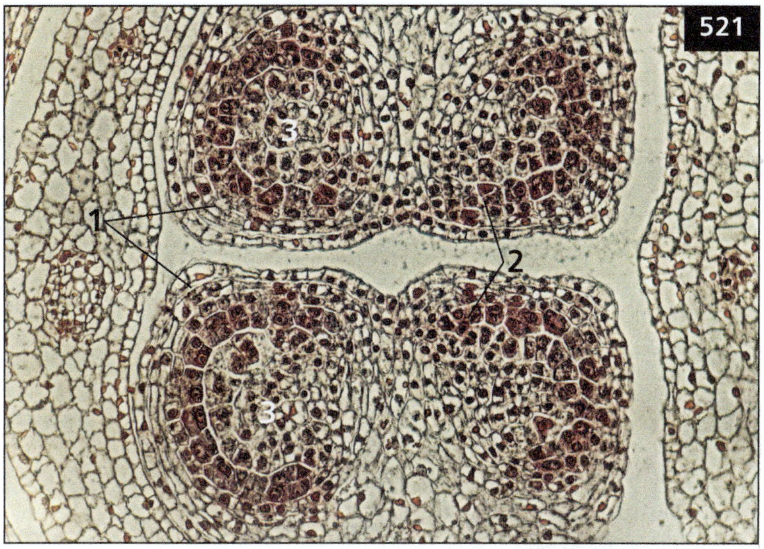

521 TS of immature pollen sacs in a flower bud of the dicot *Solanum tuberosum* (potato). Each anther lobe (cf. 513) contains two pollen sacs in which the hypodermis has undergone periclinal divisions forming vacuolated parietal tissue (1) and several layers of densely-staining tapetal cells (2). From the core of sporogenous cells (3) the haploid pollen grains are derived after meiosis. (Phase contrast LM.)

1 Parietal tissue
2 Tapetal cells
3 Sporogenous cells

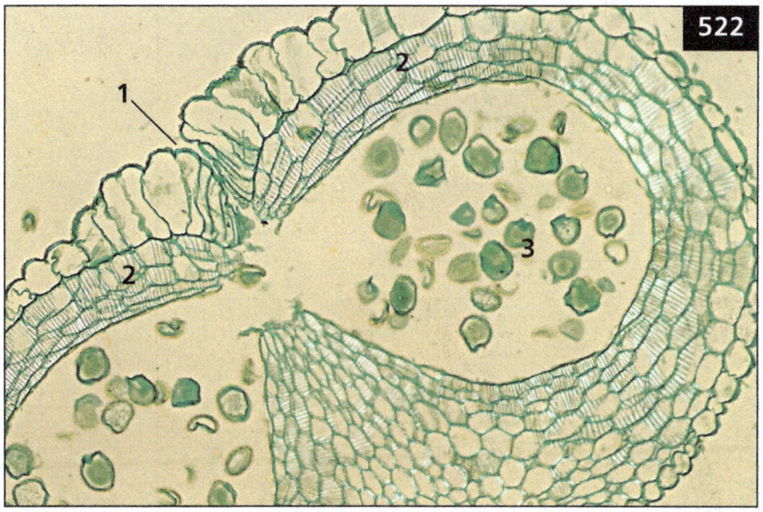

522 TS showing details of the anther wall of the monocot *Lilium* (lily). This is a mature anther (cf. 520) and shows the large but unthickened epidermal cells of the stomium (1). These contrast with the endothecium cells (2), the walls of which show anticlinal bands of cellulosic thickening. Pollen grains (3). (Polarized LM.)

1 Stomium
2 Hypodermal cells
3 Pollen grains

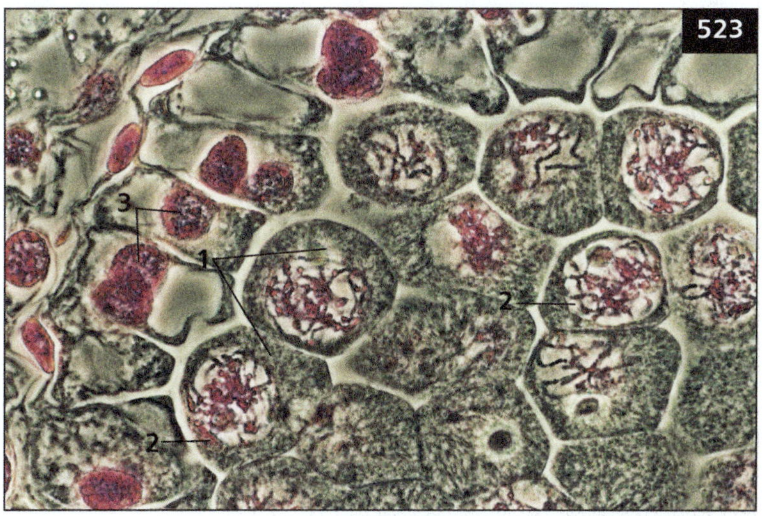

523 TS showing detail of pollen mother cells in monocot *Lilium* (lily) anther. Their large nuclei are in early prophase of the first meiotic division and the chromosomes appear as thread-like structures separated from the cytoplasm (1) by the nuclear envelope (2). Tapetal nuclei (3). (Phase contrast LM.)

1 Cytoplasm
2 Nuclear envelope
3 Tapetal nuclei

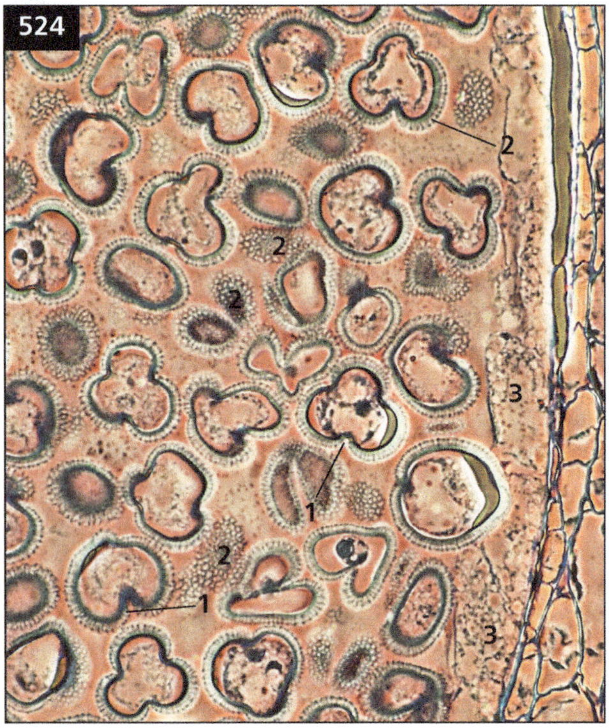

524 Sectioned anther of the dicot *Sinapis* showing pollen grains undergoing cytokinesis after meiosis. Wall formation in the pollen mother cell is of the simultaneous pattern with the wall furrows (1) developing centripetally to form four pollen grains. Note the patterned exine (2) and the degenerate remains of the tapetum (3). (G-Os, Phase contrast LM.)

1 Wall furrows
2 Patterned exine
3 Tapetum

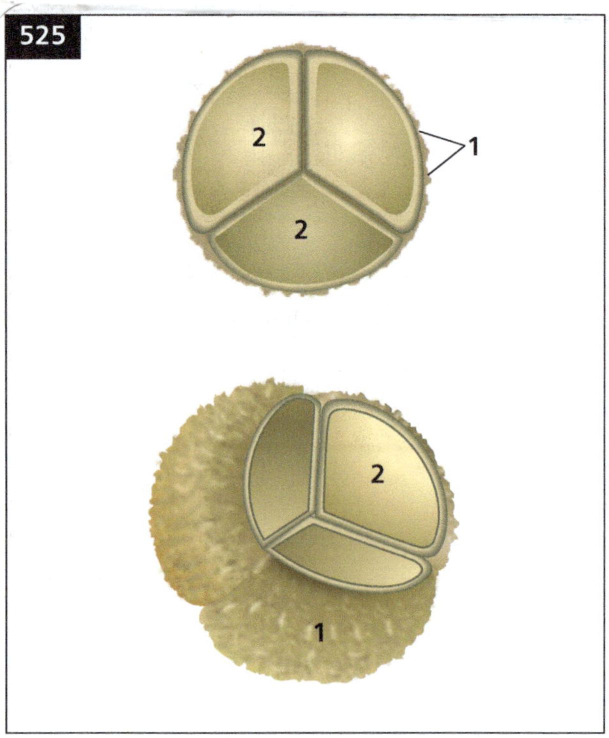

525 Diagrams of a microspore/pollen grain (top) and a newly formed spore tetrad (bottom) from which a single microspore has been removed. Exine (1), inner faces of microspore (2).

1 Exine
2 Inner faces of immature microspore

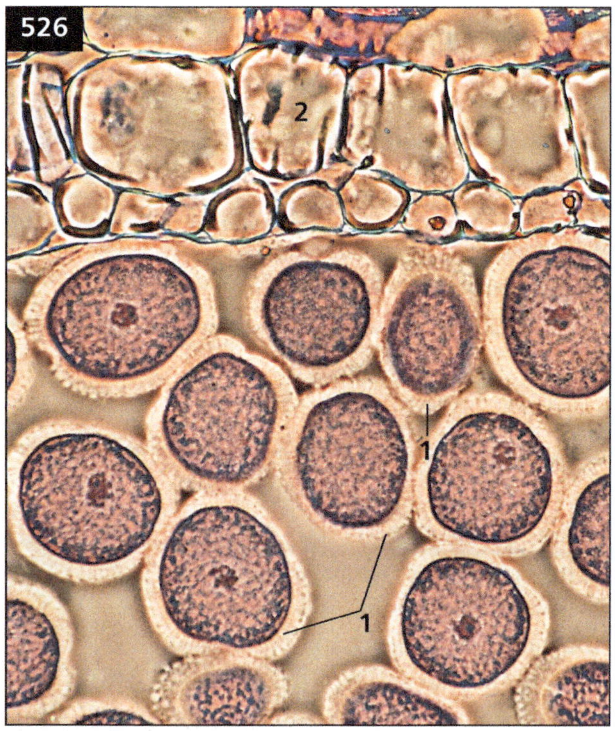

526 Anther of the dicot *Sinapis* showing individual pollen grains. These have almost separated from the tetrads (cf. 524) and are considerably enlarged; each grain is surrounded by a thick, patterned exine (1). Fibrous anther wall (2). (G-Os, Phase contrast LM.)

1 Patterned exine
2 Anther wall

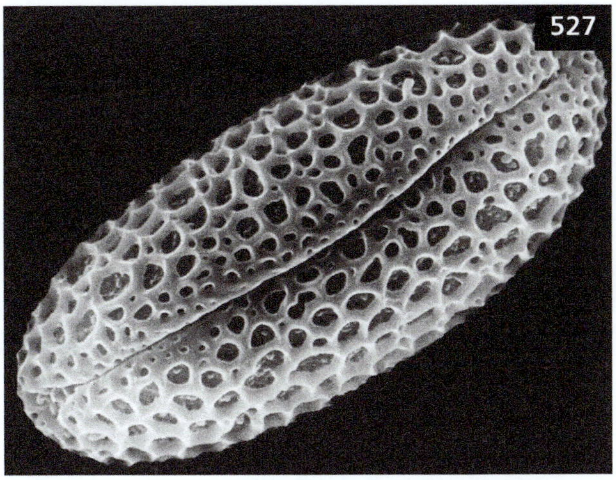

527 Mature pollen grain of the dicot *Salix* (willow). This grain has a reticulate exine and one of its three longitudinal grooves (colpi) is visible. The grains are insect-dispersed and at germination the pollen tube emerges from one of the colpi. (SEM.) (Copyright of Dr James H. Dickson.)

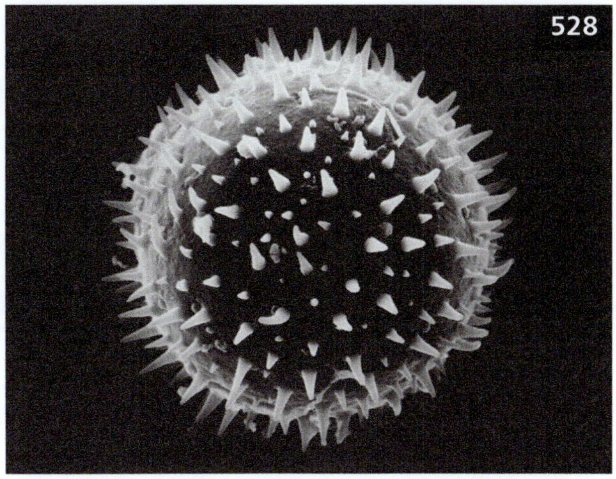

528 Mature pollen grain of the dicot *Malva* (mallow). Note that the exine is coarsely ornamented with large spikes. The grains are insect-dispersed and at germination the pollen tube emerges through one of the numerous germ pores. (SEM.) (Copyright of Dr James H. Dickson.)

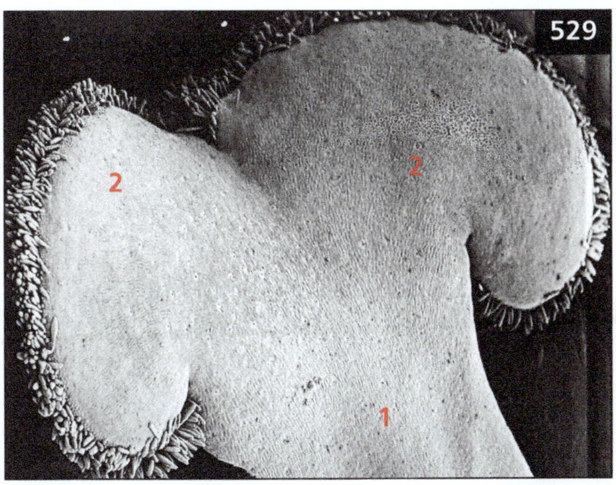

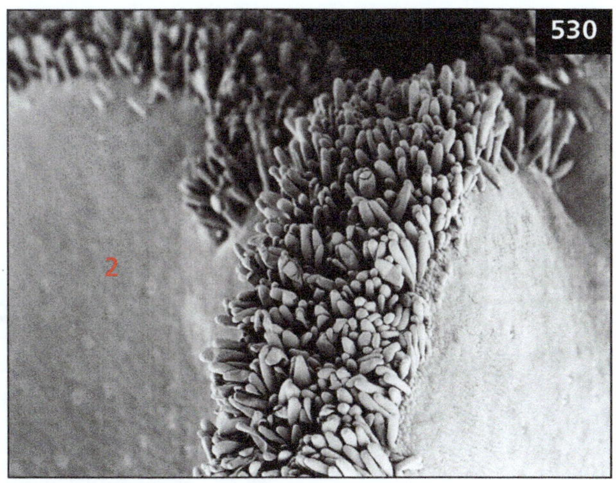

529, 530 Surface detail of the stigma of the monocot *Tulipa* (tulip). In 529, the style (1) expands into three ridges (2) reflecting the trilocular nature of its ovary. The surfaces of these ridges (530) bear closely crowded, short glandular trichomes in which pollen grains become enmeshed. (SEM.)

1 Style

2 Stigmatic ridges

531

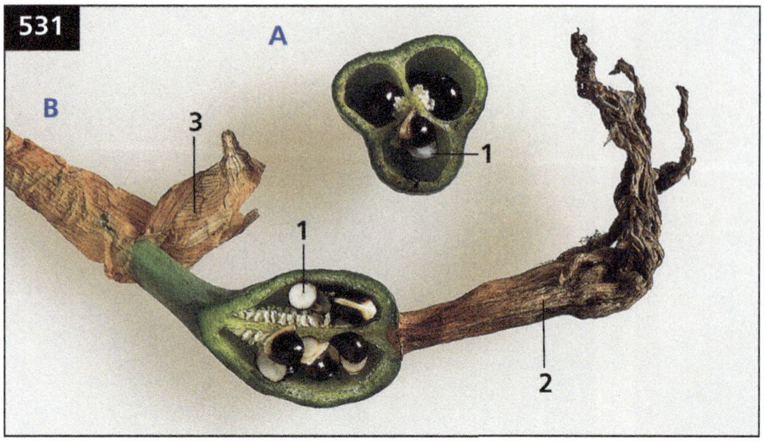

531 TS (A) and LS (B) of the fruit of the monocot *Narcissus* (daffodil). A shows its trilocular nature and the axial placentation of the seeds (1). B illustrates the inferior fruit lying at the base of the withered corolla tube (2). Bract (3).

1 Seeds
2 Corolla tube
3 Bract

532

532 This young tomato (*Lycopersicon esculentum*) fruit is developing from the flower's ovary, located between the persistent sepals (1) and the withered petals (2).

1 Sepals
2 Petals

533

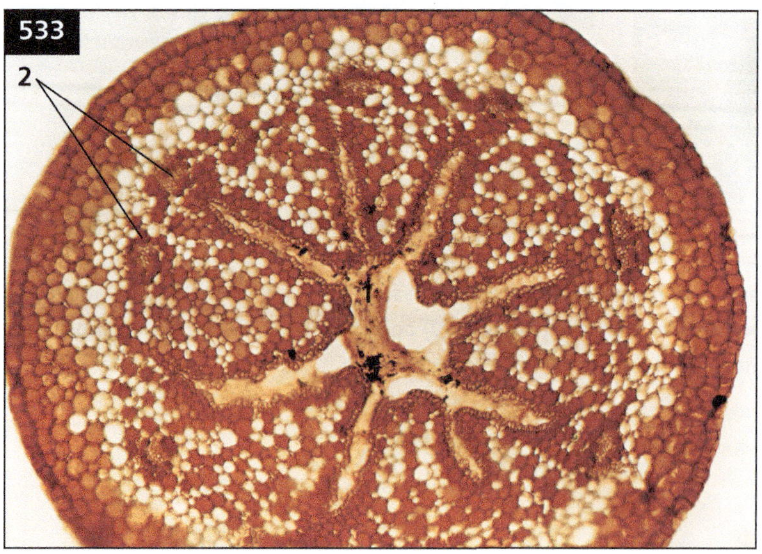

533 TS of the style of the dicot *Rhododendron*. The central channel and radiating arms contain mucilage (1) secreted by the lining epithelial cells; the growing pollen tubes are nourished by this secretion. The stylar wall contains a number of longitudinal vascular bundles (2). (LM.)

1 Mucilage
2 Vascular bundles

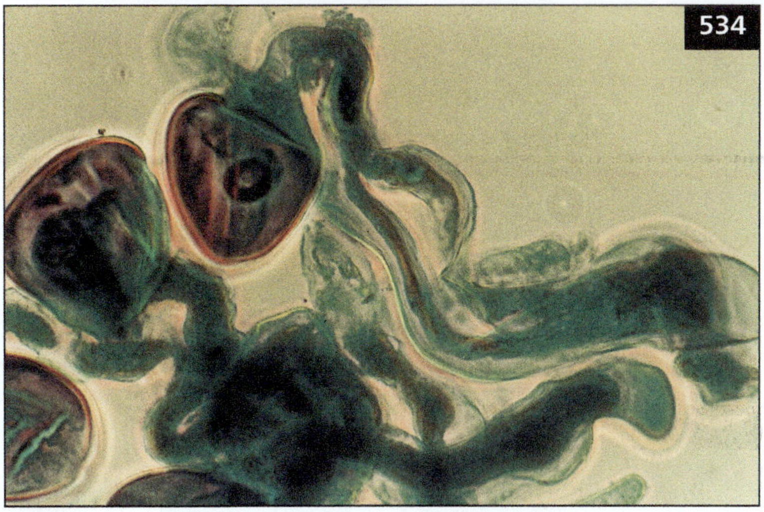

534 Details of germinating pollen of the monocot *Narcissus* (daffodil). These grains have germinated *in vitro* in agar with 7% sucrose and show well-developed, convoluted pollen tubes; several nuclei are also evident. (Phase contrast LM.)

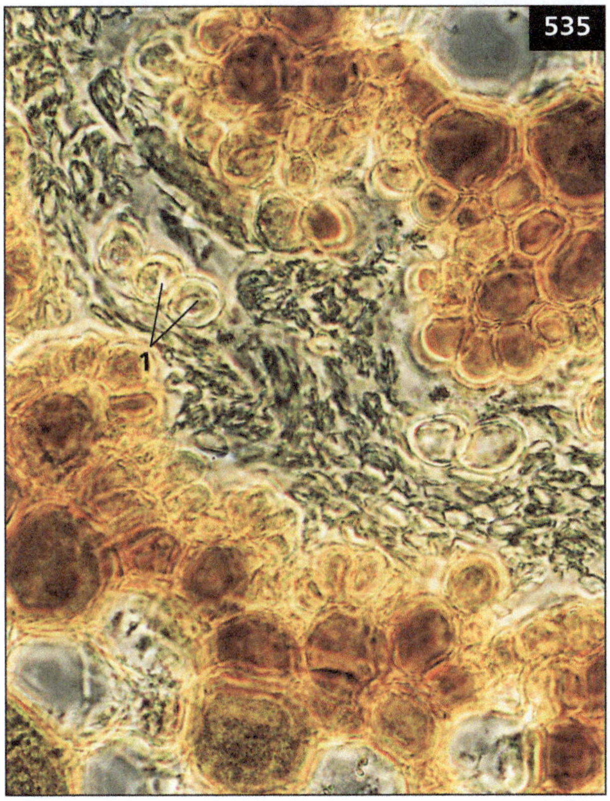

535 TS of the style of the dicot *Rhododendron* showing growing pollen tubes. Note within its mucilaginous interior (cf. 533) several transversely-sectioned pollen tubes (1). (Phase contrast LM.)

1 Pollen tubes

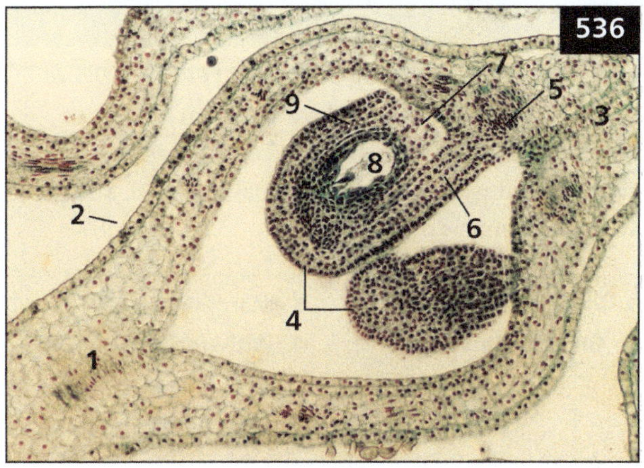

536 TS of a carpel of the dicot *Caltha palustris* (marsh marigold). This shows its leaf-like nature with the dorsal suture (1) corresponding to a mid-rib while the blade is folded with its abaxial surface (2) outermost. The two leaf margins are fused at the ventral suture (3) and the inverted ovules (4) are joined to the placenta (5) by a short funiculus (6). Micropyle (7), embryo sac (8), integuments (9). (LM.)

1 Dorsal suture
2 Abaxial surface
3 Ventral suture
4 Ovule
5 Placenta
6 Funiculus
7 Micropyle
8 Embryo sac
9 Integument

537

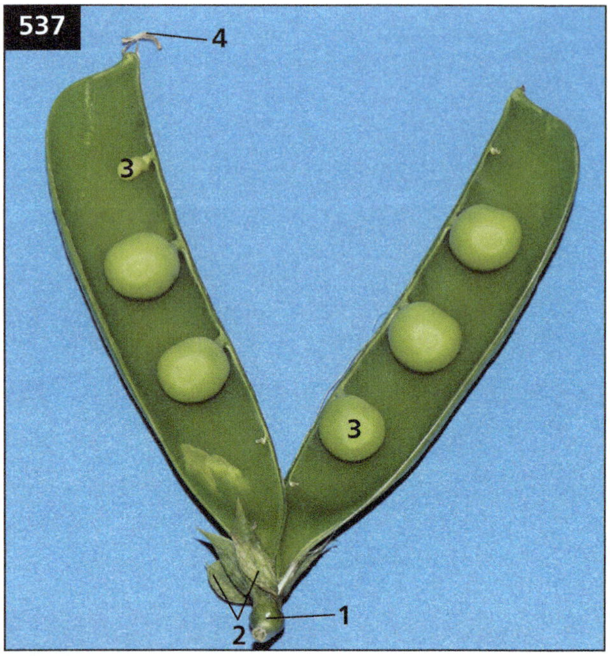

537 Young split-open pod of the dicot *Pisum sativum* (pea). In legumes the fruit develops from a superior apocarpous ovary. Note the pedicel (1), green sepals (2), swollen fruit with a row of marginal seeds (3), and withered remains of the style (4).

1 Pedicel
2 Green sepals
3 Marginal seeds
4 Withered style

538

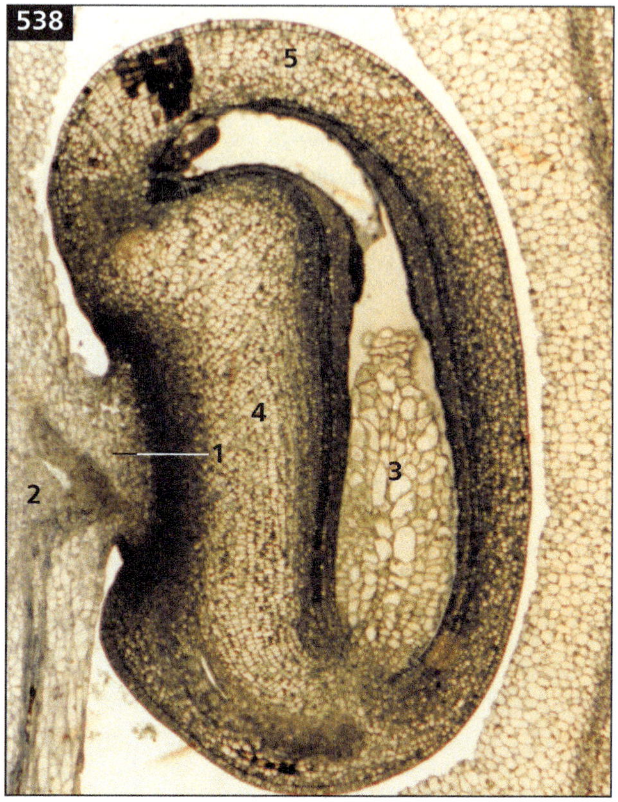

538 LS of a fertilized ovule of the dicot *Phaseolus vulgaris* (bean). This is attached by a short funiculus (1) to the placenta (2). The embryo is not visible in this section but cellular endosperm (3) is apparent within the embryo sac. Nucellus (4), integument (5). (G-Os, LM.)

1 Funiculus
2 Placenta
3 Cellular endosperm
4 Nucellus
5 Integument

539

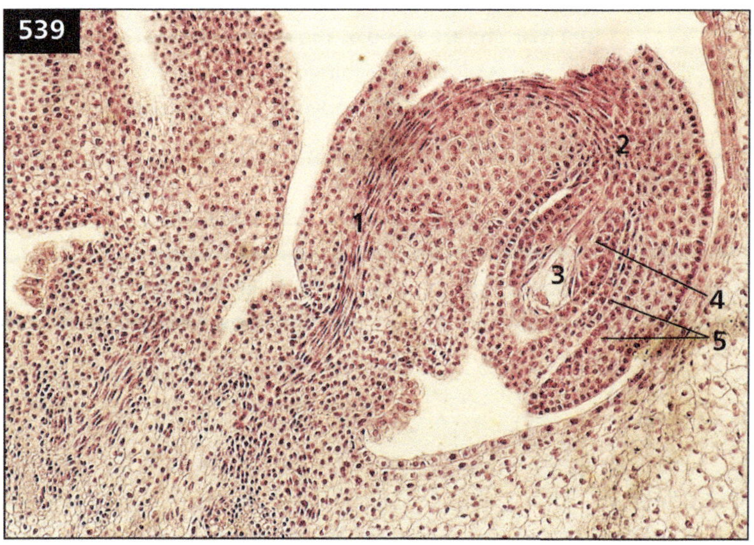

539 LS of anatropous ovules of the monocot *Iris*. The medianly sectioned ovule arises from the central axis of a trilocular ovary (cf. 531) with the funiculus, containing a vascular strand (1), running from the placenta to the chalaza (2) at the base of the embryo sac (3). Nucellus (4), integuments (5). (LM.)

1 Vascular strand
2 Chalaza
3 Embryo sac
4 Nucellus
5 Integuments

540 LS of an ovule primordium in the monocot *Lilium* (lily). At the micropylar pole of the nucellus (1) a single hypodermal cell has enlarged and its nucleus (2) is about to undergo meiosis. Integument (3), placenta (4). (LM.)

1 Nucellus
2 Hypodermal nucleus
3 Integument
4 Placenta

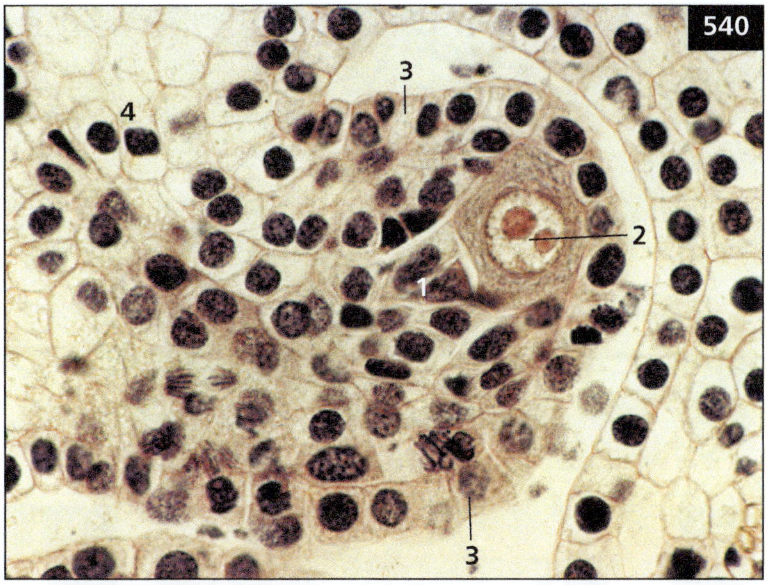

541 LS of a semi-mature ovule of the monocot *Iris*. The embryo sac contains two functional nuclei (1) but several degenerate nuclei (2) represent the nonfunctional megaspores and nucellar cells crushed by the expanding embryo sac. Integument (3), micropyle (4), nucellus (5). (LM.)

1 Functional nuclei
2 Degenerate nuclei
3 Integument
4 Micropyle
5 Nucellus

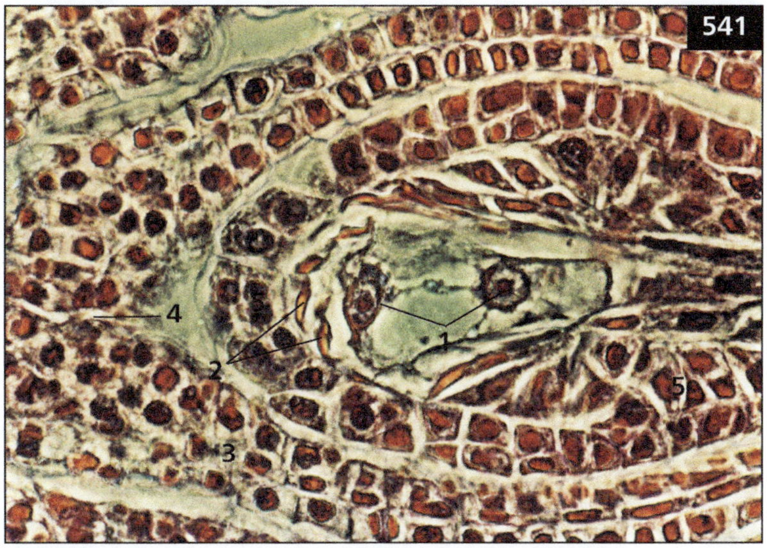

542 LS of an immature embryo sac of the monocot *Lilium* (lily). The megaspore mother cell (cf. 540) has divided meiotically to give four nuclei, but without wall formation. Three of these nuclei have fused to give a triploid nucleus (1) at the chalazal end of the embryo sac while a haploid nucleus (2) remains at the other pole. Subsequently the egg, two synergids, and one polar nucleus form from the latter, while the triploid nucleus gives rise to three antipodal and one polar nucleus. (LM.)

1 Triploid nucleus
2 Haploid nucleus

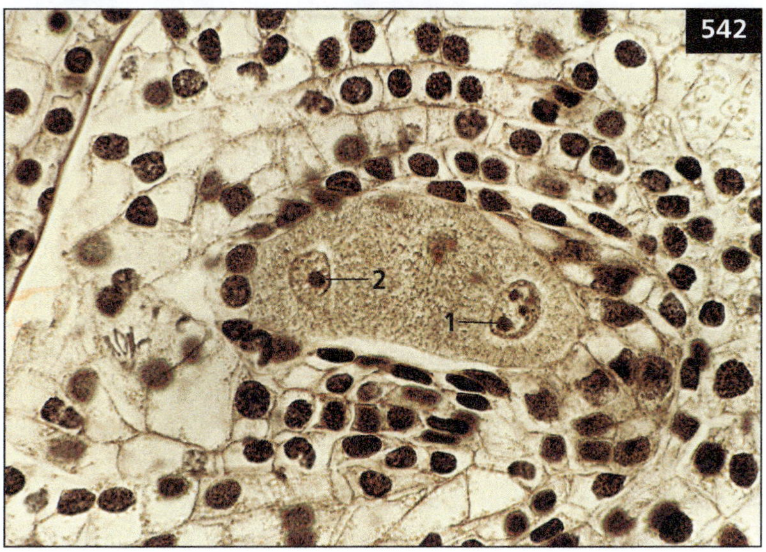

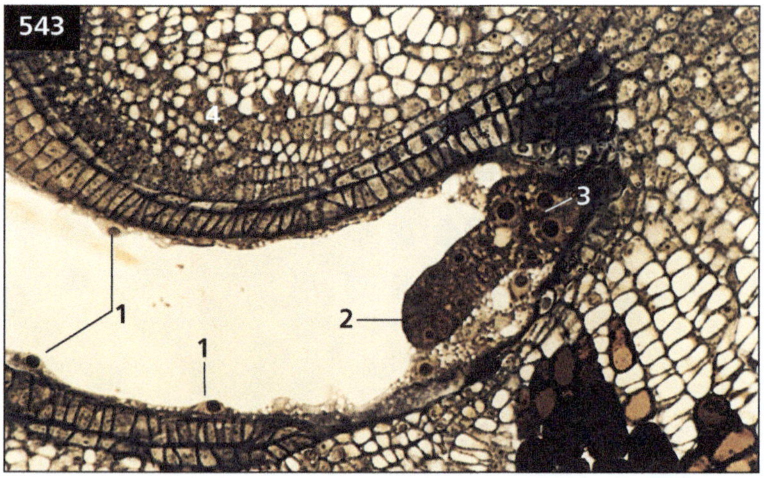

543 LS of a very young embryo of the dicot *Phaseolus vulgaris* (bean). This is located at the micropylar pole of the large embryo sac (cf. 538) and nuclei (1) of the coenocytic endosperm line the walls. The embryo is terminated by the proembryo (2) while the cylindrical suspensor (3) is distended at its micropylar end. Nucellus (4). (G-Os, LM.)

1 Endosperm nuclei
2 Proembryo
3 Suspensor
4 Nucellus

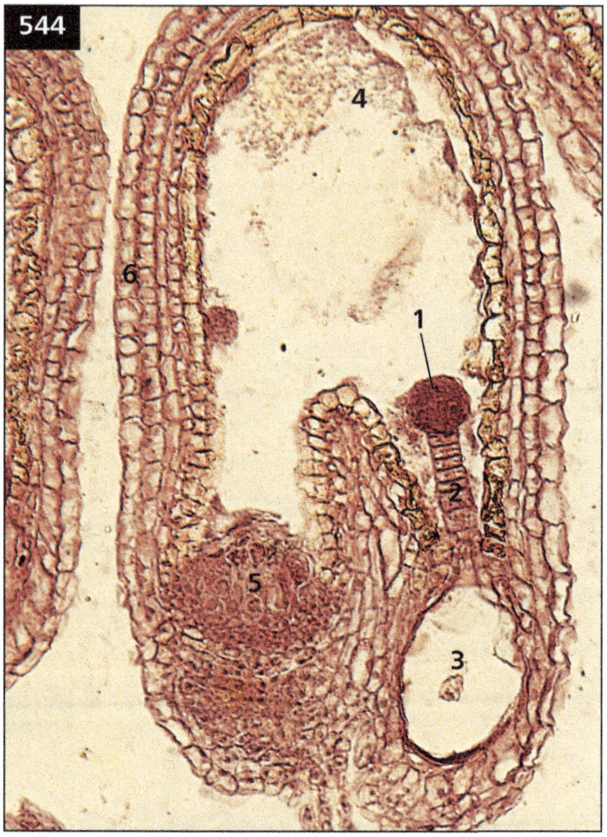

544 Immature seeds of the dicot *Capsella bursa-pastoris* (shepherd's purse). The longitudinally-sectioned seed shows a globular proembryo (1) attached to a filamentous suspensor (2), with its swollen basal cell (3) terminating at the micropylar end of the embryo sac. Coenocytic endosperm (4), hypertrophied nutritive nucellar tissue (5), testa (6). (LM.)

1 Proembryo
2 Filamentous suspensor
3 Basal cell
4 Coenocytic endosperm
5 Nucellar tissue
6 Testa

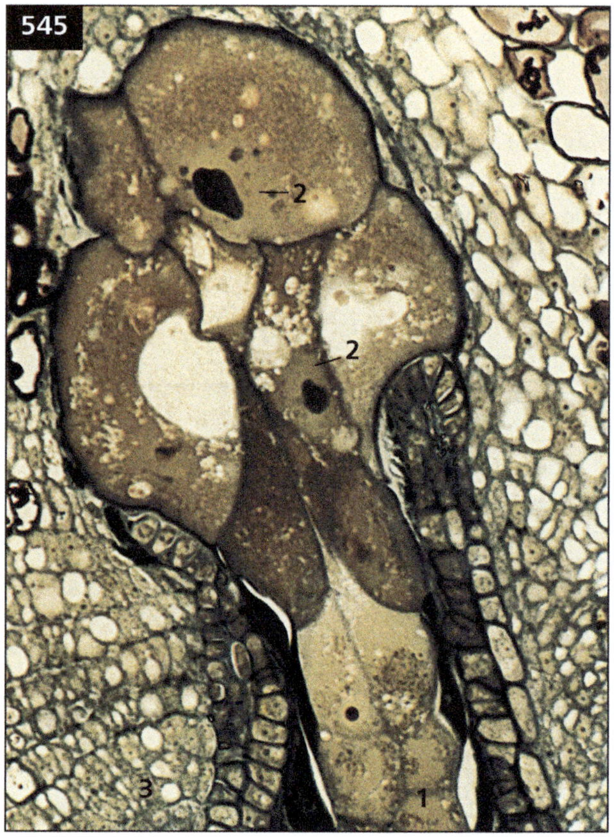

545 Detail of the micropylar pole of the embryo in the dicot *Phaseolus vulgaris* (bean). The nuclei of the basal cells of the suspensor (1) undergo numerous rounds of DNA replication, which is not accompanied by mitosis, and very large highly polytene nuclei (2) develop. Nucellus (3). (G-Os, LM.)

1 Suspensor
2 Polytene nuclei
3 Nucellus

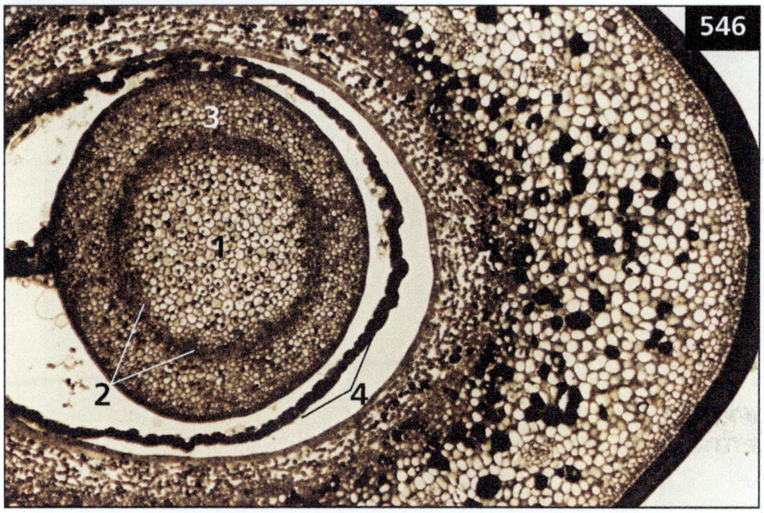

546 Immature seed of the dicot *Phaseolus vulgaris* (bean). The section cuts the embryo transversely through the hypocotyl; note the differentiation of pith (1), vascular tissue (2), and cortex (3). The peripheral endosperm layer (4) has become detached from the wall of the embryo sac. (G-Os, LM.)

1 Pith
2 Vascular tissue
3 Cortex
4 Endosperm layer

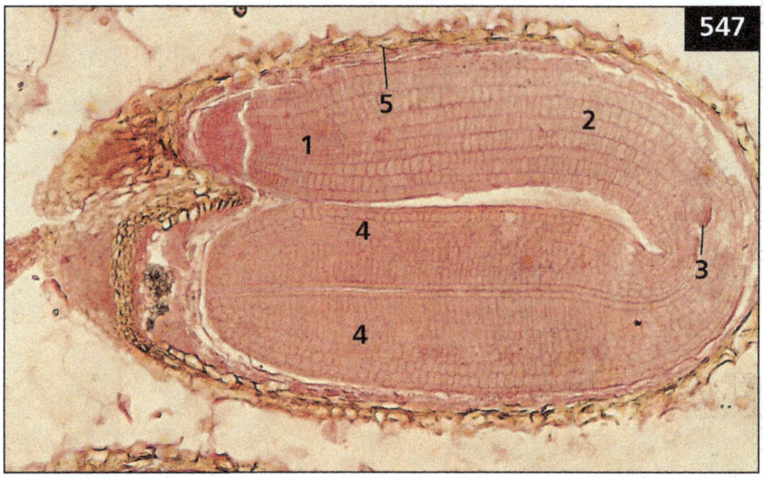

547 LS of the mature seed of the dicot *Capsella bursa-pastoris* (shepherd's purse). The embryo completely fills the embryo sac and all the endosperm has been absorbed. A wide cylindrical radicle (1, with its apex at the micropyle) is continuous with the hypocotyl (2). This is terminated by a small plumular apex (3) on either side of which arise a pair of swollen cotyledons (4) lying parallel to the radicle–hypocotyl axis. The curved nature of its embryo sac (cf. 544) results in the bent shape of the embryo. Testa (5). (LM.)

1 Radicle
2 Hypocotyl
3 Plumular apex
4 Cotyledons
5 Testa

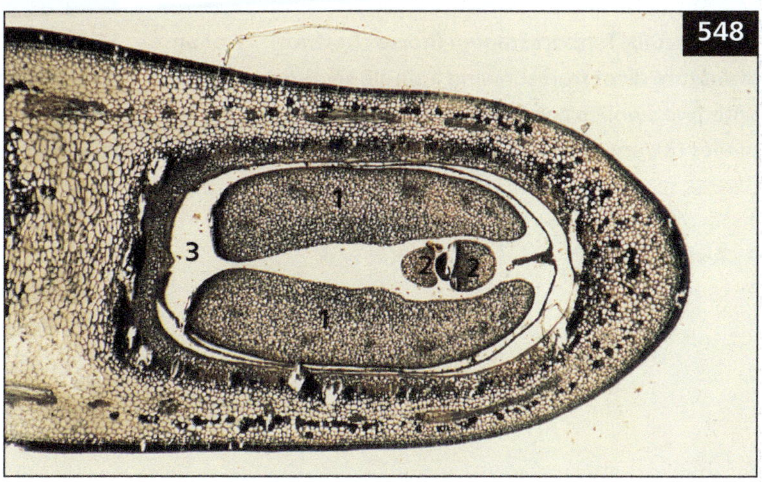

548 Semi-mature seed of the dicot *Phaseolus vulgaris* (bean). The embryo is cut obliquely and shows the paired cotyledons (1) while the primordia of the first pair of foliage leaves (2) have already been formed by the shoot apex (cf. 213). Embryo sac (3). (G-Os, LM.)

1 Cotyledons
2 Foliage leaves
3 Embryo sac

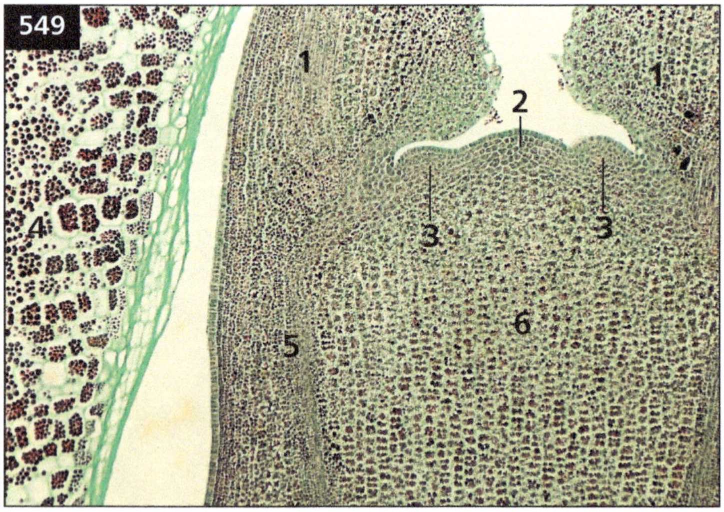

549 LS of the mature plumule in the seed of the dicot *Ricinus communis* (castor oil). Note the paired cotyledons (1) on either side of the shoot apex (2) and the first foliage leaf primordia (3). The seed contains copious supplies of endosperm (4) which nourish the germinating embryo. Procambium (5), pith (6). (LM.)

1 Cotyledon
2 Shoot apex
3 Foliage leaf primordia
4 Endosperm
5 Procambium
6 Pith

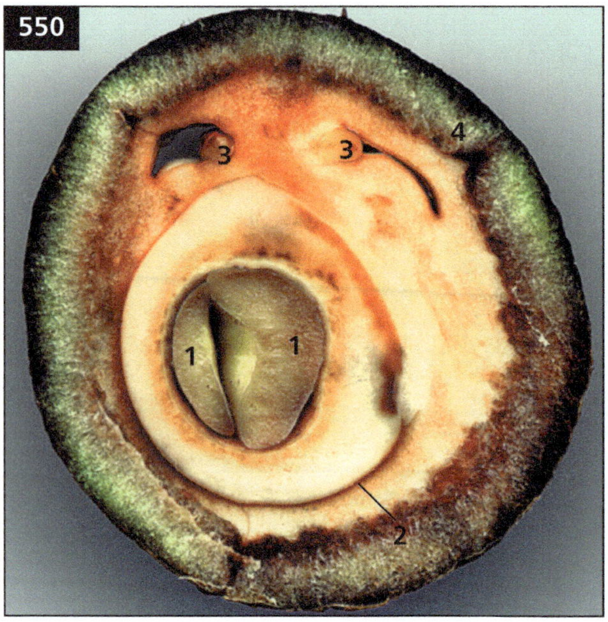

550 *Aesculus hippocastanum* (horse chestnut). TS of an immature dicot fruit showing a single semi-mature seed with two swollen cotyledons (1), testa (2), aborted ovules (3), and fruit wall (4).

1 Swollen cotyledons
2 Testa
3 Aborted ovules
4 Fruit wall

551 Interior of *Cocos nucifera* (coconut palm) revealing its thick, fleshy endosperm (1) layer (dried commercially as 'copra'); the scanty remaining liquid endosperm (coconut milk) in its central cavity (2), spilt out when the coconut was broken open. Testa (3), sclerified, hard inner layer of the fruit wall (4) (cf. 554).

1 Endosperm
2 Central cavity of seed
3 Testa
4 Inner fruit wall

552 Germinating grain of the monocot *Zea mays* (maize). Note the long radicle with its dense covering of root hairs (1) and the remains of the coleorhiza (2) at its base. The latter covered the root prior to germination, while the coleoptile (3) invests and protects the plumule. Seminal root (4). (Copyright of T. Norman Tait.)

1 Root hairs
2 Coleorhiza
3 Coleoptile
4 Seminal root

553 Fruiting crown of the monocot *Cocos nucifera* (coconut palm). These large monocots frequently grow along tropical beaches in Asia and Polynesia; they bear clusters of large one-seeded fruits that fall to the ground when ripe and are often dispersed great distances by sea currents (cf. 554).

554 Fruits of *Cocos nucifera* (coconut palm) have a smooth exocarp and a light-weight, airy fibrous mesocarp that makes the giant fruit buoyant. They can float for many weeks and cover many kilometres without suffering damage from seawater because the innermost endocarp (coconut shell; cf. 551) is extremely hard and impervious. The single seed within the fruit is enclosed by a thin brown testa (not distinguishable in this illustration) which is lined by a layer of the white cellular endosperm, which we use as food and candy.

555 LS of the fruit of the dicot *Persea americana* (avocado pear). This forms from a superior unilocular ovary bearing a single pendulous ovule that develops into the large central seed (1). The edible parenchymatous fruit wall (2) contains up to 30% of oils that initially accumulate in oil sacs. In the mature fruit the tissue degenerates and the flesh is buttery.

1 Seed
2 Parenchymatous fruit wall

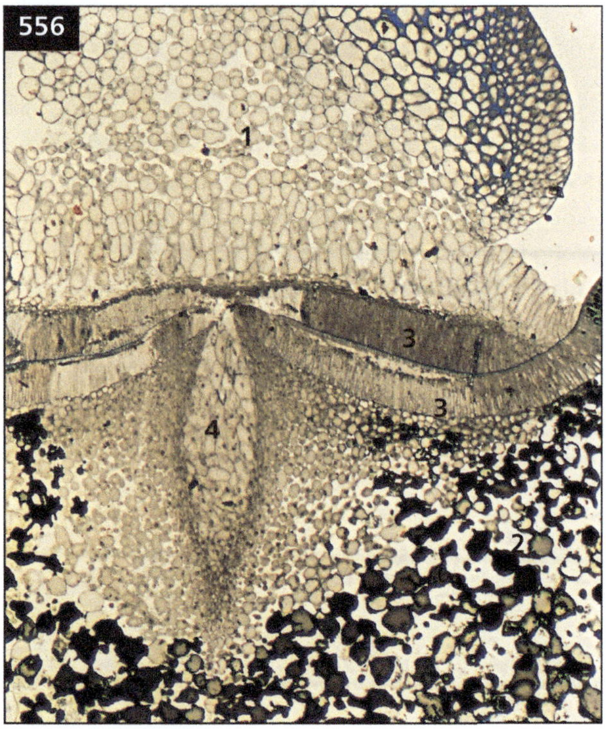

556 TS through the hilum of the semi-mature seed of the dicot *Phaseolus vulgaris* (bean). The funiculus (1) is confluent with the testa (2); at their junction two layers of columnar epidermal cells are apparent (3) that subsequently form macrosclereids (cf. 557). A compact group of tracheids later develops in the centre of the hilum from a lens-shaped group of cells (4). (G-Os, LM.)

1 Funiculus
2 Testa
3 Epidermal cells
4 Potential tracheids

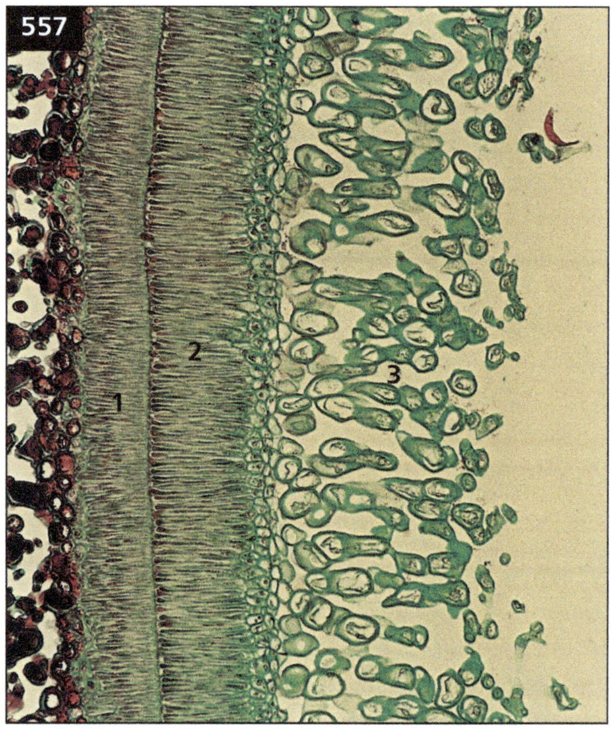

557 LS through the hilum of the mature seed of the dicot *Phaseolus vulgaris* (bean). Note the two layers of macrosclereids (cf. 556); the inner (1) is derived from the epidermis of the integument while the outer layer (2) forms from the funicular epidermis; the spongy tissue (3) is hypodermal in origin. (LM.)

1 Inner macrosclereids
2 Outer macrosclereids
3 Spongy tissue

558 Harvested cob of the monocot *Zea mays* (maize). This matures several months after fertilization from an inflorescence axis bearing numerous female flowers. The cob is invested by bracts (peeled back) that cover the closely crowded, swollen grains (fruits). Note the withered silks representing the collective remains of the pendulous, thread-like style borne by each flower; the styles may be up to 25 cm long and remain receptive to pollination for up to 2 weeks.

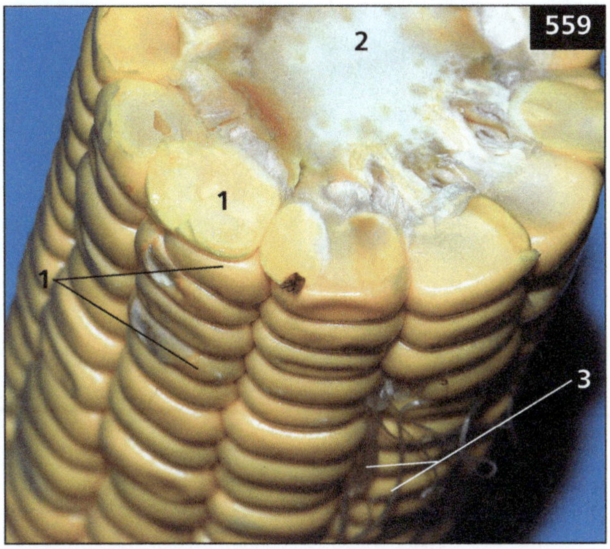

559 *Zea mays*. Cross-cut surface of mature edible cob (inflorescence axis) showing the individual fruitlets (caryopses, 1) which emanate from the central inflorescence axis (2). Remnants of styles (3).

1 Individual fruitlets
2 Central inflorescence axis
3 Remnants of styles

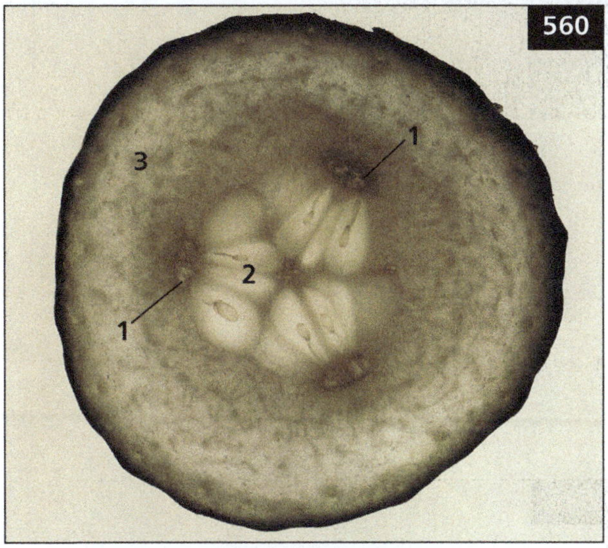

560 TS of the fruit of the dicot *Cucurbita* (cucumber). This develops from an inferior syncarpous ovary of three fused carpels with parietal placentae (1). The numerous seeds (2) are enclosed in a pulpy parenchymatous endocarp; this is surrounded by the firmer flesh of the mesocarp (3) in which a ring of bicollateral vascular bundles occurs. The exocarp forms the green skin of the fruit.

1 Parietal placentae
2 Seeds
3 Mesocarp

561

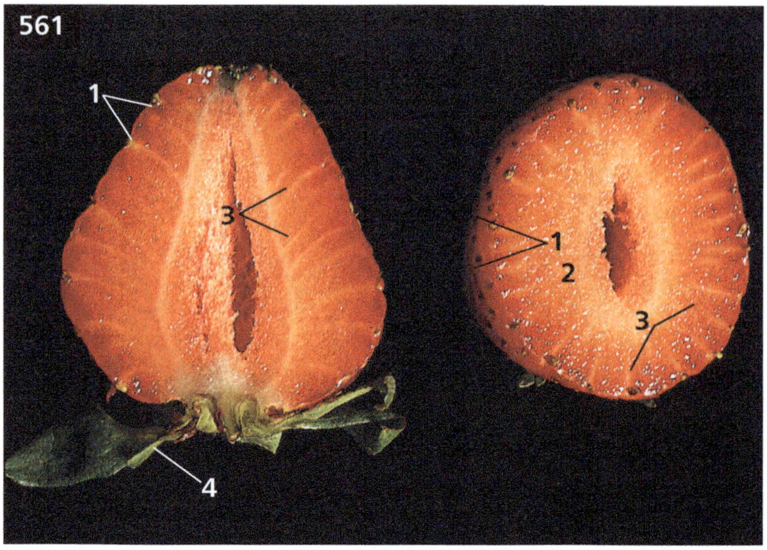

561 The aggregate fruit of the dicot *Fragaria vesca* (strawberry). This develops from the numerous free carpels of a single flower, with each superior carpel containing a single ovule. The resultant indehiscent fruitlets (pips, 1) are embedded in the hypertrophied floral receptacle (2) to form a succulent aggregate fruit. Note the white longitudinal vascular strands (3) which branch to supply the individual fruitlets. Calyx (4).

1 Fruitlets (pips)
2 Floral receptacle
3 Vascular strands
4 Calyx

562

562 Inflorescence of the dicot *Banksia* bearing closely-crowded flowers. The flowers of this indigenous Australasian genus are bird-pollinated and subsequently the ovaries become embedded in the woody inflorescence axis to form a cone-like structure (cf. 563).

563

563 Cone-like fructification of the dicot *Banksia*. This woody genus forms closely crowded sclerified fruits embedded in the swollen inflorescence axis that undergoes some secondary thickening. The fruits dehisce, usually following a bush fire, along a horizontal suture to release the seeds.

564 Each of these fruits (1) of *Ferocactus wislizenii* (barrel cactus) is actually a combination of the development of the inferior ovary and the floral shoot that surrounded it (cf. 503). Thus the outer, visible yellow tissue of each constitutes a 'false fruit' whereas the true fruit is located inside. If cut in transverse section, however, the two would be indistinguishable. The scales (arrows) are leaves, the axillary buds of which have not produced spines. Withered perianth (2).

1 Fruit
2 Withered perianth

565 Fleshy fruit of the dicot *Malus sylvestris* (apple). The core of the apple (with its tough sclereids) forms f rom an inferior syncarpous ovary with axile placentation. The parenchymatous flesh represents the greatly enlarged floral tube which surrounded the ovary. In cross-section of the fruit, four of the five ovary compartments contain seeds (1). In the inner flesh there is a ring of vascular bundles (2) that supplied the five sepals and five petals. In longitudinal view the pedicel (3) and the withered remains (4) of calyx, stamens, and styles are visible.

1 Seeds	3 Pedicel
2 Vascular bundles	4 Remains of calyx, stamens, styles

566 LS of the fruit of the dicot *Prunus* (peach). This develops from a unilocular superior ovary. The stony inner endocarp (1) encloses a single seed. The extensive mesocarp (2) is fleshy and succulent and is enclosed by the thin skin (exocarp, 3). Pedicel (4).

1 Stony endocarp	3 Exocarp (skin)
2 Fleshy mesocarp	4 Pedicel

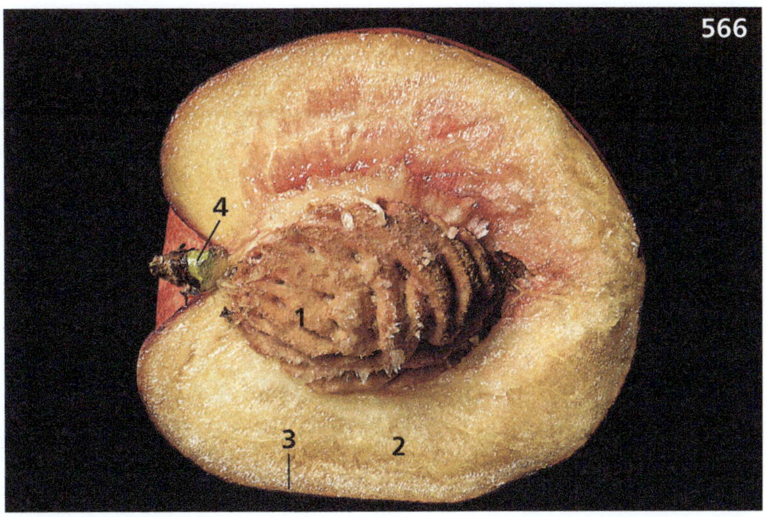

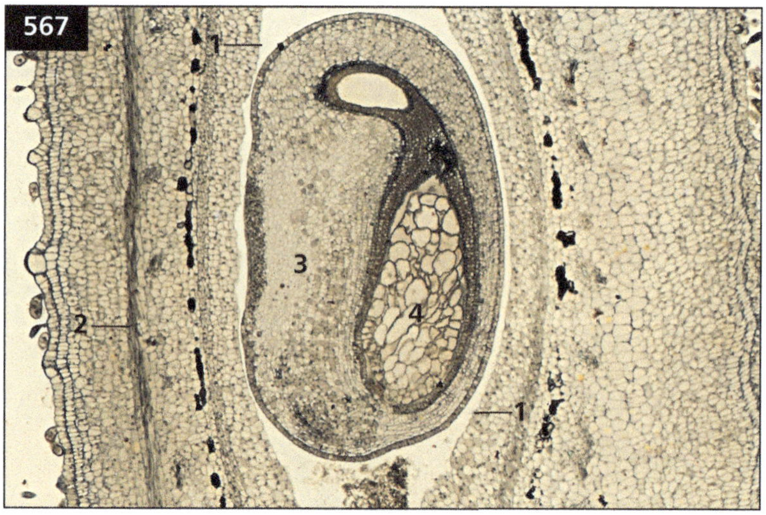

567 LS of a young fruit of the dicot *Phaseolus vulgaris* (bean). The outer epidermis of the pod bears numerous trichomes and a cuticle is present, but the inner surface (1) is hairless and lacks a cuticle. Vascular strand (2), immature seed (3), cellular endosperm (4). (G-Os, LM.)

1 Inner surface of pod
2 Vascular strand
3 Immature seed
4 Cellular endosperm

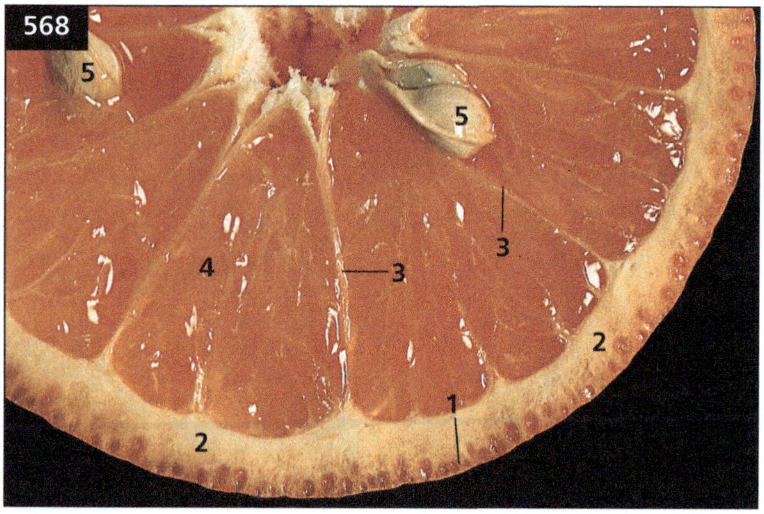

568 TS of the fruit of the dicot *Citrus sinensis* (orange). This develops from a superior ovary of ten united carpels with axile placentation. The peel of the fruit consists of the orange leathery exocarp (1, containing numerous oil glands) and white aerenchymatous mesocarp (2). The endocarp forms a thin layer internal to the mesocarp and also the radial partitions (3) between the locules. Ingrowths from the endocarp develop into the juice sacs (4) which pack the interiors of the locules. Seeds (5).

1 Exocarp with oil glands
2 Mesocarp
3 Radial partitions (endocarp)
4 Juice sacs
5 Seeds

569 Pendulous fruits of the dicot *Eucalyptus calophylla*. This indigenous Australasian genus has woody capsules that dehisce at their tips to release the seeds. The woody fruit develops from an inferior, syncarpous ovary to which the surrounding floral receptacle is united.

570 Ripening fruits of the dicot *Papaver* (poppy). The woody capsule forms from a superior syncarpous ovary. In the mature fruit the stigmatic lobes bend upwards from their previous position (cf. 508) and uncover a ring of pores (not visible) in the capsule wall. As the capsule is blown by the wind, the small seeds gradually sift through the pores.

571 Immature fruit of the dicot *Acer pseudoplatanus* (sycamore). This forms from a superior bicarpellary ovary, the walls of which develop two prominent wings, each enclosing a single seed. At maturity the fruit abscises and its wings cause the fruit to spin downwards to the ground where it separates into two nondehiscent segments, each containing a single seed. Pedicel (1), withered remnants of style (2), position of seeds within the pericarp (3).

1 Pedicel
2 Style remnants
3 Position of seeds

572 *Mecanopsis cambrica* (Welsh poppy) dehisced capsule from which the very small, mature dicot seeds escape and are scattered when the fruit is blown by the wind.

573 Dehisced fruit of the monocot *Iris*. This forms from an inferior trilocular ovary that at maturity is dry and dehisces into three valves to liberate its red seeds. (Copyright of T. Norman Tait.)

574 Dehiscing fruit of the dicot *Epilobium hirsutum* (willow herb). The inferior ovary has four locules with axile placentae bearing numerous ovules; the mature fruit dehisces from the top downwards into four valves. The plumed seeds are dispersed in the wind. (Copyright of T. Norman Tait.)

575 Dehisced pods of the dicot *Cytisus scoparius* (broom). The old pods have dehisced longitudinally along both sutures; as the dry pods split open they become twisted and violently eject the seeds.

576 *Ficus auriculata* (fig). The massed dicot fruits, which develop from cauliflorous flowers, are avidly gorged upon by various animals in the tropics.

577 *Populus trichocarpa* is known as cottonwood because it produces copious numbers of tiny, hairy seeds that blow like cotton on the wind and eventually fall to the ground in masses resembling snowdrifts.

578 *Ecballium elaterium* unexploded fruits and fruit stalks of exploded specimens. During fruit maturation the seeds are invested in a central mass of very thin-walled tissue which eventually develops an osmotic pressure nearing 27 atmospheres. This creates tension in the surrounding thick-walled tissue and, when the abscission layer at the base of the fruit finally ruptures, the central seed mass is violently ejected.

579

579 Viviparous seedlings of the dicot *Rhizophora* (mangrove). The embryo germinates within the fruit (1, note the persistent basal calyx) while the latter is still attached to the mangrove tree. The elongate seedling axis comprises a long radicle–hypocotyl axis (2) but the plumule still lies within the fruit. The seedlings reach 30–50 cm before falling from the tree and rooting in the mud flats (cf. 580). (Copyright of T. Norman Tait.)

1 Fruit
2 Radicle–hypocotyl axis

580

580 Newly established colony of the dicot *Rhizophora* (mangrove). The viviparous seedlings (cf. 579) root in the estuarine mud or shallow water and form new colonies. Note the numerous seedlings in the foreground. (Copyright T. Norman Tait.)

Bibliography

Bell AD (1991). *An Illustrated Guide to Flowering Plant Morphology*. Oxford University Press, Oxford, p. 341.

Bowes BG (1999). *A Colour Atlas of Plant Propagation and Conservation*. Manson Publishing, London, p. 224.

Bowes BG (2008). *A Colour Atlas of Trees*. Manson Publishing, London.

Burgess J (1985). *An Introduction to Plant Cell Development*. Cambridge University Press, Cambridge, p. 239.

Cresti MS, Blackmore S, van Went JL (1992). *Atlas of Sexual Reproduction in Flowering Plants*. Springer-Verlag, New York, p. 249.

Dickison WC (2000). *Integrative Plant Anatomy*. Academic Press/Thompson Learning, San Diego, California, p. 533.

Endress PK (1994). *Diversity and Evolutionary Botany of Tropical Flowers*. Cambridge University Press, Cambridge, p. 511.

Esau K (1977). *Anatomy of Seed Plants*. John Wiley & Sons, New York, p. 550.

Evert RF (2006). *Esau's Plant Anatomy: meristems, cells, and tissues of the plant body – their structure, function, and development*, 3rd edn. John Wiley & Sons, Hoboken, New Jersey, p. 601.

Fahn A (1990). *Plant Anatomy*, 4th edn. Pergamon Press, Oxford, p. 588.

Foster AS, Gifford EM (1974). *Comparative Morphology of Vascular Plants*, 2nd edn. WH Freeman & Co, San Francisco, p. 751.

Heywood VH (ed.) (1993). *Flowering Plants of the World*. BT Batsford, London, p. 335.

Hickey M, King CJ (1981). *100 Families of Flowering Plants*. Cambridge University Press, Cambridge, p. 567.

Ingrouille M (1992). *Diversity and Evolution of Land Plants*. Chapman & Hall, London, p. 340.

Körner C (2003). *Alpine Plant Life : functional plant ecology of high mountain ecosystems*. Springer Verlag, Berlin, p. 433.

Lyndon RF (1990). *Plant Development: the cellular basis*. Unwin Hyman, London, p. 342.

Mauseth JD (1988). *Plant Anatomy*. Benjamin/Cummings Publishing Company, Menlo Park, California, p. 560.

Mauseth JD (2003). *Botany – An Introduction to Plant Biology*, 3rd edn. Jones & Bartlett Publishers, Sudbury, Massachusetts, p. 848.

Mauseth JD (2006). Structure–function relationships in highly modified shoots of Cactaceae. *Annals of Botany* **98**:901–926. doi: 10.1093/aob/mc1133, available online at www.aob.oxfordjournals.org.

Raghavan V (1986). *Embryogenesis in Angiosperms*. Cambridge University Press, Cambridge, p. 303.

Rudall P (1992). *Anatomy of Flowering Plants*, 2nd edn. Cambridge University Press, Cambridge, p. 110.

Simpson MG (2006). *Plant Systematics*. Elsevier/Academic Press, Amsterdam, p. 590.

Sporne KF (1974). *The Morphology of Angiosperms*. Hutchinson & Co, London, p. 207.

Sporne KF (1974). *The Morphology of Gymnosperms*, 2nd edn. Hutchinson & Co, London, p. 216.

Taiz L, Zeiger E (1991). *Plant Physiology*. Benjamin/Cummings Publishing Company, Menlo Park, California, p. 559.

Weberling F (1989). *Morphology of Flowers and Inflorescences*. Cambridge University Press, Cambridge, p. 405.

Glossary

Abaxial surface The surface of leaf or bud primordium remote from the shoot apex or stem (cf. Adaxial). The lower surface of a bifacial leaf.

Abscission layer The layer of cells of which breakdown leads to the shedding of a deciduous organ.

Abscission zone The region at the base of a leaf, or other deciduous organ, which ruptures along the abscission layer to reveal a protective layer of cork covering the scar.

Accessory bud A bud additional to the main axillary bud.

Accessory tissue The tissues of the flower or inflorescence which may be associated with the ovary in fruit development.

Acropetal differentiation Forming from the base of an organ towards its apex.

Actinomorphic A regular, radially symmetrical flower (cf. Zygomorphic).

Adaxial surface The surface of a leaf or bud adjacent to the shoot apex or stem (cf. Abaxial).

Adventitious Plant organs arising from unusual locations; as in buds developing on roots or leaves, roots forming on leaves and stems.

Aerenchyma Parenchymatous tissue with very large intercellular spaces.

Aerobic respiration Organisms utilizing molecular oxygen for respiration.

Aggregate fruit A fruit formed from several or numerous carpels of a single flower, e.g. strawberry.

Albuminous seed A seed which at maturity contains endosperm to nourish the germinating embryo.

Alga A member of the large group of nonvascular plants which are thalloid and predominantly aquatic.

Alternation of generations In all land plants the diploid sporophyte generation is different in form to the haploid gametophyte (heteromorphic alternation). In vascular plants the sporophyte generation is dominant in contrast to the bryophytes where the gametophyte is dominant.

Amphicribral vascular bundle A vascular bundle with a xylem core surrounded by phloem.

Amphivasal vascular bundle A vascular bundle with a core of phloem surrounded by xylem.

Amyloplast A noncoloured plastid storing large quantities of starch in its stroma.

Anaphase The stage in mitosis in which the sister chromatids migrate to opposite poles of the spindle.

Androecium The collective term for the stamens of a flower.

Angiosperm A seed plant in which the ovule(s) is enclosed within an ovary at the base of the carpel and with a stigmatic surface on which pollen is deposited.

Anther The pollen-bearing terminal region of a stamen.

Anticlinal Refers to the formation of a new cell wall at right angles to the surface of the organ or duct (cf. Periclinal).

Antipodal cell One of the cells, usually three, occurring at the chalazal end of the embryo sac.

Apoplast The nonprotoplasmic region of the plant comprising cell walls, the lumina of dead cells, and intercellular spaces.

Arborescent species Woody plants which are either trees or shrubs at maturity.

Areole The smallest area of leaf mesophyll (especially in dicotyledons) which is completely invested by veins.

Artefact Something created by humans; during specimen preparation, intentional artefacts include microtoming and staining, accidental artefacts include distorting tissues by dehydration or allowing crystals to fall out of cells that have been cut open.

Auricle In grasses one of the two lateral flaps of tissue located at the junction of the lamina and leaf sheath.

Axial parenchyma The longitudinally orientated parenchyma cells of the secondary vascular tissue (cf. Ray).

Axil The angle between the stem and the adaxial insertion of the leaf; it normally bears an axillary bud.

Axile placentation In a syncarpous gynoecium the longitudinal arrangement of the placentae on the central axis of the ovary (cf. Parietal placentation).

Bark The inner bark of a tree comprises the vascular cambium and youngest secondary phloem, while the outer bark corresponds to the rhytidome.

Basal angiosperms DNA evidence indicates that before angiosperms diversified into the monocots and the eudicots, they had already produced about eight separate orders of plants; the descendants of those early groups constitute the basal angiosperms.

Basal meristem In the leaves of many monocotyledons, especially perennials, the base of the leaf remains meristematic and new leaf tissue continues to form from this source.

Bicollateral bundle A vascular strand in the shoot with phloem forming to the outside and inside of the xylem (cf. Collateral bundle).

Bifacial cambium A cambium that produces cells both along its inner side and its outer side.

Bifacial leaf A dorsiventral leaf showing palisade mesophyll on the upper (adaxial) surface and spongy mesophyll on the lower surface (cf. Isobilateral leaf).

Bisporic development The formation of an embryo sac when one of the two derivatives of the first meiotic division degenerates but the other proceeds to the second division; both haploid nuclei thus formed undergo mitosis to form the mature embryo sac (cf. Mono- and tetrasporic embryo sac).

Bryophytes A group of nonvascular land plants comprising the liverworts and mosses in which the gametophyte stage is dominant in contrast to vascular plants.

Bundle sheath Layer(s) of parenchyma or sclerenchyma cells enclosing a vein of the leaf.

Callose An amorphous polysaccharide common in the walls of sieve areas of phloem. It also forms rapidly in wounded sieve tubes where it helps to seal the sieve pores, and callose is also synthesized in developing pollen tubes.

Callus An unorganized tissue mass, initially composed mainly of parenchyma, formed at the wounded surface of plant organs and protecting the plant from infection. Later a protective layer of cork may develop within it, while adventitious roots and shoots sometimes form from meristematic nodules in the callus.

Calyptra A protective tissue derived from the gametophyte and covering the sporophyte in some mosses and liverworts. Rarely used as a technical term for the root cap (see Calyptrogen).

Calyptrogen A distinct meristematic layer in the root apex, present in the grasses and some other roots, which gives rise to the root cap.

Cambium A lateral meristem which either forms secondary xylem and phloem (vascular cambium) or cork (cork cambium).

Carbohydrate A general term for substances composed of carbon, hydrogen, and oxygen and having the general chemical formula of $C_nH_{2n}O_n$.

Carpel A component of the gynoecium which may be free or fused to other carpels and bears ovule(s) in its ovary.

Casparian band A continuous impermeable layer, composed of lignin and suberin, located in the radial and transverse (anticlinal) primary walls of the endodermis in roots and some stems.

Cavitation The formation of a bubble in the water column within a vessel or tracheid leading to the loss of their water-conducting capacity.

Cell plate The partition formed from fused Golgi vesicles, which separates the two nuclei at the end of mitosis and spreads centrifugally to divide the mother cell. The pectin interior of the plate constitutes the middle lamella and subsequently primary wall is secreted on either side of it.

Cell wall The protoplast of a plant cell is normally surrounded by a fairly rigid primary wall composed of a fibrillar cellulosic framework linked to amorphous polysaccharides and proteins. A secondary wall may be deposited internally; this usually contains a higher proportion of cellulose and is often lignified.

Cellulose A polysaccharide consisting of glucose molecules linked into long unbranched drains of up to 15,000 monomers. The chains are laterally hydrogen bonded to form microfibrils up to several micrometres long and 3–8 nm wide.

Central mother cells The terminal zone of the shoot apex from which the subjacent apical tissue is derived.

Centrifugal growth Development from the centre towards the outside as in the growth of the cell plate or differentiation of the primary xylem in the shoot.

Centripetal growth Development from the outside towards the centre as in the differentiation of the primary xylem in the root.

Cephalium The flowering zone in certain cacti; cephalia may be terminal or lateral.

Chalaza Region of the ovule where the base of the nucellus is attached to the funiculus.

Chimaera A plant or organ composed of tissues of several genotypes; as in a shoot apex in which a mutation in an initial leads to its derivative tissues being incapable of developing chloroplasts and the leaf appearing variegated.

Chitin A polymer formed from a modified sugar molecule; it is the main skeletal material in the cell wall of fungi and also occurs in insect and crustacean outer skeletons.

Chloroplast A plastid concerned with photosynthesis. The internal chlorophyll bearing membranes are very extensive and normally arranged into a complex series of stacked cisternae forming numerous grana which interconnect by stromal membranes. The surrounding stroma contains the enzymes for carbon fixation.

Chromatid The half chromosome (joined to its partner by the centromere) visible during early mitosis and also in meiosis.

Chromatin See Eu- and Heterochromatin.

Chromosome A body within the nucleus bearing genes arranged linearly. The chromosomes are normally decondensed in the interphase nucleus and not distinguishable by the light microscope, but during nuclear division they form visible thread-like bodies.

Cisternae A flattened membranous compartment bounded by a single membrane as in the endoplasmic reticulum and Golgi body.

Cladistic study A cladistic study seeks to show evolutionary relationships among a group of organisms; the resulting cladogram ('family tree') emphasizes shared, derived characters, those that were gained or lost by one evolutionary line and that thus make it distinct from other lines.

Cladode A flattened shoot that resembles a leaf, either slightly (as in *Opuntia*, **341**) or strongly (as in *Semele*, **342**).

Cleistogamy Self-pollination within an unopened flower bud.

Coated vesicle A small cytoplasmic vesicle coated with clathrin and apparently pinched off from the plasmalemma.

Coenocyte A large cell containing several to many nuclei, usually resulting from mitosis unaccompanied by cytokinesis.

Coleoptile A sheath which encloses the embryonic shoot in grass grains. During germination it forms a protective channel for the elongating shoot.

Coleorhiza A sheath enveloping the radicle of grass and certain other monocotyledonous embryos.

Collateral bundle A strand of vascular tissue with xylem and phloem on the same radius and with the latter usually lying nearest the epidermis.

Collenchyma A living supportive tissue consisting of elongate cells with unevenly thickened, nonlignified walls. It is common in the peripheral regions of the young shoot.

Companion cell A specialized parenchyma cell with extensive plasmodesmatal connections to a sieve tube member; both cells are derived from a common mother cell. The densely cytoplasmic companion cells apparently control the functioning of the enucleate sieve tubes.

Complementary tissue The cork cells underlying lenticels have abundant intercellular spaces, unlike the impermeable cork elsewhere, which allow aeration of the internal living tissues.

Compression wood In gymnosperms the wood formed in the lower sides of branches which is dense and heavily lignified (cf. Tension wood).

Conifer A cone-bearing tree belonging to the largest division of the gymnosperms; common members are pines, firs, and larches.

lamellae of adjacent cells, as in some fibres.

Isobilateral leaf A leaf in which palisade mesophyll occurs both ad- and abaxially (cf. Bifacial leaf).

Kinetochore A specialized region of the chromosome at which the two chromatids are joined and from which microtubules originate to form part of the mitotic spindle.

Kranz anatomy The radial arrangement of the mesophyll cells around each bundle sheath in the leaf which is characterisitic of plants with C4 photosynthesis.

Lamina The blade of a leaf.

Lateral root A root arising from another root (cf. Adventitious).

Laticifer A secretory cell, or series of interconnected cells, containing the milky fluid latex.

Leaf An organ arising as a lateral swelling on the shoot apex and which typically bears a bud in its axil. The leaf is in vascular continuity with the stem and is normally the major photosynthetic region of the plant.

Leaf sheath In some leaves their basal portions invest the stem to form a distinct sheath, as in many monocotyledons.

Leaflet In a compound leaf a series of individual leaflets arise from the axis of the leaf but do not subtend buds.

Lenticels Pores in the bark formed, in contrast to the adjacent compact cork, from rounded cells with intercellular spaces which allow oxygen to diffuse into the plant.

Leucoplast A colourless plastid with little starch, such as is present in the leaf epidermis.

Liane A climbing plant with a long woody stem, especially prevalent in tropical forests.

Lichen A symbiotic association between an alga and fungus living together.

Lignin A complex substance containing various phenolics which is deposited in the cellulose walls of the sclerenchyma and tracheary elements; it increases their strength and renders the walls impermeable to water.

Ligno-suberin A complex of lignin and suberin deposited in the walls of the endodermis and exodermis of the root.

Ligule In grasses a membranous projection from the adaxial leaf surface at the base of the lamina.

Lipid A group of fats and fat-like compounds which are soluble in certain organic solvents but not in water.

Liverwort A small bryophytic plant (cf. Moss); the often thallosic green plant represents the dominant gametophyte in contrast to vascular plants.

Long-shoots The ordinary shoots of most plants; this term is unnecessary whenever the plant does not have highly modified branches called short-shoots.

Lumen The central channel or space, formerly occupied by the protoplast, in dead sclerenchyma and tracheary elements.

Magnoliids A group of four orders of basal angiosperms believed to have descended from some of the earliest members of the flowering plants, those which existed before the diversification into monocotyledons and eudicotyledons.

Medullary bundles Vascular bundles located in the pith of certain dicot stems.

Megasporangium A sporangium in which each diploid megaspore mother cell gives rise by meiosis to four haploid megaspores (cf. Microsporangium).

Megaspore A large haploid spore in seed plants and certain ferns and their allies, which develops into the female gametophyte or embryo sac.

Meiosis A sequence of two specialized nuclear divisions of a diploid cell resulting in the formation of four haploid cells.

Meristem A tissue primarily concerned with growth and division in an organized manner, as in the shoot and root apex, vascular and cork cambia.

Mesocarp The middle, often fleshy layer of the fruit wall.

Mesophyll The photosynthetic parenchyma of a leaf, frequently divided into cylindrical palisade cells and irregular spongy mesophyll.

Metabolism The process in which nutritive material is synthesized into protoplasm and cell wall or in which the latter are broken down into simpler substances.

Metaphase A stage in mitosis during which the kinetochores of the chromosomes all lie at the equator of the mitotic spindle.

Metaphloem The last-formed region of the primary phloem which matures after the organ has ceased to elongate.

Metaxylem The last-formed region of the primary xylem which matures after the organ has ceased to elongate; its tracheary elements are usually scalariform, reticulate, or pitted.

Microbody An organelle demarcated by a single membrane and containing various nonhydrolytic enzymes (cf. Glyoxysome and Peroxisome).

Microfibril A series of cellulose molecules linked together by hydrogen bonding to form a fibril up to several micrometres long; microfibrils form the skeletal framework of the cell wall.

Microfilament A proteinaceous and filamentous component of the cytoskeleton of some plant cells; it is ca. 7 nm wide and narrower than a microtuble.

Micrometre (μm) A unit of length representing one thousandth of a millimetre.

Micropyle A narrow pore in the integument(s) at the apex of the ovule, via which the pollen tube frequently penetrates the embryo sac.

Microsporangium This gives rise to numerous microspores (cf. Megasporangium).

Microspore A haploid spore which develops into the male gametophyte; the pollen grain of seed plants.

Microtome A device for cutting specimens into uniformly thin sections (slices) for examination with a microscope.

Microtubule A hollow proteinaceous tubule ca. 25 nanometres wide (cf. Microfilament). Microtubules form the major component of the plant cytoskeleton and are located in the peripheral cytoplasm of nondividing cells and also form the spindle fibres of dividing nuclei.

Middle lamella A layer of mainly pectic materials, derived from the cell plate, which cements together the primary walls of adjacent cells.

Mid-rib In many simple leaves a single prominent longitudinal rib extends the length of the leaf, consisting of a large vascular bundle(s) and sheath.

Mitochondrion (pl. *mitochondria*) An organelle delimited by an envelope, the inner membrane of which is involuted into tubules or cristae; responsible for aerobic respiration.

Mitosis The division of a diploid nucleus into two diploid daughter nuclei.

Mitotic spindle The fibrillar structure formed early in mitosis, the 'fibres' of which are visible under the light microscope, and consist of fasciated microtubules. It is concerned in the segregation of the two chromatids of each chromosome to a different daughter nucleus.

Monocotyledon (monocot) One of the two groups comprising the flowering plants; the monocotyledonous embryo has a single cotyledon (cf. Eudicotyledon). A number of other features (flower parts in threes, normally absence of secondary thickening, scattered arrangement of vascular bundles in the shoot, and so on) also characterize a monocotyledon.

Monosporic development This is the typical situation in the formation of an embryo sac of a flowering plant in which only one of the four haploid cells, derived from the meiosis of the megaspore mother cell, undergoes development whilst the others abort.

Morphology The external form and development of the plant.

Moss A small, leafy bryophytic plant (cf. Liverwort); the gametophyte generation is dominant in contrast to the situation in vascular plants.

Mucigel The growing root tip secretes mucilage from its cap cells which lubricates the passage of the root between the soil particles and may be important in nourishing beneficial soil micro-organisms.

Multiple epidermis A several-layered tissue derived from the protodem by both periclinal and anticlinal divisions; only the outer layer forms a typical epidermis.

Multiple fruit This is derived from the ovaries of several to many individual flowers as in pineapple (cf. Aggregate fruit).

Mycorrhiza The symbiotic association between the roots of many plants and soil fungi (see Ecto- and Endomycorrhizal association).

Nectar A fluid secreted by a nectary; the liquid is rich in sugars and other organic substance.

Nectary A multicellular gland secreting nectar; present as floral nectaries in many plants but also occurring as extra-floral nectaries on the vegetative plant.

Nexine The inner layer of the exine in the wall of a pollen grain.

Node The region of the stem from which a leaf or leaves arise (cf. Internode).

Nucellus The inner region of the ovule surrounding the embryo sac; considered to be homologous with the megasporangium.

Nuclear envelope The double membrane enclosing the nucleoplasm; the envelope is frequently penetrated by pores and the outer membrane is linked to the endoplasmic reticulum.

Nuclear pore The outer and inner nuclear membranes are often joined to form pores in the nuclear envelope.

Nucleolus A densely-staining granular body, commonly spherical, which occurs in the interphase nucleus; it is composed of RNA and protein and is the site of ribosome synthesis.

Nucleus A large organelle bounded by a double membrane and containing the chromosomes, nucleolus, and nucleoplasm; commonly only a single nucleus is present per cell but some cells are coenocytic.

Organelle A cytoplasmic body with a specialized function.

Osteosclereid An elongate sclereid with enlarged ends.

Ovary The basal region of a carpel which contains one to many ovules; after fertilization the ovary enlarges and differentiates to form the fruit.

Ovule In seed plants the female gametophyte (embro sac) is enclosed within the nucellus and integument(s). These, together with the funiculus, comprise the ovule which later develops into the seed.

P-protein The proteinaceous material occurring in sieve tubes which in damaged tissue forms a plug blocking the sieve plate.

Palisade mesophyll The parenchymatous tissue of which the cylindrical cells lie with their long axes perpendicular to the epidermis; this compact photosynthetic layer occurs on the adaxial surface of bifacial leaves.

Parenchyma cell An unspecialized, highly vacuolated cell with typically only a primary wall of uniform thickness; it occurs as extensive regions of tissue in the pith, cortex, and mesophyll of the plant body.

Parietal placentation Occurs in an ovary in which the ovules are attached to peripheral placenta(e).

Parthenocarpy Development of a fruit without fertilization of the ovules.

Perennial A plant the vegetative body of which persists for many years.

Perforation plate The end wall of a vessel element, either a single large hole (simple plate), commonly lying in a transverse wall, or several pores forming a compound plate on an oblique end wall.

Perianth The collective name for the sterile outer parts of a flower which are often differentiated into the outer sepals and inner petals.

Pericarp A synonym for fruit wall.

Periclinal Refers to a cell wall forming parallel to the surface of an organ (cf. Anticlinal).

Pericycle In a root the tissue, usually parenchymatous, lying between the endodermis and vascular tisses.

Periderm The secondary protective tissue (bark) replacing the epidermis; it comprises the phellem, phellogen, and phelloderm, plus any cortex or phloem lying internal to a deeply situated cork cambium.

Perinuclear space The region lying between the two membranes of the nuclear envelope.

Periplasmodium The coenocytic mass, formed from fused tapetal protoplasts, which occurs around the developing pollen grains in some anthers.

Perisperm A nutritive storage tissue formed from the nucellus present in the seeds of several dicotyledonous families.

Peroxisome A single membrane-bounded microbody lying adjacent to a chloroplast and involved in the metabolism of glycolic acid associated with photorespiration.

Petal An inner perianth member which is distinct in form from a sepal.

Petiole The narrow stalk which attaches the leaves of many plants to the stem.

Phanerogam A plant that reproduces by means of seeds; examples are cycads, conifers, and angiosperms (cf. Cryptogam).

Phellem The nonliving outer layer (cork) of the periderm which is impermeable due to the deposition of suberin within the walls of its constituent cells; formed from the phellogen.

Phelloderm A parenchymatous tissue which is formed centripetally from the phellogen of some plants.

Phellogen A lateral meristem (cork cambium) which by regular periclinal divisions forms phelloderm to its exterior; in some plants a phelloderm is also formed to the interior.

Phloem The main food transporting tissue of vascular plants; consisting of the conducting sieve elements, various types of parenchyma, and sclerenchyma.

Phragmosome The layer of cytoplasm stretching across a vacuolated cell in which the nucleus is situated during division, and which demarcates the plane of the newly-forming cell plate.

Phylloclade Interpreted as a petiole which has become flattened and replaces the lamina of a leaf.

Phyllotaxy The pattern in which the leaves are arranged on the stem.

Pit A region of the cell wall in which the primary wall remains uncovered by the deposition of secondary wall; the recess in the wall may be of uniform width (simple pit) or the pit is bordered at its outer margin.

Pit field A thin region of primary wall with numerous plasmodesmata; if secondary wall is later deposited, a pit develops over this region.

Pith The central ground tissue (usually parenchymatous) of the stem and some roots.

Placenta The regions of the ovary to which the ovules are attached.

Plasmalemma The single membrane which demarcates the cell protoplast from the externally lying wall; the plasmalemmae of contiguous cells are in continuity via their plasmodesmata.

Plasmodesma (pl. *plasmodesmata*) A pore in the cell wall linking adjacent protoplasts; it is lined by plasmalemma and contains an axial desmotubule linked to endoplasmic reticulum at either end.

Plastid The generic name for a varied group of organelles, bounded by a double membrane, which are derived from a proplastid; common examples are chloroplasts and amyloplasts.

Plastoglobulus (pl. *plastoglobuli*) A small densely-staining lipidic vesicle occurring within the stroma of chloroplasts and other plastids.

Plicate mesophyll cell A parenchyma cell, the primary walls of which are enfolded into the protoplast, as in pine leaves.

Plumule The region lying above the cotyledon(s) in the embryo and which forms the young shoot in the seedling (cf. Radicle).

Pneumatophore A negatively geotropic root projecting from the substratum; produced by trees living in swamp conditions and serving for aeration of the underground root system.

Polar nucleus In the ovule of a flowering plant one of the (normally) two nuclei occurring in the central cell; the endosperm results from the fusion by a male gamete with these nuclei.

Pollen grain In seed plants the term used for microspore.

Pollen tube The tube developing from a germinated pollen grain, in which the male gametes are transported to the embryo sac.

Polyarch The roots of monocotyledons have numerous protoxylem poles and are termed polyarch.

Polyploid Referring to a plant or cell possessing a multiple of the normal diploid set of chromosomes, e.g. a tetraploid has a double set.

Polysaccharide A carbohydrate composed of many monosaccharides linked together in a chain, as cellulose and starch.

Polysome A complex of ribosomes concerned with protein synthesis.

Primary cell wall The wall formed by the protoplast up to the end of expansion growth; the cellulose microfibrils are often randomly orientated and are less abundant than in the secondary wall.

Primary thickening meristem This occurs in the sub-apical region of plants with greatly thickened primary stems; divisions throughout the incipient cortex, procambium, and pith lead to rapid radial growth of the axis, as in cycads and monocotyledons.

Procambium A meristematic tissue arising directly from the apical meristem; in the primary plant body it differentiates into the primary vascular tissues; in dicotyledons and gymnosperms it also forms the fascicular cambium.

Proembryo The embryo before the onset of organ and tissue differentiation.

Prolamellar body The star-shaped complex of membranous tubules occurring in an etioplast; on exposure to light this body is transformed into grana and frets.

Proleptic growth Rhythmic growth of a perennial plant (cf. Sylleptic).

Prop root An adventitious root formed on the stem above the soil surface and helping to anchor the plant.

Prophase The early stage of nuclear division; characterized by the appearance of the chromosomes, the breakdown of the nuclear envelope, and the development of the spindle apparatus.

Proplastid A small and undifferentiated plastid occurring in meristematic tissues; the progenator of all other plastid types.

Protein A large and complex molecule composed of various amino acids.

Protoderm Meristematic tissue which gives rise to the epidermis.

Protophloem The first phloem to differentiate from the procambium and usually consists of sieve elements only. These are short-lived and usually crushed in the developing shoot (cf. Metaphloem).

Protoplast The protoplasm confined within the walls of an individual cell.

Protoxylem The first xylem to differentiate from the procambium, usually consisting of annular or spirally-thickened tracheary elements (cf. Metaxylem).

Proximal Nearest the point of attachment or origin (cf. Distal).

Pulvinus A joint-like thickening of the leaf petiole (or of a petiolule) in which the central vascular strand is surrounded by a broad expanse of parenchyma. Loss of turgor in this tissue causes the leaf to droop.

Quiescent centre The terminal region of the root apex in which cell divisions are absent, or occur very infrequently relative to the adjacent meristematic cells.

Radicle The embryonic root situated beneath the hypocotyl in the seed and forming the main root of the seedling (cf. Plumule).

Raphide A slender, needle-like crystal of calcium oxalate; raphides occur in bundles in the central vacuole of some cells in some plants.

Ray A panel of parenchyma extending radially across the secondary vascular tissues; a ray is formed from an initial in the vascular cambium and is of variable width and height.

Ray initial A squat, semi-cuboidal, cell of the vascular cambium giving rise to the ray parenchyma of the secondary vascular tissues (cf. Fusiform initial).

Receptacle The terminal region of the flower stalk to which the floral parts are attached.

Resin canal A long duct lined with epithelial cells which secrete the sticky resin common in conifers.

Rhizome An elongate horizontal stem growing beneath the soil; a common organ of perennation in monocotyledons.

Rhizosphere The region of soil immediately surrounding the root.

Rhytidome The outer bark inclusive of the periderm and any cortical and phloem tissues isolated from the functional phloem by a deep-sited phellogen.

Rib meristem The sub-terminal axial region evident in some shoot apices; its derivatives divide predominantly transverse to the long axis of the young stem and give rise to the pith.

Ribosome A small organelle composed of RNA and protein which is concerned with protein synthesis; ribosomes may be aggregated into polysomes.

Ring porous wood Secondary xylem with the vessels of the spring wood much wider and more numerous than in later wood; this pattern of vessel formation leads to rings being visible in transverse section.

Root A plant organ which is linked to the shoot and is typically subterranean; roots are primarily concerned with absorption of water and mineral salts, anchorage, and nutrient storage.

Root cap A cap of cells enclosing the root apex.

Root hair A simple cylindrical bulge from an epidermal cell of the young root which extends laterally between the adjacent soil particles and extends the absorptive surface of the root.

Root pressure The water pressure in the xylem resulting from the active transport of mineral salts into the vascular cylinder by the endodermis, thus causing intake of water from the cortex.

Rosette A shoot with very short internodes but bearing fully expanded leaves. Also a group of cellulose-synthesizing enzymes located at the plasmalemma.

Scale leaf A nonfoliage leaf often investing dormant buds or found in underground stems.

Schizogeny The separation of cells along their middle lamellae to form an intercellular space.

Sclereid A type of sclerenchyma cell characterized by its very thick lignified walls and numerous pits; the shape is variable but it is generally much shorter than a fibre.

Sclerenchyma A supporting tissue, the cells of which are commonly dead at maturity and possess thick, lignified secondary walls, as in fibres and sclereids.

Scutellum The highly modified cotyledon present in grasses which supplies nutrients from the endosperm to the germinating embryo.

Secondary cell wall The wall formed by the protoplast at the end of expansion growth; the cellulose microfibrils are closely crowded and, in any one layer, lie parallel to each other (cf. Primary cell wall). Secondary walls often become lignified, as in sclerenchyma and tracheary elements.

Secondary thickening This occurs in gymnosperms and most dicotyledons and some anomalous monocotyledons. The stem and root increase in diameter due to the formation of secondary vascular tissues by the vascular cambium (or by the secondary thickening meristem), while the epidermis is normally replaced by cork formed from the cork cambium.

Secondary thickening meristem In some arborescent monocotyledons (e.g. *Dracaena, Cordyline*) an anomalous form of secondary thickening occurs from a meristem which arises in the outer cortex and cuts off discrete vascular bundles, plus parenchyma, centripetally.

Seed The structure which develops from the fertilized ovule; it contains the embryo and a food supply to support early seedling growth.

Sepal An outer perianth member which is distinct in form from a petal.

Septate fibre A fibre with thin cross-walls which develop after the longitudinal walls have become thickened.

Sexine The outermost region of the wall of pollen grains ectine (cf. Nexine).

Shoot The nonroot region of the plant; it is usually aerial and composed of the stem bearing numerous photosynthetic foliage leaves (cf. Root).

Short-shoots In certain plants, highly modified axillary buds and branches that grow as dwarf shoots with short internodes. Short-shoots may have ordinary leaves as in apple and larch, or modified leaves as in cacti (cactus spines are leaves of short-shoots).

Sieve area Modified pit fields in the side and oblique end walls of sieve cells or tubes; the plasmodesmata have been transformed into narrow sieve pores and lateral translocation probably occurs via them.

Sieve cell The enucleate translocating element in gymnosperms and lower vascular plants, possessing sieve areas on all walls.

Sieve plate The transverse or somewhat oblique wall occurring in sieve tubes; it contains either a single series of large pores or is compound with several series of pores (cf. Sieve area).

Sieve pore The hole in a sieve area or plate through which cytoplasmic continuity occurs from one sieve element to another; the wall surrounding the pore is commonly impregnated with callose.

Sieve tube An elongate element comprising several to many enucleate cells interconnected via the sieve plates (former cross-walls). Sieve tubes are confined to angiosperms (cf. Sieve cell).

Simple fruit Formed from the single ovary of an individual flower.

Softwood The wood of a conifer which generally lacks thick-walled, lignified fibres; it is therefore easier to saw than most hardwoods.

Spongy mesophyll Very irregular green parenchyma cells with large intercellular spaces between them; in bifacial leaves this tissue occurs abaxially (cf. Palisade mesophyll).

Sporangium A structure in which spores are produced; in most ferns the spores are uniform in size but in seed plants different sized spores are produced in mega- and microsporangia.

Spores Haploid cells formed as derivatives of the meiotic division of a diploid spore mother cell within a sporangium. Each spore germinates to form the gametophyte.

Sporophyte The diploid phase of the life cycle; in vascular plants the sporophyte is dominant (cf. Gametophyte).

Sporopollenin The substance composing the exine of pollen grains; it is formed from cyclic alcohols and is highly resistant to microbial decay.

Stamen The male organ of the flower composed of the terminal anther-bearing pollen and the basal sterile filament.

Starch The chief food storage polysaccharide of plants composed of several hundred hexose sugars; it is insoluble and accumulates within the stroma of various plastids.

Starch sheath In many primary dicotyledonous stems the inner layer of the cortical parenchyma forms a sheath with rich deposits of starch in its cells.

Stigma The receptive zone of a carpel at the tip of the style, upon which pollen is deposited and germinates.

Stipules Projections of tissue on either side of the base of the leaf, which in dicotyledons are sometimes large and vasculated.

Stoma (pl. *stomata*) A complex consisting of a pore in the shoot epidermis which is surrounded by two specialized guard cells; their turgidity causes the opening and closing of the stomatal pore and thus controls gaseous exchange with the external atmosphere.

Stone cell A small, thick-walled and more-or-less isodiametric sclereid.

Stroma The nonmembranous ground substance of a plastid.

Style The region of a carpel lying between the stigma and ovary.

Suberin A fatty, hydrophobic deposit in the cell walls of cork, associated with lignin in the thickened walls of the root endodermis and exodermis.

Subsidiary cells In some plants these occur adjacent to the guard cells of a stoma; subsidiary cells are morphologically distinct from the general epidermal cells.

Succulent A plant with fleshy leaves and stems containing many large, water-storing parenchyma cells.

Superior ovary An ovary which is inserted into the receptacle above the level of the other floral parts.

Suspensor A multicellular structure, usually filamentous, which is anchored at one end near the micropyle of the ovule and at the other to the radicle pole of the embryo; extension of the suspensor pushes the growing embryo into the endosperm.

Sylleptic growth Continuous growth of a perennial plant without rest phases (cf. Proleptic growth).

Symbiosis The mutually beneficially association of two different kinds of living organisms (e.g. lichens, nitrogen-fixing root nodules).

Symplast The combined protoplasts of the plant body; these are all linked by their plasmodesmata (cf. Apoplast).

Synergids In the embryo sac of flowering plants the two cells adjacent to the egg, which apparently play an essential role in the transmission of the male gametes to the egg and polar nuclei.

Tapetum The layer of nutritive cells lining the pollen sac; the tissue is absorbed as the pollen grains mature.

Tap root The main root of many dicotyledons which is directly derived from the persistent radicle.

Telophase The last stage of mitosis in which the two daughter nuclei are reorganized at the poles of the mitotic spindle.

Tension wood Forms on the upper sides of branches in arborescent dicotyledons and is characterized by the occurrence of numerous fibres, the walls of which are nonlignified and highly hydrated (cf. Compression wood).

Testa The investing layer of a seed formed from the modified integuments of the ovule.

Tetrasporic embryo sac An embryo sac which develops when all four haploid nuclei, formed by the meiosis of the megaspore mother cell, survive and contribute to the mature embryo sac.

Thallus The nonleafy, dorsiventrally flattened gametophyte of many liverworts.

Thylakoids The photosynthetic internal membranes of a chloroplast consisting of grana and frets.

Tissue A group of cells forming a discrete functional unit; in simple tissues the cells are all alike, whereas in a complex tissue varied cell types occur.

Tonoplast The single membrane which encloses a vacuole.

Torus The bi-convex thickened disc forming the central part of the primary wall of a bordered pit in conifers.

Totipotent A differentiated plant cell or tissue (e.g. parenchyma) which retains all of the genetic material present in the embryo; due to wounding or hormonal influence such cells may undergo dedifferentiation, regain their meristematic capacity, and give rise to adventitious organs and embryos.

Tracheary element A collective term for the vessels and tracheids of the xylem. These dead, water-conducting elements show various patterns of secondary wall thickening (annular, spiral, scalariform, reticulate, and pitted).

Tracheid An elongated imperforate tracheary element with various patterns of secondary wall deposition (cf. Tracheary element).

Transfer cell This shows labyrinthine ingrowth of its walls which greatly increase the surface area of plasmalemma; these cells function in the large-scale transport of solutes over short distances.

Transfusion tracheids Specialized xylem, confined to gymnosperm leaves, in which the tracheids are short and with nontapering ends.

Translocation The movement of sugars and other organic substances throughout the vascular plant body via the sieve elements of the phloem.

Transmitting tissue The specialized tracts of tissue in the style through which the pollen tubes grow towards the ovary.

Transpiration The movement of water from the root to the shoot in the tracheary elements of the xylem and the subsequent loss of water vapour from the leaf surface via the stomata.

Trichome Any outgrowth from an epidermal cell; it may be unicellular or multicellular and is often glandular.

Tuber A much-enlarged underground stem or root from which the plant perennates.

Tunica The outermost layer(s) of the shoot apex in flowering plants characterized by divisions which are only anticlinal (cf. Corpus).

Tylose An outgrowth of a xylem parenchyma cell which penetrates through the pit of a tracheary element (commonly a vessel) and enlargens within its lumen to block it.

Unifacial cambium A cambium that produces cells on only one of its sides, not both.

Vacuole An organelle bounded by the tonoplast and containing a watery fluid. During differentiation the many small vacuoles of the meristematic cell become confluent and expand to form a large central vacuole.

Variegated leaf A leaf showing a distinct pattern of green and lighter or white regions in the leaf blade; commonly formed from a varigated chimaeral shoot apex.

Vascular bundle A strand of tissue composed of primary xylem and phloem (and in dicotyledons, cambium) running lengthwise in the shoot.

Vascular cambium see Cambium.

Vascular tissue A general term referring to both the xylem and phloem.

Vegetative cell In flowering plants this is the largest of the two cells in the young pollen grain.

Velamen The water-absorptive multiple epidermis in the roots of tropical epiphytic orchids and on some aroids.

Venation The pattern formed by the veins supplying the leaf blade.

Vessel A long, perforate tracheary element, formed from a longitudinal series of cells by the breakdown of their end wall (cf. Perforation plate). The pattern of secondary wall deposition is variable (cf. Tracheary element).

Wood The secondary xylem of arborescent dicotyledons and of gymnosperms.

Xeromorphic Morphological features typical of xerophytes.

Xerophyte A plant of dry regions with various xeromorphic features such as hairy and inrolled leaves, thick cuticles, sunken stomata, or spiny reduced leaves with succulent photosynthetic stems.

Xylem A complex tissue composed of the water-conducting tracheary elements, sclerenchyma, and parenchyma.

Zygomorphic flower A flower with bilateral symmetry.

Zygote The diploid cell resulting from the fertilization of a haploid egg cell by a male gamete; subsequent division of the zygote leads to the formation of the proembryo.

Index

For Product Safety Concerns and Information please contact our EU representative GPSR@taylorandfrancis.com Taylor & Francis Verlag GmbH, Kaufingerstraße 24, 80331 München, Germany

Batch number: 10732774

Printed by Printforce, the Netherlands